WENNER-GREN CENTER
INTERNATIONAL SYMPOSIUM SERIES

VOLUME 41

SOMATOSENSORY MECHANISMS

SOMATOSENSORY MECHANISMS

Proceedings of an International Symposium held at
The Wenner-Gren Center, Stockholm, June 8-10, 1983

Edited by

Curt von Euler
Department of Neurophysiology, Karolinska Institutet, S-104 01 Stockholm, Sweden

Ove Franzén
Department of Psychology, University of Uppsala, S-751 04 Uppsala, Sweden

Ulf Lindblom
Department of Neurology, Karolinska Hospital, S-104 01 Stockholm, Sweden

David Ottoson
Department of Physiology, Karolinska Institutet, S-104 01 Stockholm, Sweden

PLENUM PRESS • NEW YORK AND LONDON

First published 1984 by
PLENUM PRESS, NEW YORK
A Division of Plenum Publishing Corporation
233 Spring Street, New York, N.Y. 10013

ISBN 0–306–41842–8

Printed in Great Britain

CONTENTS

PERIPHERAL AND INTEGRATIVE MECHANISMS OF SOMATOSENSATION

SOMATOSENSORY DYSFUNCTIONS

LIST OF CONTRIBUTORS AND INVITED PARTICIPANTS

Brian Cooper
Department of Neuroscience
University of Florida
GAINESVILLE
Florida 32610 USA

Marshall Devor
Life Sciences Institute
Hebrew University of Jerusalem
JERUSALEM
Israel

Curt von Euler
Department of Neurophysiology
Karolinska Institutet
S-104 01 STOCKHOLM
Sweden

Ove Franzén
Department of Psychology
University of Uppsala
S-751 04 UPPSALA
Sweden

Esther Gardner
Department of Physiology & Biophysics
New York University
School of Medicine
550 First Avenue
NEW YORK
N.Y. 10016 USA

Gisèle Gilbaud
Unité de Recherches de Neurophysiologie Parmacologique
INSERM
2, Rue d'Alésia
F-75014 PARIS
France

Ragnar Granit
Eriksbergsgatan 14
S-114 30 STOCKHOLM
Sweden

Rolf Hallin
Department of Clinical Neurophysiology
Huddinge Hospital
S-141 86 HUDDINGE
Sweden

Ainsley Iggo
Department of Veterinary Physiology
University of Edinburgh
EDINBURGH
U.K.

Roland Johansson
Department of Physiology
University of Umeå
S-901 87 UMEÅ
Sweden

Kenneth Johnson
Department of Neuroscience
The Johns Hopkins University
School of Medicine
725 North Wolfe Street
BALTIMORE
Maryland 21205 USA

Jon Kaas
Department of Psychology
Vanderbilt University
134 Wesley Hall
NASHVILLE
Tennessee 37240 USA

Dan Kenshalo
Department of Psychology
Florida State University
TALLAHASSEE
Florida 32306 USA

Robert LaMotte
Department of Anesthesiology
Yale University
School of Medicine
333 Cedar Street
NEW HAVEN
Connecticut 06510 USA

Lea Leinonen
Department of Physiology
University of Helsinki
Siltavuorenpenger 20 J
SF-00170 HELSINKI
Finland

Ulf Lindblom
Department of Neurology
Karolinska Hospital
S-104 01 STOCKHOLM
Sweden

Vernon Mountcastle
Department of Neuroscience
The Johns Hopkins University
School of Medicine
725 North Wolfe Street
BALTIMORE
Maryland 21205 USA

Ulf Norrsell
Department of Physiology
University of Göteborg
S-300 33 GÖTEBORG
Sweden

José Ochoa
Department of Neurology
Dartmouth Medical School
HANOVER
New Hampshire 03756 USA

David Ottoson
Department of Physiology
Karolinska Institutet
S-104 01 STOCKHOLM
Sweden

Edward Perl
Department of Physiology
51 Medical Research Building
The University of North Carolina at Chapel Hill
CHAPEL HILL
North Carolina 27514 USA

Per E. Roland
Department of Neurology
Rigshospitalet
Blegdamsvej 9
DK-2100
COPENHAGEN
Denmark

Wolfgang Schady
Department of Clinical Neurophysiology
Akademiska Hospital
S-750 14 UPPSALA
Sweden

Floyd Thompson
Department of Neuroscience
University of Florida
GAINESVILLE
Florida 32610 USA

Erik Torebjörk
Department of Clinical
Neurophysiology
Akademiska Hospital
S-750 14 UPPSALA
Sweden

Patrick Wall
Department of Anatomy
University College London
Gower Street
LONDON WC1E 6BT
England

Åke Vallbo
Department of Neurophysiology
Karolinska Institutet
S-104 01 STOCKHOLM
Sweden

Barry Whitsel
Department of Physiology
University of North Carolina
at Chapel Hill
CHAPEL HILL
North Carolina 27514 USA

Charles Vierck
Department of Neuroscience
University of Florida
GAINESVILLE
Florida 32610 USA

Zsuzsanna Wiesenfeld
Department of Clinical
Neurophysiology
Huddinge Hospital
S-141 86 HUDDINGE
Sweden

William Willis
Marine Biomedical Institute
University of Texas Medical
Branch
GALVESTON
Texas 77550 USA

Clinton Woolsey
Department of Neurophysiology
University of Wisconsin
627 Waisman Center
MADISON
Wisconsin 53706 USA

Manfred Zimmermann
Department of Physiology
University of Heidelberg
Neuenheimer Feld 326
D-6900 HEIDELBERG
FRG

Professor Yngve Zotterman, M.D., V.M.D.h.c., F.R.S.

PREFACE

It is a great honour and a personal pleasure for me to introduce this symposium volume on "Somatosensory Mechanisms" dedicated to the memory of Yngve Zotterman. It was actually Yngve Zotterman himself who in the last year of his life began to plan for this meeting, by which he wanted to commemorate the discovery of the punctuate nature of skin sensations made by the Swedish physiologist, Magnus Blix in 1983. Yngve wanted by this Symposium to enlighten the progress in recent years in various fields of somatosensory physiology and to expose new conceptual aspects which have emerged as a result of these advances. We, the organizers of the Symposium, conceived it our clear duty to try to pursue the plans outlined by Yngve. For us it is a particular pleasure that so many of Yngve's close friends have been able to accept our invitation.

The field of somatosensations has expanded enormously in the last decade and discoveries and new ideas have greatly deepened our understanding of a multitude of phenomena at all levels from transfer functions of sensory receptors to integrated cortical functions. Many of the advances made in recent years have been achieved thanks to new and powerful techniques and ingenious experimental design. We are therefore at present in a much better position to successfully tackle many of the problems which only a decade ago were out of reach for experimental approach. It has not been possible within the scope of this symposium to cover all the aspects of progress in recent years but we hope that this meeting will reflect some of the most exciting achievements and thereby form a meaningful attempt to present the state of art in somatosensory physiology.

Yngve Zotterman during his scientific career made many important discoveries which brought new light to different aspects of somatosensory functions. Together with Adrian he solved the code of the sensory message in single axons, he was the first to record the activity from single non-myelinated pain fibres, he made important contributions to the understanding of thermosensitivity, taste, vestibular functions and receptor functions. In all his work he was driven by curiosity. For him science was fun but he was, as we all know, also a man who enjoyed life in all its aspects. As sensory physiologist he had the expert knowledge to appreciate all the nuances of sensations that life offered to him and he was a man who generously shared his joy with his friends.

We are all here today not only as scientists to bring our contributions to various fields of sensory physiology but also because of the feelings of personal affection for Yngve that we all share. However, I don't think Yngve would have liked this meeting to be solemn but rather a lively gathering with exciting presentations and animated discussions. Let us therefore make it a symposium in the true spirit of Yngve Zotterman.

David Ottoson
Karolinska Institutet

INTRODUCTORY LECTURE

NEURAL MECHANISMS IN SOMESTHESIS: RECENT PROGRESS AND FUTURE PROBLEMS

VERNON B. MOUNTCASTLE

Bard Laboratories of Neurophysiology, Department of Neuroscience, The Johns Hopkins University, Baltimore, MD, USA

INTRODUCTION

Professor Ottoson, friends and colleagues: I am pleased to be invited to attend this symposium, and to participate in this opening program. Firstly, because it honours Professor Yngve Zotterman, whom I admired greatly for his many contributions to the subject we address: the neural mechanisms in somesthesis. And, for his warm and effervescent personality, filled with collegiality for Neurophysiologists from every land. Secondly, because I have for a dozen years worked in a different field, the central neural mechanisms of spatial perception and directed visual attention. I hope this temporal detachment will lend some perspective to what I wish to say. That is, to summarize what I perceive to have been the main advances of the last decades; and, to suggest where opportunity lies in the years ahead.

A heuristic change has been the realization that the study of sensory performance called Psychophysics, and of the neural events set in motion by sensory stimuli, called Sensory Neurophysiology, are simply different experimental approaches to the same set of problems. What more evidence of this union do we need than that 14 of the members of this symposium have made investigations in which they measured both sensory performance and neural events, and sought explanations of the former in terms of the latter. Thus the CNS physiologist now aims to characterize brain events evoked by natural stimuli in a waking, sensing subject; in a behavioral context allowing measure of the sensations evoked by them in terms of detection, discriminations and ratings. The experiment has evolved through several stages, so that now we move to study somesthetic mechanisms in the combined experiment that has proved so successful in studies of the motor system. We may now observe the dynamic action of the sensing brain.

THE PERIPHERY

There is no longer any doubt of the modality specificity of mechanoreceptive afferent fibers. They are differentially sensitive to mechanical stimuli as opposed to thermal or noxious ones. Moreover, among the larger mechanoreceptive afferents those terminating peripherally in a particular class of receptors are differentially sensitive to a particular feature or features of the set of all possible stimulus attributes. A similar specificity obtains for the warm, the cool, and the two sets of nociceptive afferents as well. Thus many of the uncertainties on this subject and the validity of such alternatives as the pattern theory it would have been necessary to discuss 20 years ago, now appear anachronistic.

Work of great interest has been carried out on the primate hand by many of you here today - by Drs. Iggo and Perl and LaMotte and Lindblom, Kenshalo and Kenneth Johnson, as well as by others not with us, Darian-Smith and Werner. Quantitative studies including both neural and Psychophysical measures of sensory performance have established a principle. Namely, that along several continua our quantitative relations to the external world are set at the interface between the relevant receptor sheet and the environment. This is of some importance in the conceptualization of the central mechanisms involved. Our knowledge of the quantitative relations between sensory stimuli and sensory experience comes from 123 years of Psychophysics. I believe we owe more to S.S. Stevens than to any other in that history; the field has since been advanced by the contributions of Professors Franzen, Lindbloom, LaMotte and others.

The work on the primary afferents innervating the monkey hand has now been followed by an explosion of discovery concerning the first-order afferents and efferents in the peripheral nerves of humans. And, this has been an almost uniquely Scandinavian accomplishment! I need not detail it to this audience, for much of the work has been done by those present: Professors Valbo and Hagbarth, and Johansson and Hallin and Torebjork.

Dr. Ochoa has allowed me to read in manuscript form a description of the work he and Dr. Torebjork have just completed using intradermal microstimulation. They were able to stimulate a single afferent fiber, prove its singularity by proximal recording, show that excitation of a single fiber evokes a modality specific sensory experience that fits the functional properties of the fiber set determined separately. The projected receptive field of the conscious sensation coincides exactly with the receptive field determined by mechanical stimulation and recording from the same fiber. These discoveries lay to rest ideas concerning those perceptual mechanisms that arose in the minds of some from the results obtained by direct stimulation of the surface of the postcentral gyrus.

Coding

Johnson and Darian-Smith have made significant advances in the field of coding, especially of the afferent signals for fine form and texture discrimination. Problems of coding earlier evoked much interest, but have been neglected in the last decade; they deserve renewed study. For example, of the question I believe Dr. Johnson will raise, whether the peripheral afferent signals - the codes - for certain complex stimuli may lie in the dynamic relation between the patterns of stimulus evoked activity in two or in several of the different sets of primary afferent fibers. This introduces the population problem, which I shall return to in discussion of the cerebral cortex.

CENTRAL NERVOUS MECHANISMS IN SOMESTHESIS

Introduction

An observer of studies of the CNS mechanisms in somesthesis is struck by the fact that research of the last decades has been largely anatomical in nature. That is, if you classify as anatomical, as I do, descriptions of connectivity and cytoarchitecture, mapping or other studies of what is called "functional organization", even though the latter are made with electrophysiological methods. In a recent review, I consulted more than 1000 papers published on our subject in this period. While I have no exact tabulation, I estimate that less than 10% dealt with the dynamic operation of the nervous system in somatic sensing; i.e., with physiology! The reasons are not hard to find:

(1) The ease and beauty of the tracer methods introduced a decade or so ago;
(2) The productivity of the micromapping method introduced by Welker;
(3) The realization that the study of dynamic operations in anesthetized or otherwise reduced preparations is at best a hazardous affair; and,
(4) The long delay in the adoption by sensory physiologists of the combined experiment in waking monkeys.

The era of anatomical work has produced a base of information for the future: a detailed connectivity hitherto only dimly perceived; the plasticity of the microconnectivity of the system; the multiplicity of somatic sensory fields in the neocortex; the sources, courses, and destinations of ascending pathways; in sum, a base for what I am confident will be a surge of effort to understand how the brain works in sensing, as well as where and in what pattern.

Prethalamic Components

Work of the last decade has emphasized the role in somatic

sensibility of the dorsal ascending systems of the spinal cord and their combined forebrain projections through the medial lemniscus, particularly in primates. We now know that the lemniscus collects ascending axons from a complex of nuclei, including gracilis and cuneate, x and z, and part of the lateral cuneate. These I group together as the dorsal column nuclear complex. This system is essential for complex sensory tasks like those requiring discrimination of movement and direction, as Dr. Vierck has clearly shown; capacities that depend upon the cardinal properties of the dorsal ascending systems: precise somatotopy, modality specificity, and a strong synaptic security.

A major discovery is that neurons located in the middle laminae of the dorsal horns give rise to ascending systems that run in each of the funiculi of the spinal cord, and that these systems are important for mechanoreceptive sensibility. These are the spinocervicothalamic system, highly developed in furred quadripeds, comparatively less so in primates, and still of unknown function; the postsynaptic dorsal column system that projects upon the DCN; second order elements of the dorsolateral columns that project upon the DCN complex, including x and z; and the spinothalamic tract, more highly developed in primates than in carnivores, which gains in the former a clear projection upon the ventrobasal complex, as well as the intralaminar nuclei.

The systems of dorsal horn origin other than that in the dorsolateral funiculus contain some elements that receive convergent input from two or more classes of first-order fibers; some can be activated by light mechanical, noxious, and thermal stimuli. The primate spinothalamic tract plays a dominant if not exclusive role in pain sensibility, but this is not at all certain in carnivores, in whom ascending elements of the dorsolateral columns may contribute to pain. What role the spinothalamic system may play in mechanoreceptive sensibility remains uncertain and I hope that Drs. Wall and Willis and Perl will clarify this for us.

Several new discoveries have led to new concepts of the prethalamic pathways. It has been known for a long time that systems descending from the forebrain project directly upon the origins of prethalamic systems. For the spinocervicothalamic system one function of these descending systems is to maintain in a dynamic way the modality and place specificity of the cells or origin of ascending axons, as A.G. Brown has shown. What local network operations may effect such a control is unknown, but Brown's discovery raises the likelihood that modality and to a degree place specificity in the somatic system may, to different degrees for different components, be dynamically maintained by descending control systems.

The important discovery by Whitsel and Werner, of the resorting mechanisms of the dorsal columns in the monkey for both place and modality provides an explanation for the transition from a segmental to a somatotopic representation of the body form in the lemniscal system, and for the segregation of certain mechanoreceptive afferent classes to the dorsal columns, and of others via second order projections into the dorsolateral columns. These two systems re-combine in the projection through the DCN complex to the VB complex of the thalamus.

This discovery of the dorsal column shuffle dispelled much of the confusion that clouded lesion work on the spinal afferent systems in the 60's and 70's, particularly those experiments in which the remaining capacities of monkeys with dorsal column lesions were tested in simple somesthetic tasks; or, those like position sensibility that depend upon the re-shuffled projections - for the hindlimb - through the dorsolateral column. Properly chosen and more complex tasks promptly revealed the predicted deficiencies.

Finally, recent experiments by Dykes and by Jones have shown that in the lemniscal system there is a bundling together of elements with common modality and receptive field properties, and that this bundling begins in the peripheral nerves and ascending spinal systems. They project upon modularized elements of the DCN complex, and they upon the modules or rods of the VB complex, and they upon columns of the postcentral gyrus. The implication is that the modular organization of the somatic system, as regards its static properties of place and modality, is generated at its first-order input stage.

The lemniscal system is thus the major autobahn for transmitting signals from peripheral mechanoreceptors to the forebrain with specificity as regards modality and precision in the temporal domain. It is the forebrain projection pathway for the several medullary nuclei grouped together in the dorsal column nuclear complex, which receives projections from systems in both the dorsal and dorsolateral funiculi of the cord. And, I add, descending projections from the neocortex, about which virtually nothing is known in terms of physiological mechanisms.

THE VENTROBASAL COMPLEX

I pass quickly over the large volume of tracer research defining the thalamic targets of ascending somatic systems. The major but not the only target is the ventrobasal complex, which receives projections from the lemniscus, the spinothalamic tract, and the primate remnant of the lateral cervical system, all overlapping almost completely in the gross spatial sense. The lemniscus and STT project also upon POm, and the STT upon VPLo and the intralaminar nuclei, as well.

Fibers of each system are bundled by place and modality,

and each bundle pre-empts a rod of VB neurons 1-200 um in diameter and about 3 times that long. Remarkably, the overlap between systems, or between modalities within a system, upon VB rods or upon single neurons is very small. Thus each mode is represented with VB in a distributed set of modules and each modular set composes in a sense a separate somatotopic representation, even though members of a single set are separated from one another by interdigitations with other sets. This interdigitation is incomplete, producing a differential distribution for modality.

Is there a single or are there several somatotopic representations in VB? The matter is largely one of interpretation, but the partially interlocked modular sets of different classes compose, at a larger level, a single overall somatotopic pattern.

Each of the rods of VB cells projects upon a very local zone of the postcentral cortex, and receives in return a locally focussed corticothalamic projection. VB receives also a recurrent input from a portion of the thalamic reticular nucleus, and a projection from the central core systems of the brain stem. Obviously, VB has the potential for complicated interactions, which may control its relay-projection function. However, no information on this matter obtained in normal, sensing animals is available. I emphasize the importance of study of these problems in waking monkeys, especially of what is clearly a very local and powerful reverberating network between VB and the postcentral cortex. One hypothesis to be explored is that the input to the postcentral gyrus is controlled and shaped by intracortical mechanisms reflecting back upon VB, and themselves open to other mechanisms, like those for state controls such as directed attention.

FUNCTIONAL ORGANIZATION OF THE SOMATIC SENSORY AREAS OF THE NEOCORTEX

It has been known for a century that the body surface is represented in an orderly array in the postcentral gyrus, on evidence from lesions, local epilepsy, and electrical stimulation in man. A new era opened exactly fifty years ago, in 1933, with the publication of a 3-page paper by Gerard, Marshall and Saul describing the evoked potential method. There quickly followed a number of preliminary studies, culminating in the first evoked potential map, that by Woolsey, Marshall and Bard, of the monkey, in 1938 and 1942.

It is largely due to the studies of C.N. Woolsey and his many colleagues and students, in a research program carried out over forty years, that maps have been made in several dozen mammalian species, including man himself. Several principles evolved: the multiplicity of cortical representations, partially shifted overlap, the relation of cortical space in a map to peripheral innervation density and somesthetic specialization of a part; and many others. And now, the intensive application of the method of single neuron analysis and the micromapping method of Welker has revealed complexities undoubtedly smeared in the evoked potential method. Principally, that there is a separate representation of the body form in each of the four cytoarchitectural areas of the postcentral gyrus; and moreover, that while those precise patterns reveal at any given moment a precise neuronal connectivity, they depend for maintenance upon dynamic and plastic mechanisms that can be modified by sensory experience.

Structure and connectivity of somatic sensory areas

A somatic sensory area is one that receives direct somatic inflow, whose neurons respond to somatic stimuli, and which can be shown on other grounds to play a role in somesthesis. SI and SII surely qualify; the motor cortex does not, for its removal leaves somesthetic capacities intact. Parietal homotypical areas like the supplementary sensory area, whose neurons respond to somatic stimuli and whose removal produces complex perceptual disorders, are in another category, for their somatic inputs are from other cortical areas.

Each of the four postcentral areas has a distinctive cytoarchitecture and pattern of extrinsic connections; each receives its thalamic input from largely separate sets of VB modules. This is undoubtedly the basis for the strong but incomplete segregation of cortical columns of different modality properties between the four areas. At least such a restricted relation between thalamus and cortex is the implication of the recent micro-tracer experiments of Jones and Hand and their colleagues. Somatic I and II receive a number of other ascending projections, from the basal forebrain, from the intralaminar nuclei, and from the monaminergic nuclei of the brain stem. Little is known of the function of these confluent and less specific projections; the speculation that they regulate cortical excitability and play a role in central control states awaits experimental proof.

Corticocortical connectivity

Major anatomical discoveries have revealed a more numerous, detailed and specific connectivity of somatic cortical areas than earlier supposed. The generalizations are these:

(1) Area 3-b projects upon areas 1 and 2, but receives no projection from any other ipsilateral field.
(2) Successively more posterior parietal areas are reciprocally linked with successively more anterior frontal fields.
(3) The supplementary sensory area receives from all SI fields, projects to 4 and 6, is reciprocally linked to the supplementary

motor area.
(4) SII is reciprocally linked with areas 1 and 2 of SI, and projects upon area 4 and some posterior parietal fields.
(5) Callosal connections spare the zones of representation of distal body parts, are strictly homologous except for a projection of SI to both SI and SII of the contralateral side.

It was a wholly unpredictable discovery apparently true for all cortical areas including the somatic that the cells of origin and the reciprocal fibers of termination of an ipsilateral or contralateral connection are arranged to columns of 0.5-1.0 mm dimensions, separated by zones linked only sparsely in that particular connection. And, the sources and sinks of different connections appear to be interdigitated in a cortical area. The generalization that appears to be emerging is that the total set of modules of an otherwise unitary cortical area is fractionated into a number of subsets, and that each subset entertains a sample, not all, of the extrinsic connections of the area. Each area is thus a member of a number of distributed systems.

Functional Organization of the Postcentral Gyrus, S-I

Two discoveries by Merzenich and his colleagues initiated two major themes. The first was that the postcentral gyrus contains more than one somatotopic pattern; the second, that the details of representation can be changed by altering peripheral input. This led to a series of studies by Merzenich, Kaas, Sur, Nelson and their colleagues that have established that there are four separate representations in S-I. The representations in 3b and 1 are roughly parallel, with mirror reversals of the pattern at the border. Each is a somatotopic composite, and the map of the periphery cannot be related to the map of the cortex by any simple transformation. Although the majority of the neurons of area 2 are related to deep receptors, the proportion of cutaneous neurons, particularly in the hand area, is sufficient to establish a third representation there, with a second mirror reversal at the 1-2 border. Neurons of 3-a are almost all activated from deep afferents, many from muscle, thus no detailed map is available, but the mediolateral representation there in general parallels that of 3b, immediately adjacent posteriorly.

Place properties of postcentral neurons. Single neuron studies of receptive fields have lent support to many of the discoveries of Woolsey, especially a quantitative measure of the innervation density - cortical area rule, in terms of magnification factors defined by Sur. That for 3b is about a third larger than that for area 1. Receptive field studies of postcentral neurons are of course familiar to you all, as well as the continuous distribution of the excitatory and inhibitory zones; when the latter are larger, surround inhibition results, but in-field inhibition is always much stronger. The virtual identity of RF centers for cells arranged vertically across the cortical layers is a defining property for a cortical column; some scatter and variation in size of course occurs, especially for the deeper layers.

The Representation of Modality in the Postcentral Gyrus: Specificity and Differential Distribution. Postcentral neurons are specific for modality, like those of the DCN and VB complexes; they respond differentially to the features of peripheral stimuli that selectively activate one or another set of primary afferent fibers. That is the definition of modality in Neurophysiological research, and should not be confused with the definition that obtains in sensory Psychophysics. The two are frequently but not always congruent. The evidence for modality specificity of postcentral neurons is now beyond any reasonable doubt; yet it may depend upon dynamic mechanisms, and disturbance of those mechanisms may reveal a suppressed convergence. The evidence is most complete for neurons related to the glabrous skin of the hand, still incomplete for those related to the hairy skin.

This constancy of modality for neurons in translaminar arrays in the cortex is the second static defining property of columnar organization in the postcentral gyrus, together with that of place.

The differential distribution of modality classes in the postcentral cortex has now been observed in primates in a wide variety of experimental conditions. Area 3-a contains mainly modules of neurons activated by stimulation of deep afferents, including those from muscle. Areas 3-b and 1 contain predominantly cutaneous neurons; area 2 contains mainly deep neurons, notably those related to joint afferents, as well as muscle. It is important to emphasize that this segregation is not absolute, and that the degree of segregation varies with mediolateral position on the postcentral gyrus, and thus with body part. Detailed studies of 3-b and 1 show an intermittently recursive mapping of different classes of cutaneous elements within each local somatotopic zone.

Parallel and Serial Processing Channels in Somesthesis: Implications for a Model

Many facts support the idea that the cortical mechanisms in somesthesis consist of interdigitated parallel and serial processes, both between and within large structures of the system. A pervasive characteristic is the segregation of neurons with different signatures for place and modality into separate processing channels. The implication is that different stimulus attributes are transduced peripherally by different sets of primary afferent fibers, and those signals projected and processed in a quasi-isolated manner through the initial cortical levels of the system. Such a parallel processing model has implications for a later convergent reconstruction of a central neural transform of the total sensory input to form the neural mechanics of somatic perception.

This is simply a restatement of the old idea of "association", which leads to the assumption that this convergent reconstruction occurs in what are sometimes called the association areas of the posterior parietal lobe. I do not believe that the association model is correct or any longer heuristic, and the first decade of work on the parietal homotypical areas has provided little support for it.

Other facts of equal certainty support a serial processing model, particularly the feed-forward nature of the connections from 3b to other somatic areas. Of this we know little, and it is easy to predict that a major subject of study in waking monkeys in the years ahead will be the difference in dynamic operations in the four different postcentral areas.

Taken together, the facts available indicate that both serial and parallel processes operate, and that they in this way compose distributed processing systems, a model that awaits experimental test.

An Alternative View of the Pattern of Representation in S-1

An alternative view of the representation of the body in the postcentral gyrus is vigorously defended. Werner, and Whitsel and his colleagues, have used the method of single neuron analysis, in contrast to the multi-neuron method of Welker used by Merzenich and Kaas, to study the same set of problems. They have interpreted their results to define a single projection of the body upon the postcentral gyrus, one complex pattern covering all four of its cytoarchitectural areas. These authors have argued that the aggregate of RF's of first-order fibers combine to ordered sets in terms of dermatomal trajectories; and, that aggregates of cortical neurons similarly compose ordered sets, each related to a particular dermatomal trajectory. Consequently, the mapping of body to brain is considered by them to be a homeomorphism.

It is my own conclusion that the differences between these two major studies can largely but not completely be explained by differences in method and interpretation. I shall not detail them here, but say that the unequivocal demonstration of the mirror image reversals of the patterns at the 3b/1 and 1/2 borders, in the experiments of Merzenich and Kaas, constitutes very strong evidence for the multiple representation hypothesis. It is important to emphasize, however, that the somatotopic maps of Merzenich and Kaas are strictly limited to the entry level of the cortex, layer IV, and should not be generalized to include the results of intracortical processing. Thus it may be that Whitsel and his colleagues, using a different method under different experimental conditions, may have already seen some of the results of that processing.

Plasticity of Microconnectivity in the Somatic System

It was Wall who in 1971 first found that a peripheral nerve transection induced changes in the pattern of representation in the somatic afferent system. This was soon followed by the papers of Paul, et al., and by a number of studies in neonatal and adult animals, by Wall and his colleagues, Killackey, by Van der Loos and T.A. Woolsey and a number of others, all tending to show that a certain plasticity exists in the microconnectivity of the system. It can be changed in both young and adult animals by denervation or amputation of a part. Merzenich, Kaas et al. have recently studied the problem in the monkey. They have used the micromapping method repeatedly in the same brain, before and at intervals after nerve section, etc. Their results show that there is a remarkable variability in the detailed maps of normal and otherwise similar individuals of the same species. Further, that a series of re-arrangements of the detailed map occurs after removal of afferent input; the simplest is the occupation of the denervated region by the representation of adjacent body parts. The time course of this change is still uncertain, but it begins in a matter of hours and continues for many days and weeks. Some observations of others indicate that changes may appear immediately after section of a peripheral nerve. Recently, Merzenich and his colleagues have been able to produce map changes by changes in sensory experience, without denervation or amputation. The shifts in the maps that have been documented appear to be limited to a few hundred micrometers, about the diameter of the axonal terminal field of a single thalamocortical fiber. Many explanations have been proposed for these changes; they range from the idea that peripheral denervation simply reveals a convergence normally suppressed, to the proposition that a growth of new connections occurs. The mechanisms of these changes are completely unknown.

These findings are of great importance in the present context; namely, that the connectivity with the system and thus the properties of central neurons depend in part upon a dynamic process that can be modified by sensory experience.

DYNAMIC NEURAL OPERATIONS IN SOMESTHESIS

The term dynamic describes the temporal and spatial distributions of activity in central neural populations on a time scale from tenths to hundredths of a sec. It includes how that activity is modified in each CNS target of first-order input, subjected as it is at all levels to modification by local processing mechanisms, by the action of descending systems, and by central control states. And, how those projected neural ensembles invade cortical sensory and homotypical areas, leading in some still unimaginable way to perception, to storage, and at choice to neural commands for responsive action. It is my contention that these matters can now be studied in combined experiments, with some further developments I shall describe later.

Flutter-Vibration

Studies of flutter-vibration have provided some examples of the dynamic operation of the somatic system, especially for stimuli delivered to the glabrous skin of the primate hand. The general plan of study is to measure the sensory performance of man and monkey under identical circumstances. Observations of neural events are made in the monkey during sensory performance. When the capacities of man and monkey are identical, one can infer that neural operation like those in the monkey brain may occur in the brains of humans under similar circumstances. One can then design further tests of the inference. An important objective is to discover which aspects of sensory performance; e.g., the neural codes for detection, discrimination, pattern recognition, frequency and amplitude discriminations, etc., fit with the neural codes observed.

The sense of flutter-vibration is by its very nature an excellent somesthetic mode for the study of neural dynamics. The detection functions for man and monkey are exactly identical over the full frequency range in which they can detect oscillations, from about 5 to 4-500 Hz. They both detect frequency differences of 2-3 Hz at 40 Hz, and about 5-10% differences in stimulus amplitudes.

Studies of the postcentral gyrus in unanesthetized, immobilized and later in waking, sensing monkeys revealed classes of neurons that mimic in their dynamic as well as static properties those of the large mechanoreceptive afferents that innervate the glabrous skin of the hand. Two classes of cortical neurons mimic those of the Meissner and Pacinian afferents, which respectively serve flutter and vibration. The cortical "Meissners" are, like the linked set of first order afferents, especially sensitive to low-frequency mechanical sinusoids. There is, however, a code change. The one impulse/cycle tuned peripheral discharge is replaced at the cortical level by a form of frequency modulation imposed upon the ongoing spontaneous activity of the cortical neurons. It is an example of a sequential order code, for the periodicity is destroyed by a random shuffle of the interval sequence. These facts lead to the hypothesis that the relevant cortical signal for the identification and discrimination between mechanical sinusoids of different frequencies is the appearance of a rhythmic periodic signal. And, that there must exist an intracortical neuronal mechanism for estimating differences in period lengths.

The periodicity hypothesis was tested in a cross-species experiment. The capacity of humans to make frequency discriminations was tested in the context of decision theory, and the underlying sensory events were assumed to be the observer's estimates of the neuronal period lengths. The variability of those estimates, the d', was then shown to co-vary exactly with the variability of the periodic discharge in monkey postcentral neurons, with changes in stimulus amplitude and frequency. That co-variance strongly supports but gives no direct proof of the periodicity hypothesis.

Other examples of studies of neuronal dynamics in somesthesis could be described, but perhaps this suffices to show that the subject is of great importance for understanding central neural mechanisms in somesthesis.

Feature Extraction

There have evolved in many sensory systems mechanisms for increasing the information content and thus the efficiency of central signals of sensory events. One is feature extraction, by which is meant the selective sensitivity of a set of central neurons to a particular combination of afferent signals reaching it which depicts a stimulus attribute not specifically transduced and coded for as such in any single set of primary afferent fibers. The extraction of a neural signal for the direction of movement of a cutaneous stimulus, an attribute not transduced at the level of single afferent fibers but of course present in the population, has now been studied in the postcentral gyrus by Whitsel & Werner, Hyvarinen, and by Gardner. Directionally sensitive neurons occur in both supra- and infragranular layers but not in layer IV, are less common in 3-b than in areas 1 and 2, and in the latter are found in columnar sets for which the defining parameters are not only place and mode, but the dynamic one of best direction, as well. This directional property is produced by an active inhibitory process eccentrically imposed, thus producing null and on directional sensitivities.

WHAT MIGHT BE AN IDEAL EXPERIMENT?

It must be obvious to you all that even these early studies of cortical neuronal dynamics and feature extraction have revealed the limitation of the method of single neuron analysis, which for the 57 years since the papers of Adrian and Zotterman in 1926 has been the most productive method for the study of somesthesis. This is so because many of the dynamic properties of interest are coded in the ensemble characteristics of neuronal populations, and in the precise temporal relations between the discharges of different members of the active population. Such patterns cannot be reconstructed post hoc, after study of members one by one.

Imagine if you will an experiment in which hundreds of micro-leads have been implanted in the brain of a non-human primate, recording there the activity of literally thousands of neurons, simultaneously. Signals of neural and behavioral events are continuously telemetered to the observer's station. The animal lives in a quasi-natural environment, yet one perturbable by the investigator - in the present context, by somesthetic tasks.

The information obtained is displayed and analyzed in such a way that the observer comes for a time to live in the brain of his subject - to perceive the world as he perceives it - to react to events as he does, or sees him do in terms of his brain activity.

In such a setting we may approach the study of the dynamic activity of large populations of neurons. And, in such a setting Neurophysiology transits from little science to big science, with all the social and technical and support problems that implies.

All are difficult; all can be solved.

And, what could be more worthwhile, for the study of brain mechanisms in somesthesis is simply one way of approaching the general problem of the relation of brain to behavior, and ultimately how human brains control human behavior. That is, I believe, the perfect union of art and science, in the search for the beauty and order inherent in the function of the brain.

Sessions I and II

CORTICAL MECHANISMS OF SOMATOSENSATIONS

COMPARATIVE EVOKED POTENTIAL STUDIES ON SOMATOSENSORY CORTEX OF MAMMALS

CLINTON N. WOOLSEY

Department of Neurophysiology, Waisman Center, University of Wisconsin, Madison, WI 53706, USA

Dr. Mountcastle, Dr. Ottoson, Dr. Franzén, distinguished colleagues! It is an honor and privilege to be invited to take part in this symposium honoring the memory of Dr. Yngve Zotterman.

In the time allotted to me I propose to survey work done since 1937, chiefly with or by colleagues, on mapping somatotopic representation in the postcentral gyrus and its homologues in a variety of mammals. Studies have been made on 50 different species. While the results have been reported at various scientific meetings, several major studies have yet to be published in full.

The evoked potential method was first applied to mapping the postcentral somatosensory cortex of the macaque monkey by Marshall et al. (1937a, b, and 1941). The cortex was explored in 1 mm steps with a saline-moistened cotton thread in a metal hypodermic tube, moved over the cortex by a micromanipulator. The thread made contact with an area approximately 0.5 mm in diameter. When microelectrodes were introduced, this came to be known as a "gross" electrode. An indifferent electrode was clipped to the scalp. The stimulus consisted of a fine camel's hair brush moved electromagnetically about 1 mm against the hairs of the body, which had been clipped to a length of approximately 2.5 mm. Non-hairy surfaces were stimulated with a cat's vibrissa attached at a right angle to the moving armature of the tactile stimulator. Stimuli were applied every one to two seconds. For each cortical site examined, the cutaneous area, whose stimulation evoked responses at the cortical site, was determined. Figure 1 (from Woolsey, et al., 1942, Fig. 1) shows the nature of the surface positive responses (positive up)

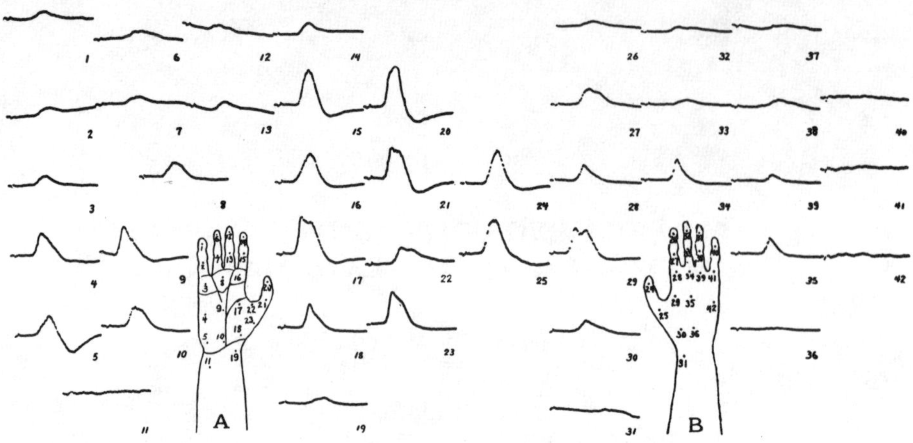

Fig. 1. EPs evoked at a single site in the hand area of M. mulatta (Woolsey, Marshall and Bard, 1942)

and the way in which they varied in amplitude according to the site stimulated under deep nembutal anesthesia sufficient to eliminate spontaneous electrical brain waves. Under lighter anesthesia, the responses at a given cortical site became more variable, since they were modified by immediately preceding spontaneous activity.

The cutaneous area for each cortical site examined was illustrated by a figurine and the figurines for all sites were then assembled as a figurine map. A composite map of the whole postcentral gyrus, including the medial wall of the hemisphere and the caudal bank of the central sulcus, was then prepared showing dorsal and ventral views of the body parts. This map was used by Dr. Bard in his Harvey Lecture (1938) and in Macleod's Physiology in Modern Medicine (1938). The final paper on this study, with figurine maps for each experiment, was published by Woolsey et al. in 1942.

In the composite map of the postcentral gyrus, we see some of the novel features of cortical sensory representation, which this study revealed for the first time (See Fig. 5). These include, after representation most medially of the tail and genitalia, (1) the postaxial leg from hip to heel and lateral aspect of the foot. Then (2) the foot and toes were represented from lateral to medial aspects of the foot, with the hallux on the dorsal surface of the hemisphere at the medial end of the central sulcus. Lateral to the hallux, (3) centers for the preaxial leg from medial foot to anterior hip and trunk were

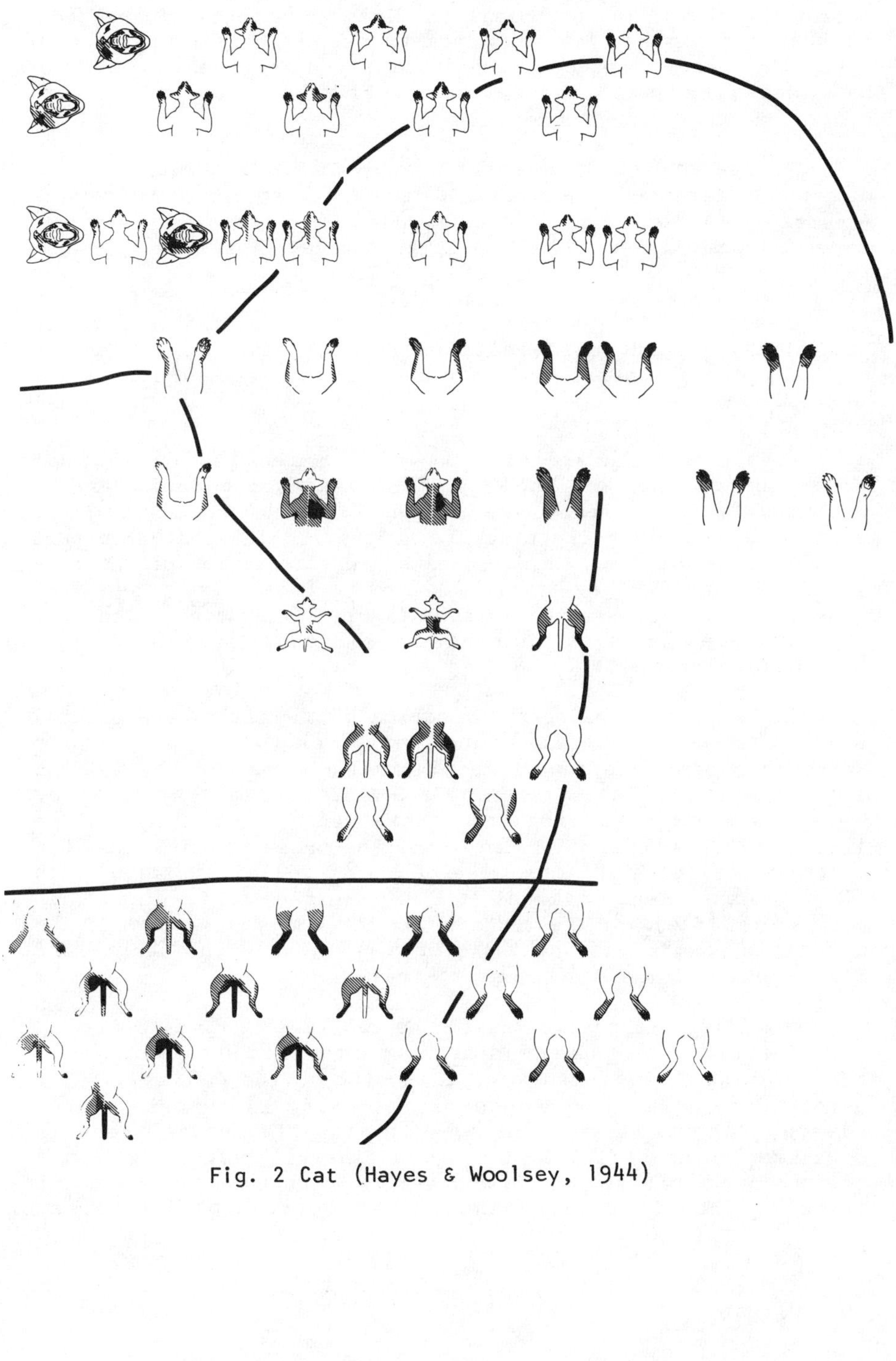

Fig. 2 Cat (Hayes & Woolsey, 1944)

found. Another unexpected finding was (4) the presence of representation of the occipital and lateral aspects of the head adjacent to centers for the upper trunk. From this upper head area, the sensory sequence continued for upper arm, lower arm, hand and fingers to the face area at the lower end of the central sulcus.

Our second evoked potential study was on the cat, carried out with George J. Hayes and reported to the American Physiological Society in 1944. A full report of this study has never been published, only a general diagram (Woolsey, 1952 and 1958).

In 1940 and 1941, Adrian reported results on the cat, which included a second representation for the paws. This was to become known as the second somatic sensory area (Woolsey and Fairman, 1946).

Figure 2 illustrates some of the results on the cat obtained by Hayes and Woolsey. The area mapped includes the medial wall of the hemisphere (at the bottom of the figure), the sigmoid gyrus behind the cruciate sulcus and the posterior part of the coronal gyrus. This shows the representation of the body from tail through leg, trunk and arm to face. Two aspects of body parts are represented in each figurine. Studies on the cat were extended later in two other unpublished studies, which I shall illustrate shortly.

In 1945, the somatosensory cortex of the rabbit was studied with G. H. Wang (Woolsey and Wang, 1945). A previously unpublished figurine map of this animal's areas S I and S II is shown in Fig. 3. The face area of S I is nearly four times as large as the arm and leg areas combined. The map of S II is incomplete, as is also the area for the trunk in S I. The plan of organization of the face area of S I is well illustrated with the top of the head adjacent to S II, the side of face and chin more medially toward the centers for the arm and leg of S I, the upper lip rostral to the top of the head, with the tongue and the interior of the mouth most rostrally.

Attention was next given to the rat in a study made with D. H. LeMessurier of Australia and reported in Fed. Proc., 1948. Figure 4 shows the results of this study (from Woolsey, 1952; Bard, 1956). The importance of the study is that it showed clearly for the first time a basic plan of cortical sensory localization as a distorted image of the body surface with its various parts related to one another very much as in the actual animal. The figure also shows the location of the second

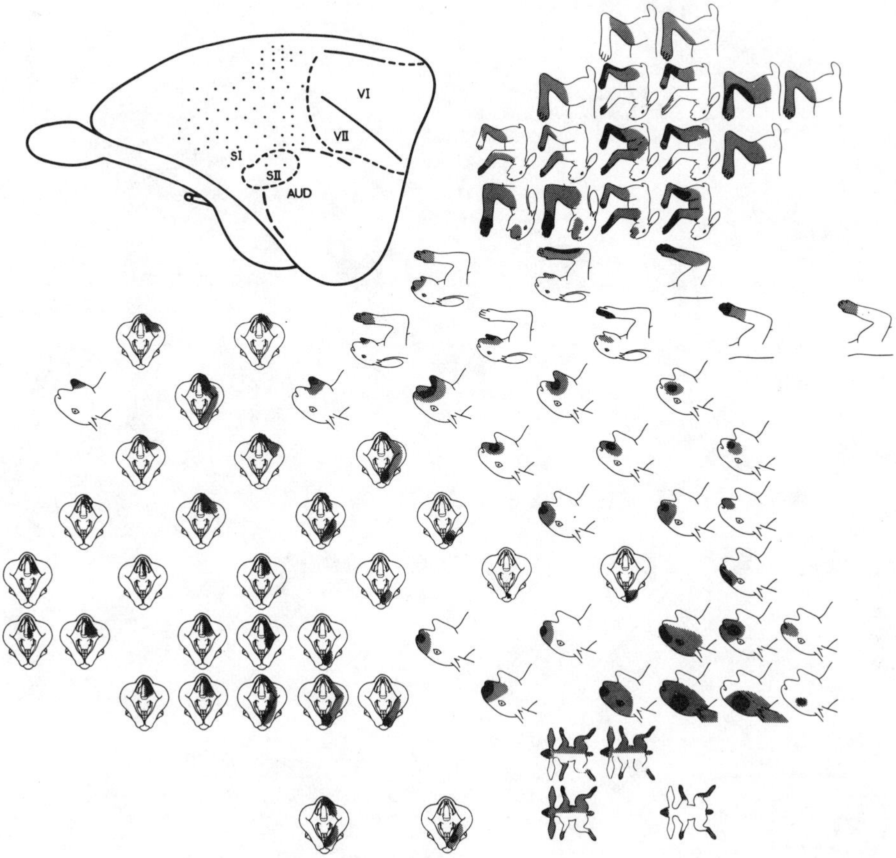

Fig. 3 Rabbit (Woolsey and Wang, 1945)

somatic sensory area (S II) in the rat.

These studies on the rabbit and rat led us to redraw the figurine map of the somatosensory cortex of the monkey. The results are shown in Fig. 5, which is reproduced from Fig. 24 of Woolsey (1958).

Further studies on the cat were made at Wisconsin by Woolsey, Cranston and Luethy in 1956 and by Dr. Ronald Tasker in 1960. Figure 6 shows the results of mapping the face area of S I on the coronal gyrus and the second somatic area (S II) of the anterior ectosylvian gyrus. Note that the face in S I is

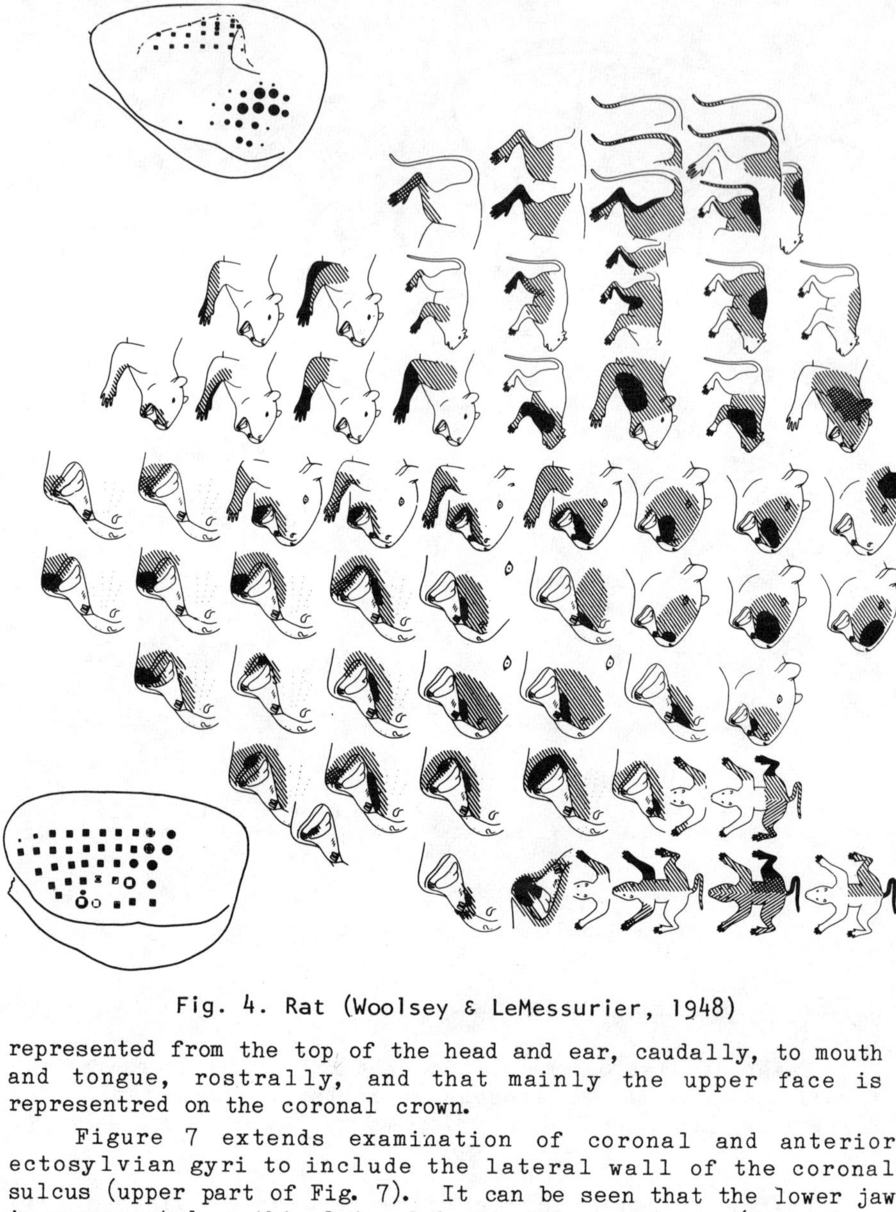

Fig. 4. Rat (Woolsey & LeMessurier, 1948)

represented from the top of the head and ear, caudally, to mouth and tongue, rostrally, and that mainly the upper face is representred on the coronal crown.

Figure 7 extends examination of coronal and anterior ectosylvian gyri to include the lateral wall of the coronal sulcus (upper part of Fig. 7). It can be seen that the lower jaw is represented on this lateral bank. These two maps (Figs. 6 and 7) were prepared by Tasker from protocols of experiments of Woolsey, Cranston and Luethy (1956). Comparison of the face area of the cat with those for rabbit and rat shows a similar plan of

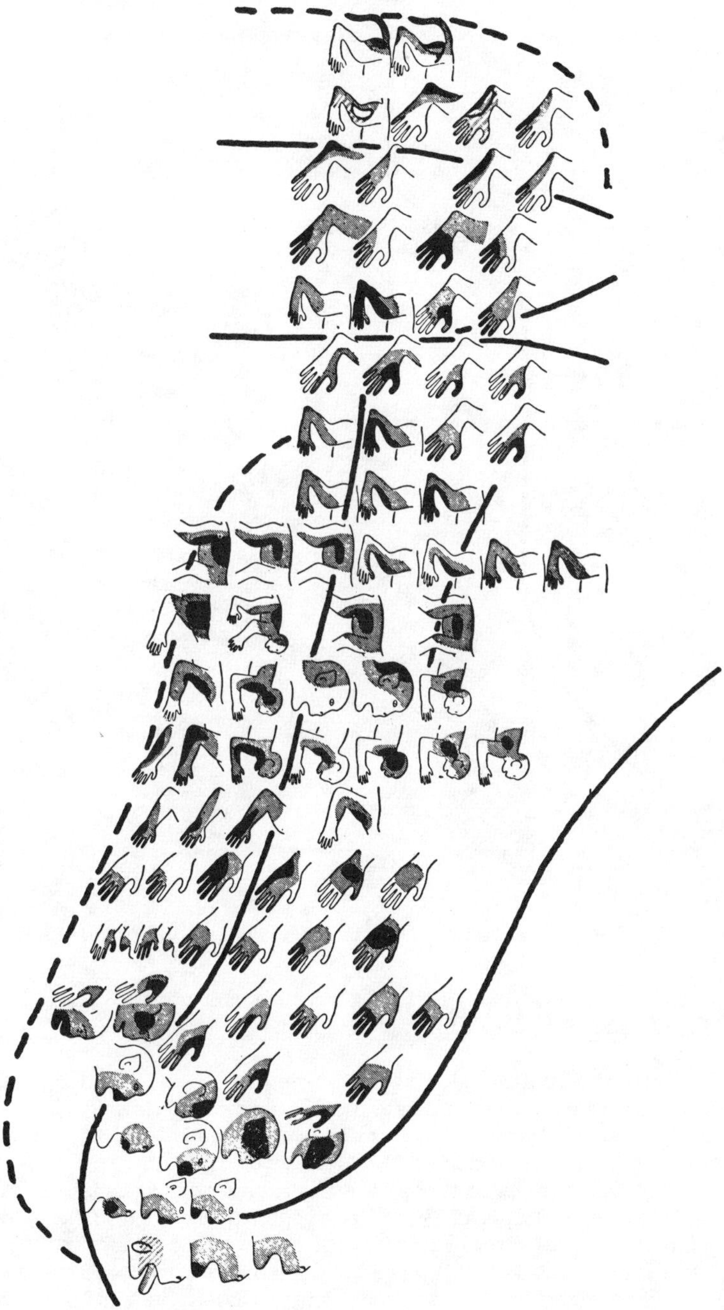

Fig. 5 Macaca mulatta (Woolsey, 1958)

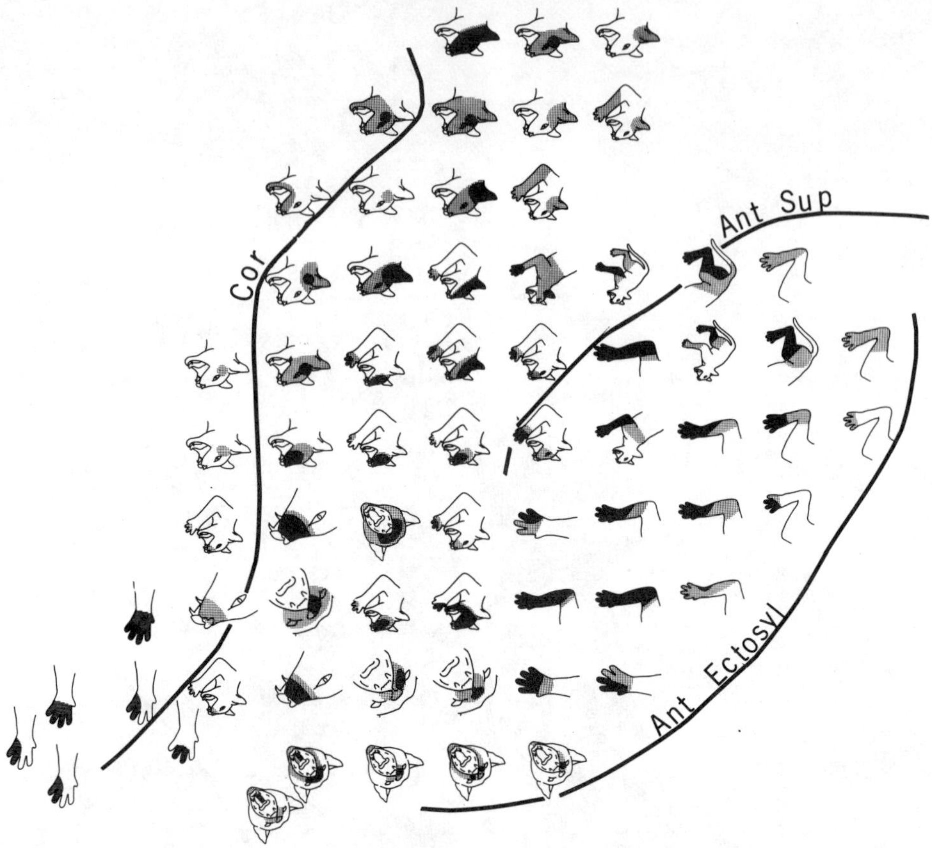

Fig. 6 Cat (Woolsey, Cranston, Luethy, 1956)

organization for each. This is also true of the dog (Pinto Hamuy et al., 1956; Fig. 9) and of the monkey (Fig. 5).

Figure 8 (A and B) is a composite unpublished map of SI and S II of the cat prepared by Dr. Ronald Tasker (1960). This map shows an extension of tactile cortex from the cruciate sulcus to well behind the ansate sulcus. Particularly noteworthy is the existence of two representations of the trunk in Fig. 8A, one laterally on the posterior sigmoid gyrus and a second on the medial wall. Tasker concluded that the trunk area of the medial wall was a part of a third somatosensory area (S III), which continues across the cortex behind S I to the anterior portion of the suprasylvian gyrus, where the head is represented, in an area later studied by Darian-Smith et al, (1966), and earlier reported by Marshall (1949). See also Tanji et al, (1978), and Landgren and Olsson (1980).

I believe the medial trunk area belongs to the cat's homologue of the supplementary sensory area, as suggested by Tasker (1960), which was first described for man by Penfield and Rasmussen (1950) and experimentally demonstrated in the squirrel monkey by Blomquist and Ambrogi-Lorenzini (1965). Data from this area in M. mulatta were reported in 1981 by Murray and Coulter. See also Fig. 23 of Woolsey et al (1979). Whether all parts of Tasker's area III belong to a single area needs to be further studied, since the arm area is disproportionately large

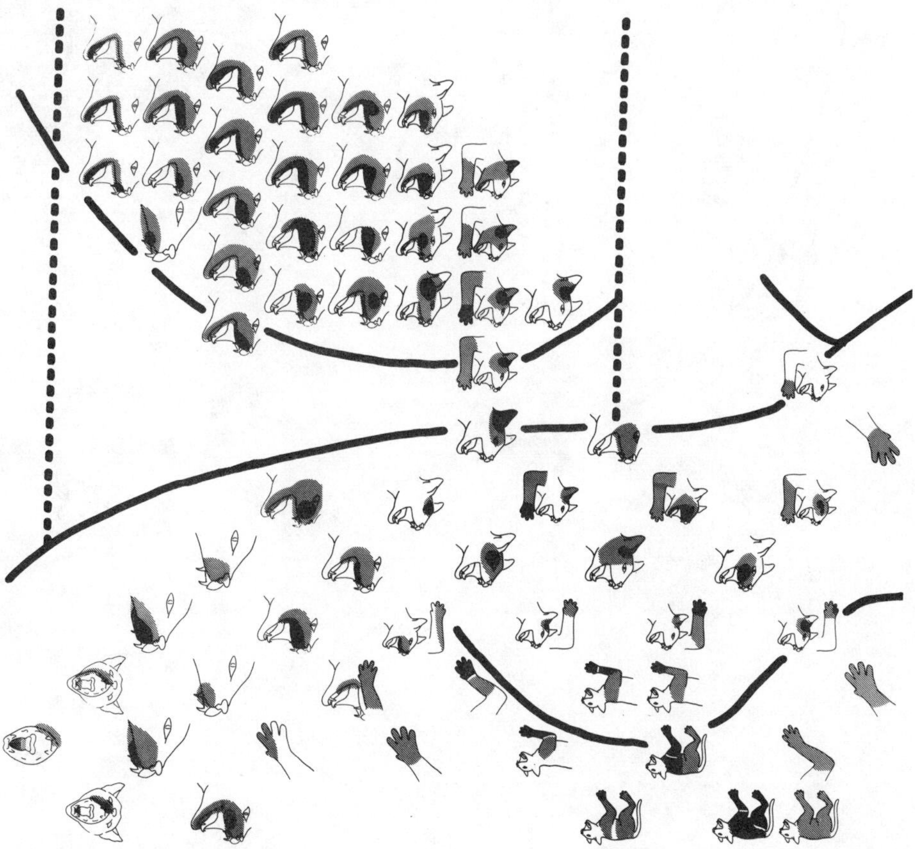

Fig. 7 Cat (Woolsey, Cranston, Luethy, 1956)

Fig. 8A. Cat. Composite map of medial somatosensory cortex - S I and S III

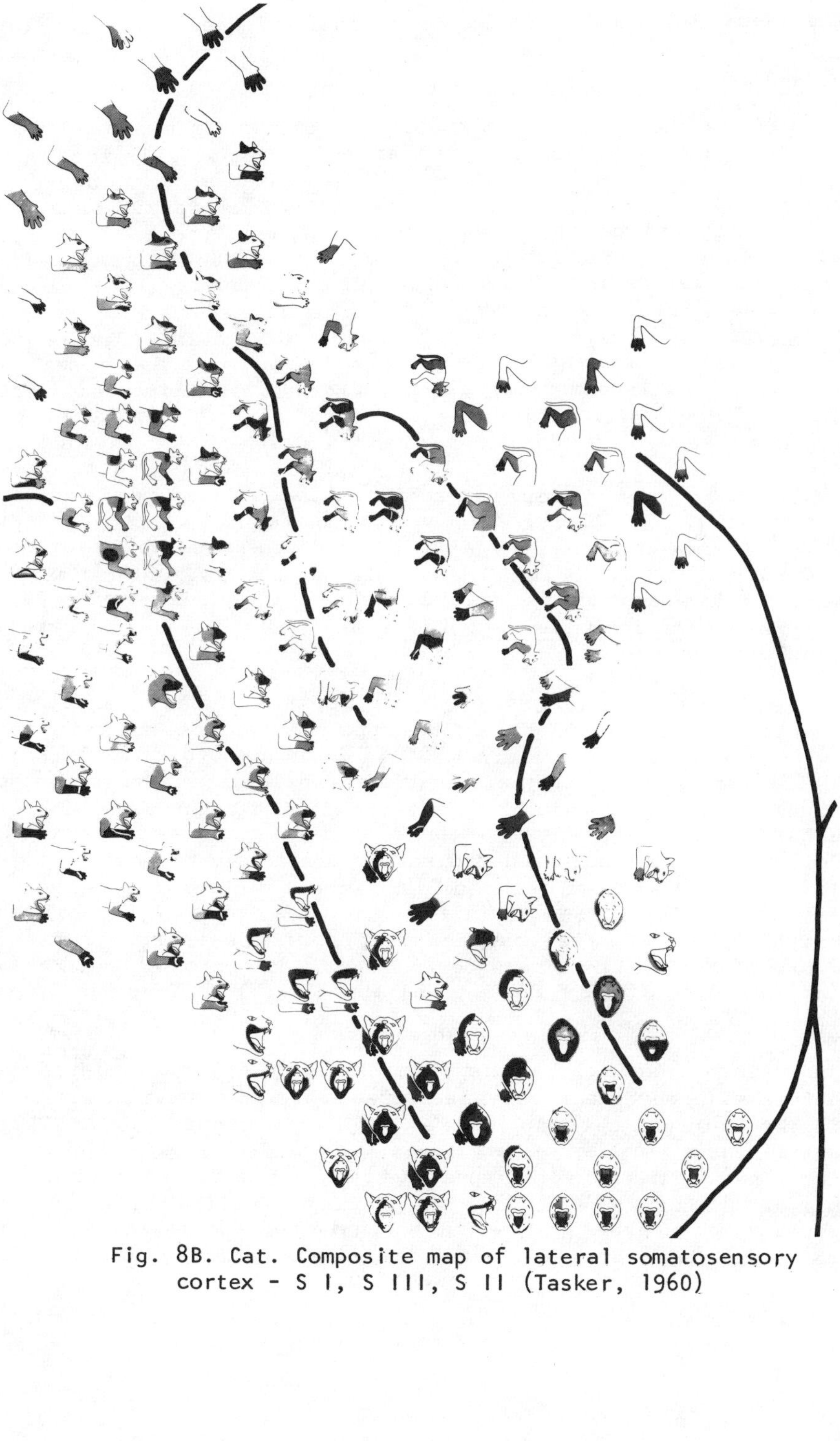

Fig. 8B. Cat. Composite map of lateral somatosensory cortex - S I, S III, S II (Tasker, 1960)

and there is no clear cut leg area distinguishable from the leg area of S I, unless it is situated within the cruciate sulcus medially.

Figure 8B shows a continuation of Tasker's map of the cat's somatic sensory areas to their lateral aspect. This includes the arm area of the lateral part of the posterior sigmoid gyrus, and details of representation in the anterior suprasylvian gyrus with continuation onto the coronal gyrus. The face is not completely represented in the anterior suprasylvian gyrus of this map, unless the area extends into the suprasylvian sulcus (But see Darian-Smith et al, 1966). Also shown is a fairly complete map of S II on the anterior ectosylvian gyrus. Note the position of centers for the trunk along the anterior ectosylvian sulcus and for distal fore and hind limbs adjacent to S I face area. This map of S II fits the pattern of organization for S II, first clearly defined for the dog by Pinto Hamuy et al. (1956) (Fig. 9) and by Lende and Woolsey (1956) for the porcupine (Fig. 10). This pattern of organization of S II for the cat has been questioned by Haight (1972) and by Burton et al (1982). It parallels, however, the pattern defined for S II of the squirrel monkey by Benjamin and Welker (1957) and for the macaque monkey by Whitsel et al (1969) and for M. fascicularis by Robinson and Burton (1980) and Burton and Robinson (1981).

Recently Clemo and Stein (1982) have described a new somatosensory area (S IV) on the rostral bank of the anterior ectosylvian sulcus in the cat. This is an area to which Celesia (1963) demonstrated projection of dorsal roots innervating the trunk in the cat, which suggested that the trunk area of S II extended onto the anterior bank of the ectosylvian sulcus. Celesia also demonstrated projection of these same dorsal roots to the trunk area of S I, but he did not record responses to stimulation of these roots from the upper part of S II, where Haight (1972) located representation of the trunk. Further detailed studies will need to be done to determine conclusively the layout of S II in the cat and the possible existence of an area S IV.

Another animal, whose somatosensory area was mapped in detail with surface steel electrodes 0.3 to 0.5 mm in diameter, is the American raccoon (Welker and Seidenstein, 1959). This animal has a very large hand area, reflecting its skill in using the hands. Dr. Welker has provided me with a detailed figurine map of this animal's S I cortex, which was not included in the paper cited above. Figure 11A illustrates the medial half of this area beginning with the tail and postaxial leg on the

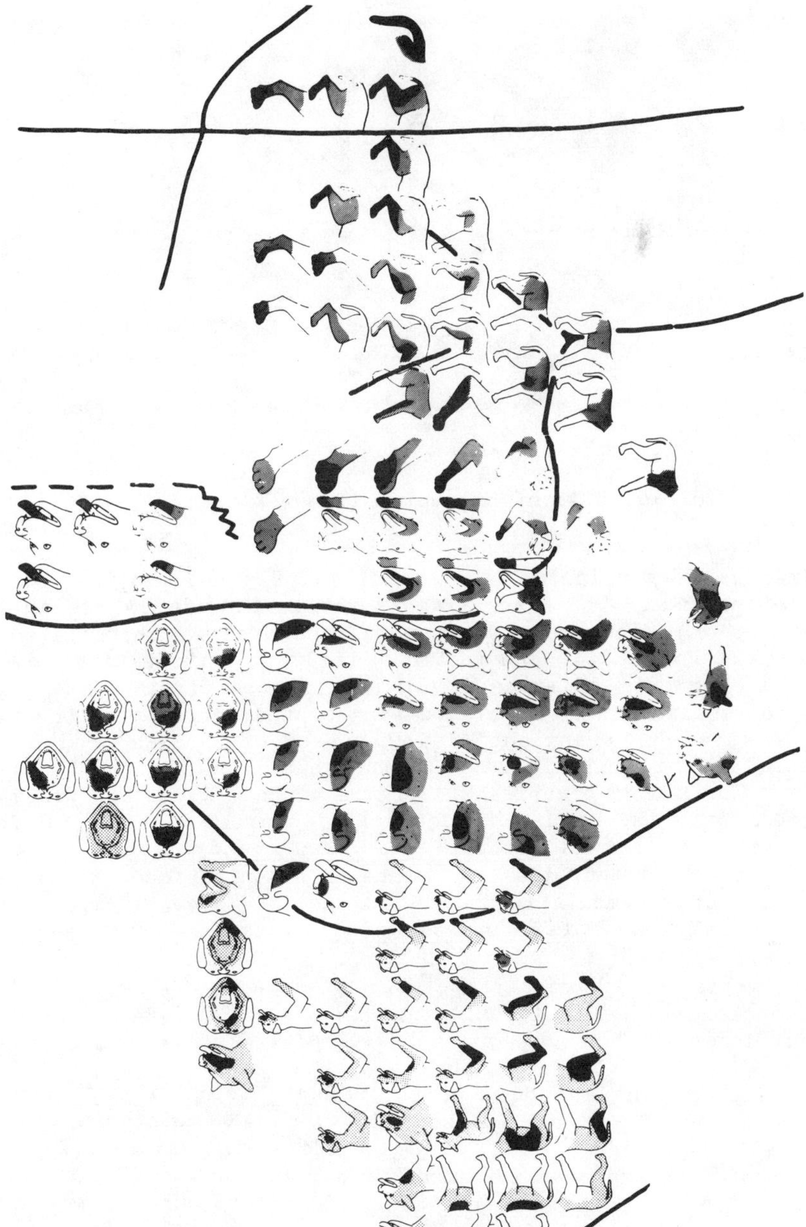

Fig. 9. Dog (Pinto Hamuy, Bromiley and Woolsey, 1956)

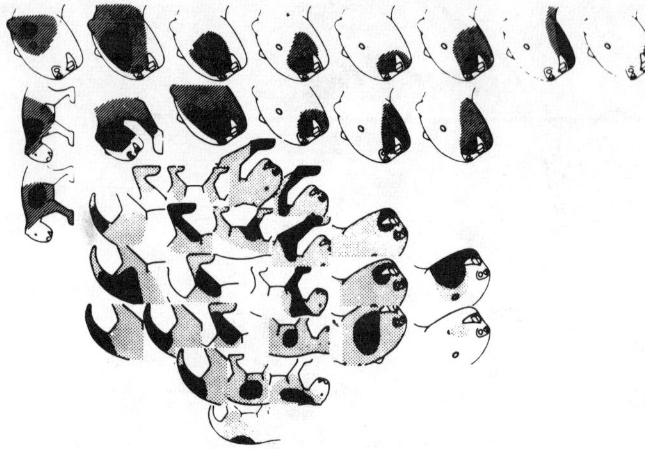

Fig. 10. S II of porcupine (Lende & Woolsey, 1956)

medial wall, continuing to the preaxial leg and the foot on the dorsal surface. The trunk area is situated caudally, as in the dog (Fig. 9). The greatly expanded hand area is illustrated to the right in Fig. 11A and in the medial two-thirds of Fig. 11B. Note the extents of areas devoted to single digits. The upper arm representation is separated by the hand representation, with the rostral shoulder and radial aspect of the arm projecting laterally and caudally next to the caudal portion of the face area. The latter occupies the entire coronal gyrus and probably extends onto the walls of adjacent fissures. Again, the face area is organized as in other animals already described. Note the ipsilateral and bilateral representation of face, tongue and interior of the mouth rostrally in the face area. Compare this with ipsilateral face representation in the pig (Woolsey and Fairman, 1946).

Papers on other subprimates, which have been studied with the evoked potential techniques, are listed at the end of this paper under ADDENDA and by Johnson et al., (1982).

The original studies on the monkey (Woolsey et al, 1942) and the cat (Hayes and Woolsey, 1944) used a moistened cotton-thread electrode. All other studies in which a "gross" electrode was employed were made with insulated metal electrodes 0.2 to 0.5 mm in diameter. Dr. Welker and associates have compared receptive fields defined by using microelectrodes with those obtained with a "gross" electrode 0.2 mm in diameter (Krishnamurti et al., 1976). These relationships are shown in Fig. 12 (Fig. 5 of Krishamurti et al.). The larger electrode essentially summates several of the smaller receptive fields defined with the microelectrode.

In several primates in addition to Macaca mulatta, the somatosensory areas of the cerebral cortex have been mapped. These include the tree shrew -Tupaia glis (Lende, 1970; Sur et al., 1980b, 1981a, 1981b),the marmoset-Hapale jacchus (Woolsey, 1952, 1954), the slow loris - Nycticebus coucang concang (Krishnamurti et al., 1976), the potto - Perodicticus potto edwarsi (Boisacq-Schepens et al ., 1977), the lesser bush baby - Galago senegalensis (Kanagasuntheram, 1963; Jameson et al, 1963; Sur et al., (1980, 1980a), Galago crassicaudatus (Carlson and Welt, 1980, 1981; Sur et al, 1980a), the squirrel monkey - Saimiri sciureus (Benjamin and Welker, 1957; Blomquist and Ambrogi-Lorenzini, 1965, Sur et al, 1982), the owl monkey - Aotus trivirgatus (Merzenich et al., 1978, 1981), the cebus monkey - Cebus capucinus (Hirsch and Coxe, 1958), the cynomolgus monkey - M. fascicularis (Robinson and Burton, 1980; Burton and Robinson, 1981; Friedman, 1981), the spider monkey - Ateles ater (Chang et al., 1947; Pubols and Pubols, 1971), the gibbon - Hylobates lar (Woolsey et al., 1960; Welt, 1962), the chimpanzee - Pan troglodytes (Woolsey et al., 1960; Welt, 1962), and man (Penfield and Rasmussen, 1950; Penfield and Jasper, 1954; Woolsey et al., 1979).

Figure 13 is a figurine map of the sensory areas of the marmoset. S II is at the lower right. The remainder of the map illustrates the organization of S I. Tail, genitalia, posterior hip and lower back are represented on the medial wall of the hemisphere. Postaxial leg and preaxial leg tend to be separated rostrally by the area for the foot and toes. The back is not well represented, but the dorsal neck and occipital aspect of the head fall between the leg and the arm areas. The face area is organized very much as in cat, dog and raccoon, with the top of the head projecting caudally; central face, lower lip and chin are next to the thumb area; the upper lip is adjacent to S II laterally, while the tongue is surrounded by the lips. The most rostral lateral points are related to the ipsilateral aspect of the tongue and face. Face, arm and leg are nearly equally represented at 21, 24 and 30 sites respectively. (This map has not been published).

Figure 14, reproduced from Kanagasuntheram (1963), illustrates somatosensory areas I and II of Galago senegalensis senegalensis. S II is situated on the upper bank of the Sylvian fissure. (In the marmoset S II was on the exposed surface of the cortex). In Galago the map of the face area is less clear and detailed than in the marmoset. The arm area is somewhat

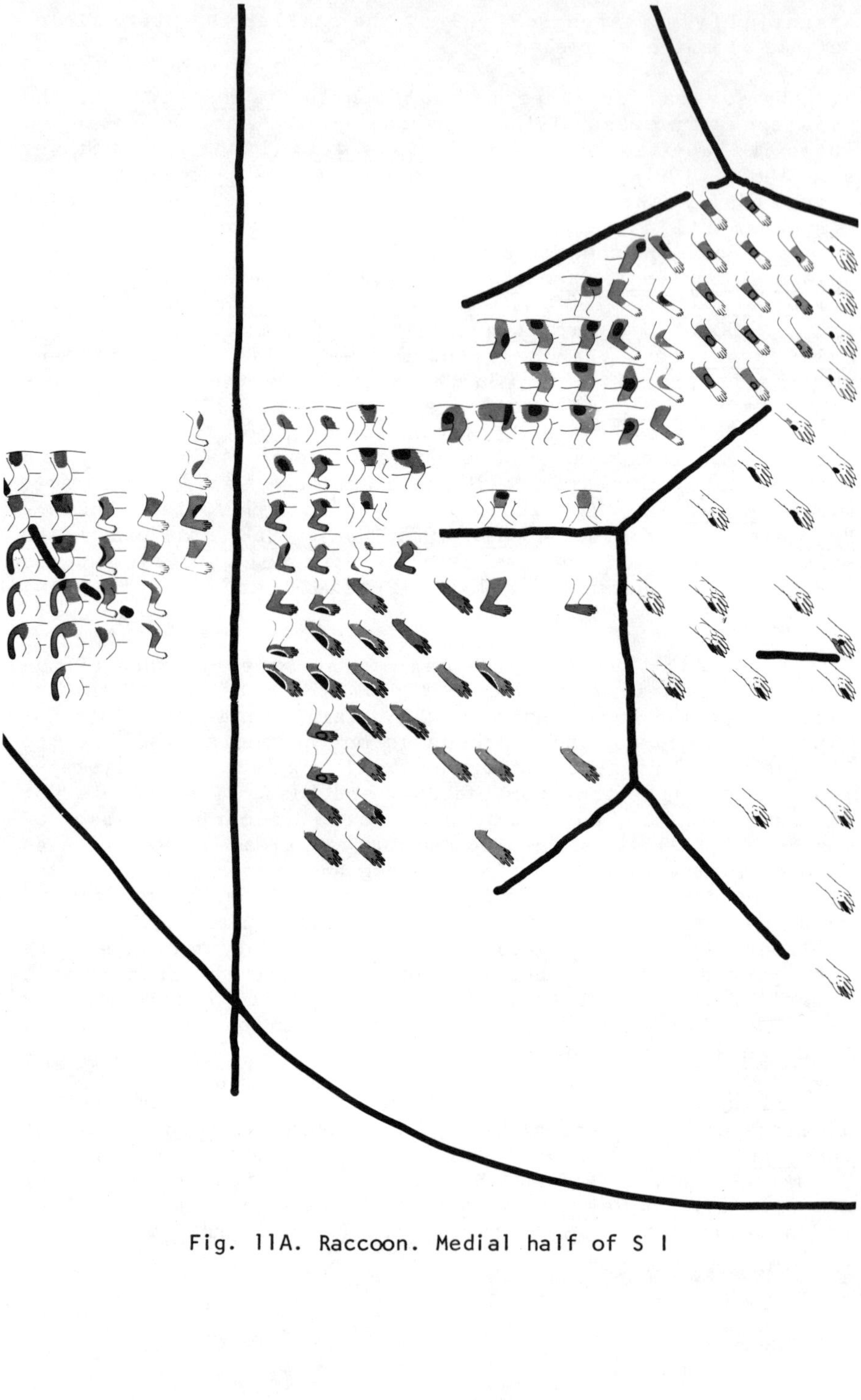

Fig. 11A. Raccoon. Medial half of S I

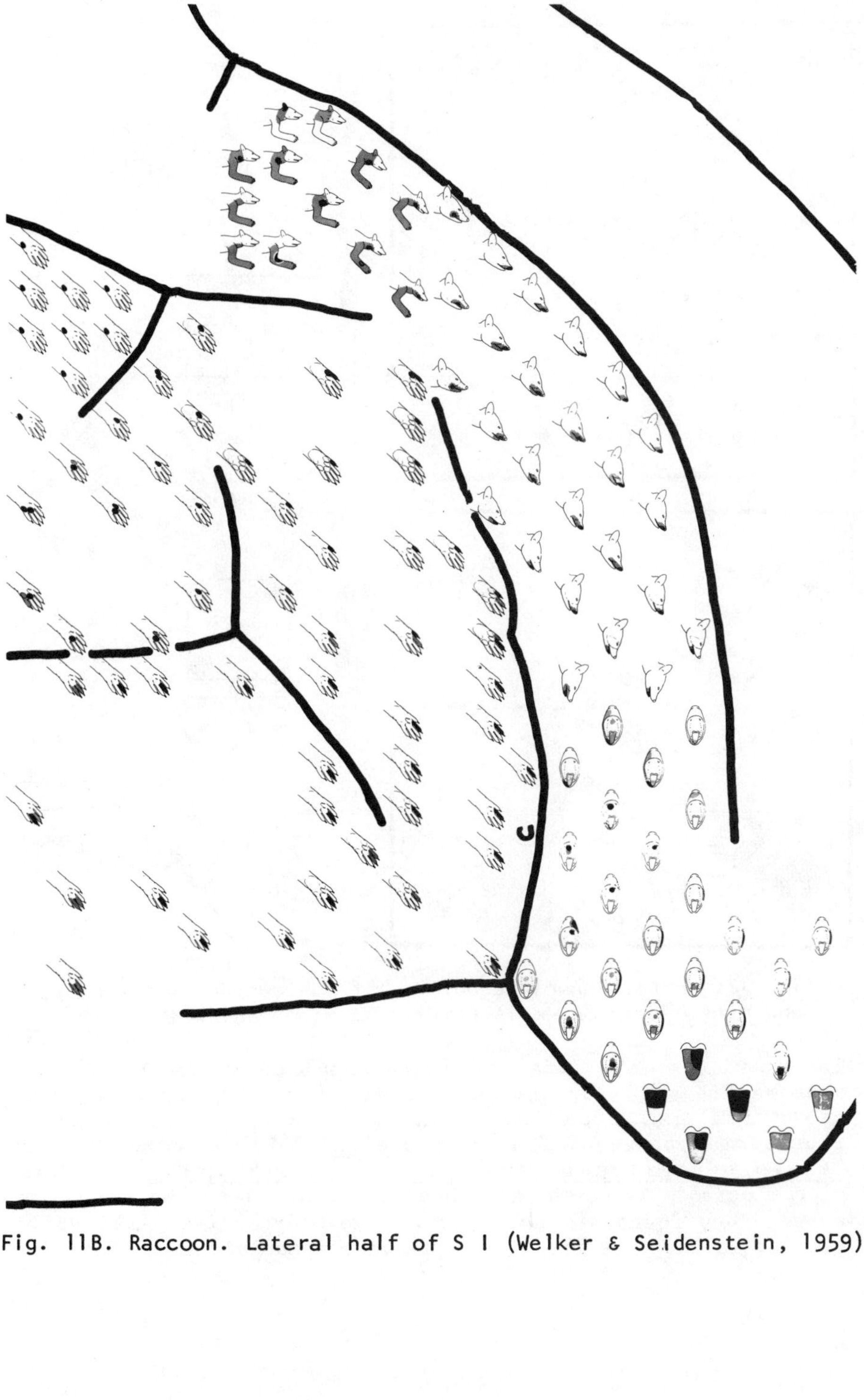

Fig. 11B. Raccoon. Lateral half of S I (Welker & Seidenstein, 1959)

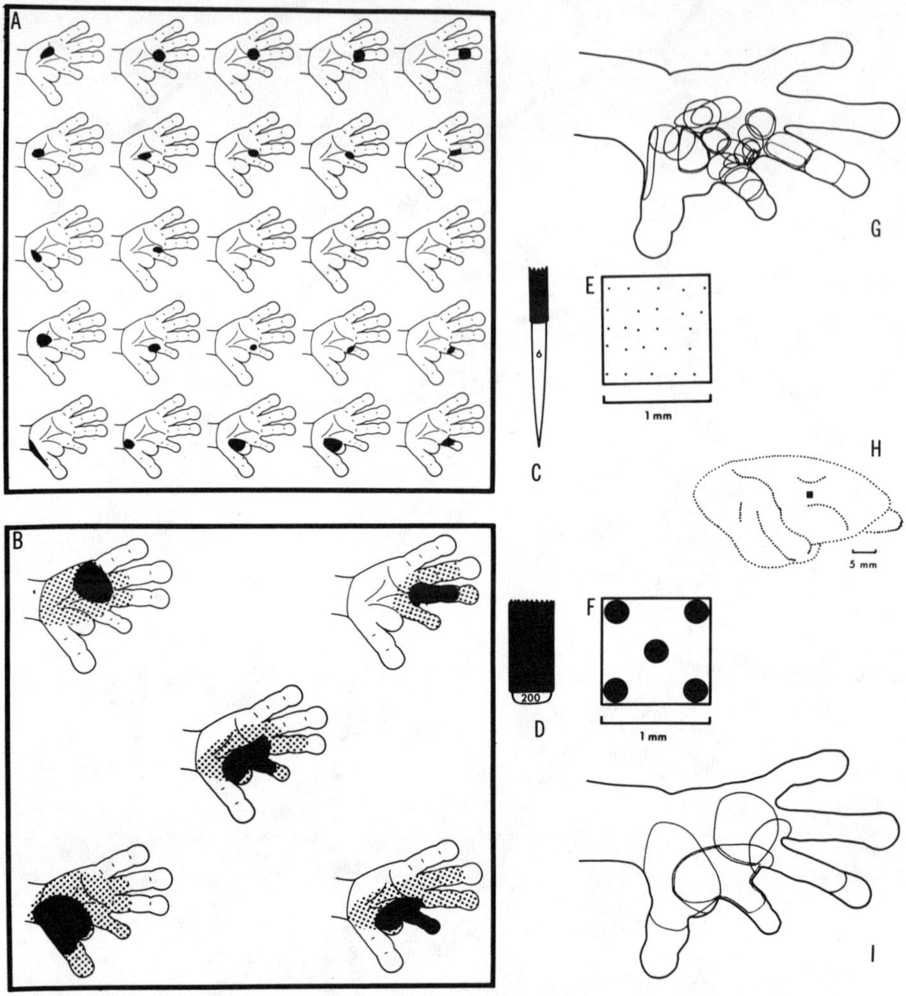

Fig. 12. Comparison of receptive fields recorded with "gross" and microelectrode (Krishnamurti, Sanides and Welker, 1976)

larger with a more detailed representation of the hand. The same may be said for the foot, which more sharply separates postaxial and preaxial leg representations. No back representation was defined. Sur _et al_. (1980) mapped S I in _Galago senegalensis_ and _Galago crassicaudatus_ with the microelectrode technique and located a trunk area less than 1 mm wide. They found a single map of the body surface basically similar to that described for non-primate mammals.

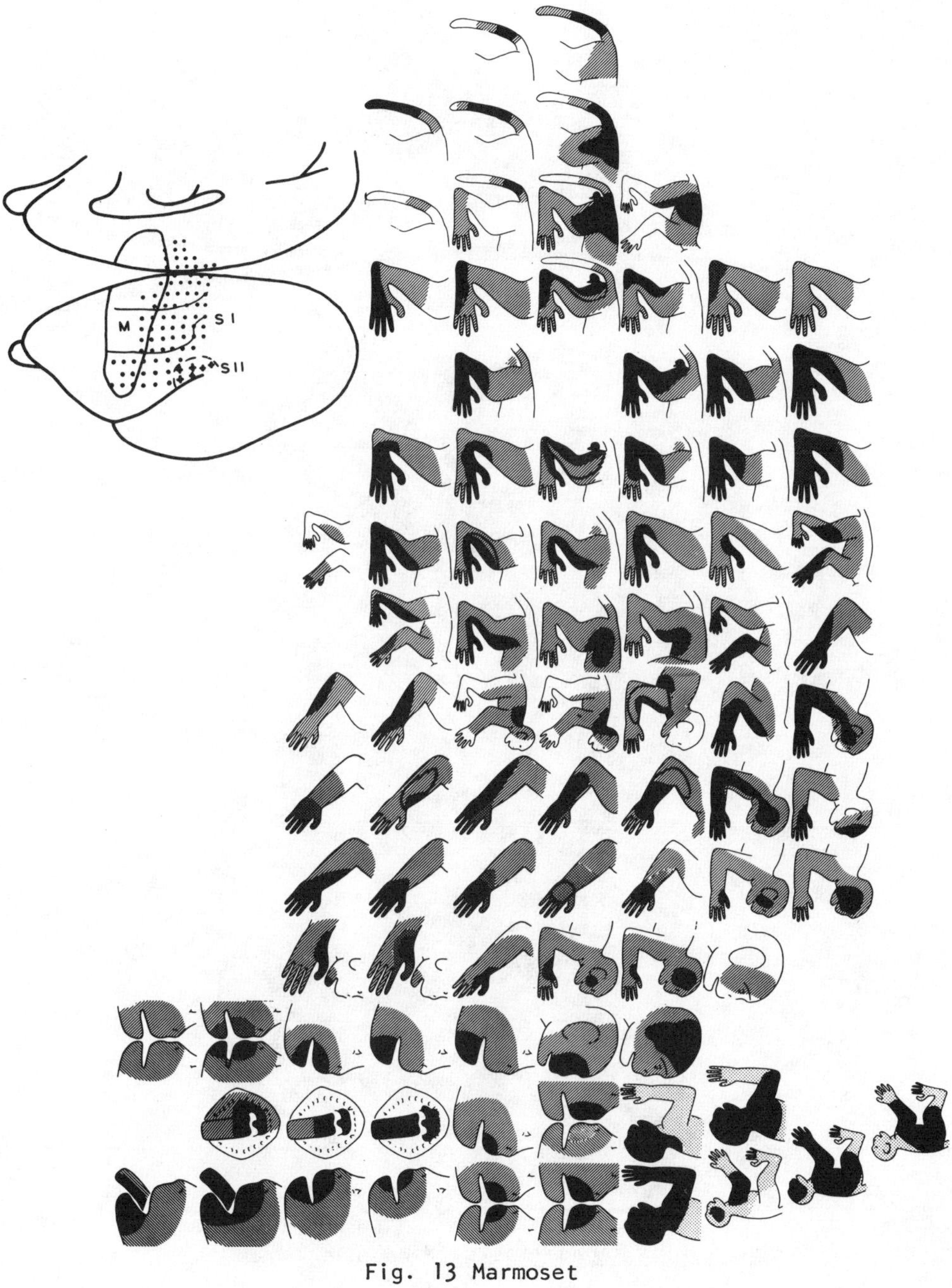

Fig. 13 Marmoset

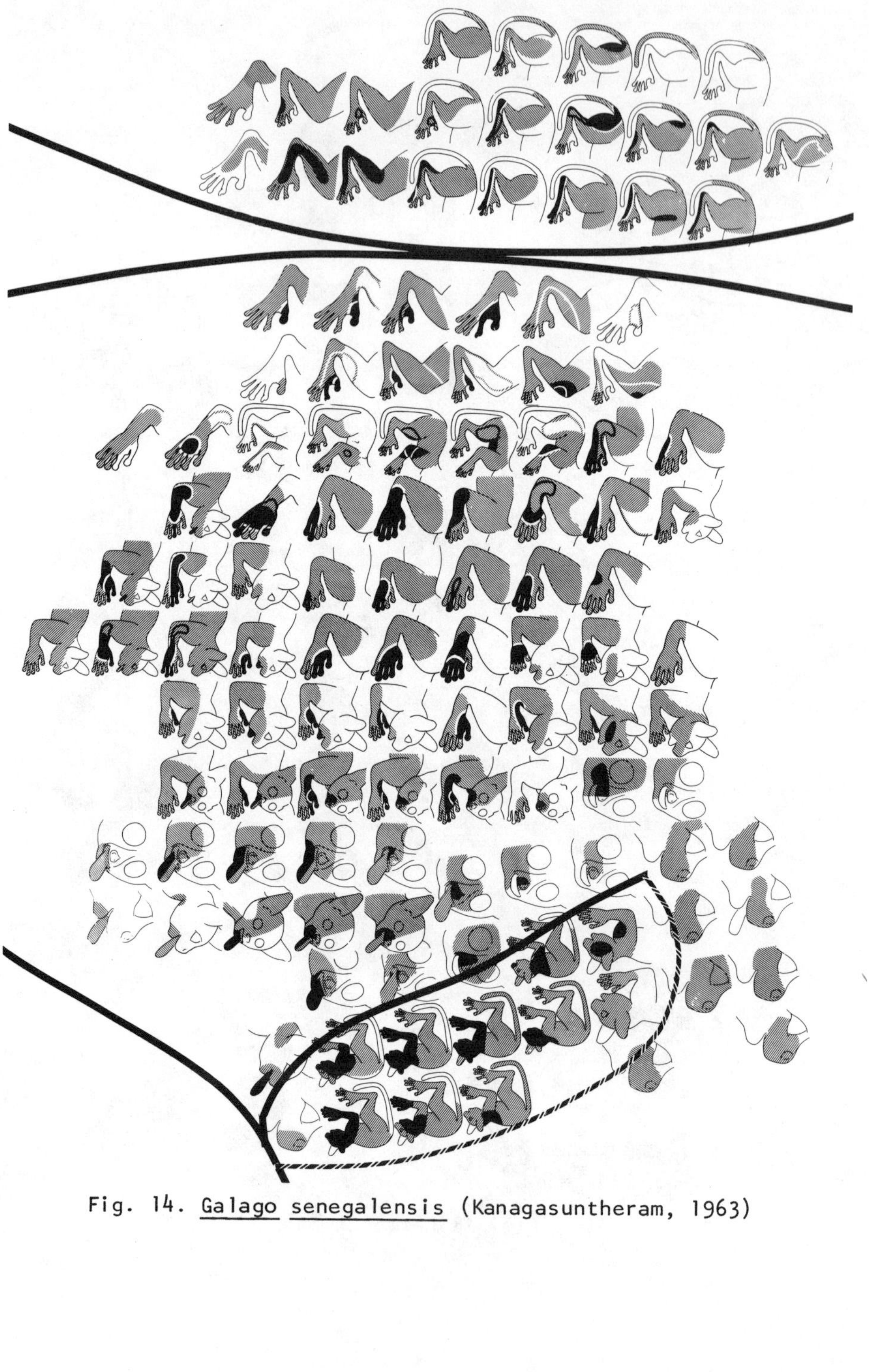

Fig. 14. Galago senegalensis (Kanagasuntheram, 1963)

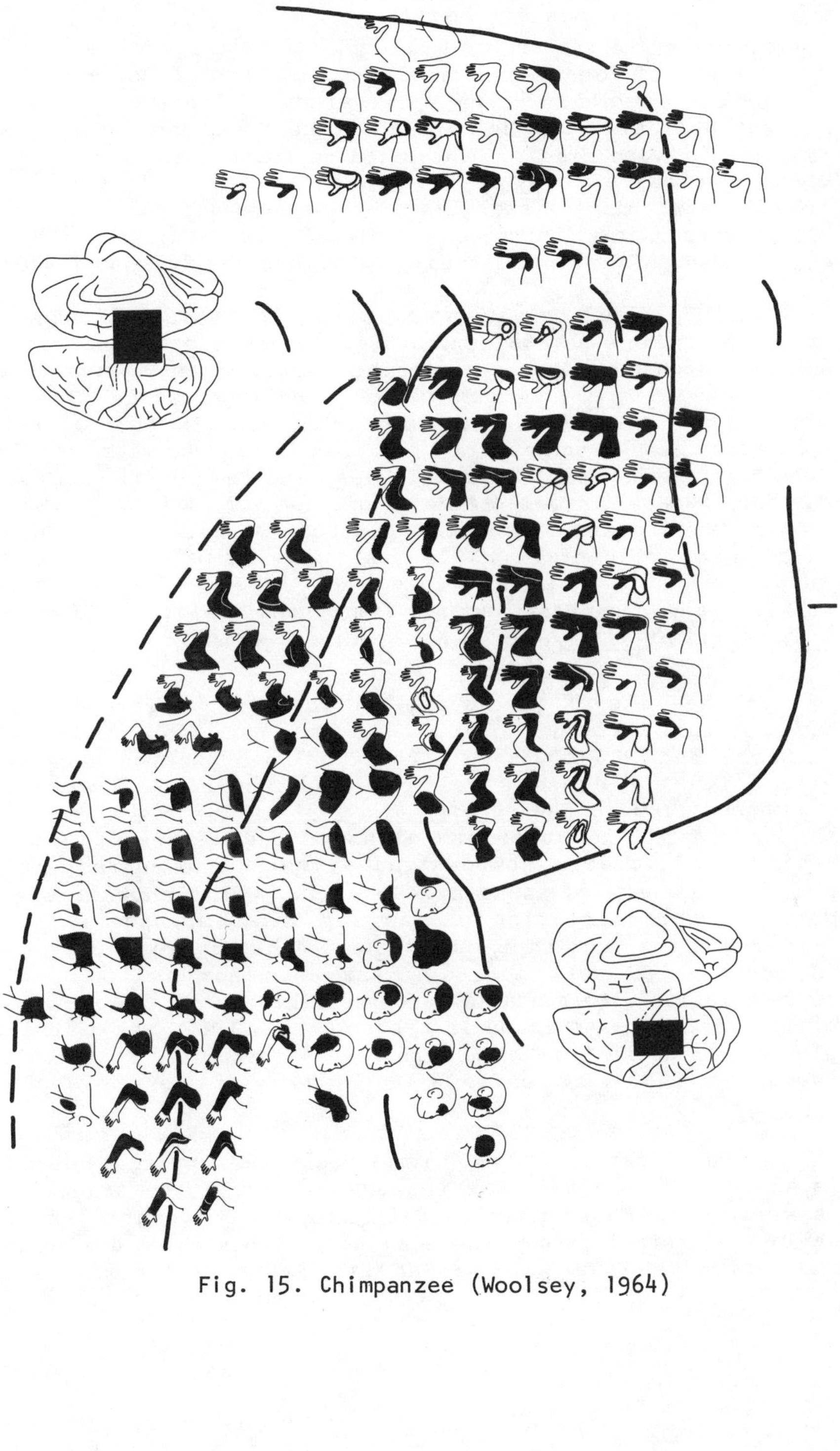

Fig. 15. Chimpanzee (Woolsey, 1964)

In the two studies made of the spider monkey (Chang *et al.*, 1947; Pubols and Pubols, 1971), the tail area was found to be greatly enlarged, extending well out onto the dorsal surface of the hemisphere (See Fig. 7 of Hirsch and Coxe, 1958, and Fig. 1 of Pubols and Pubols, 1971). The study of Chang *et al.* (1947) showed that parts of the hand and face areas had been rolled into the Sylvian fissure, and extended rostral to the lower end of the central sulcus, as the postcentral gyrus and parietal lobe had been pushed lateralward by the expanded tail area, with elimination of the intraparetal sulcus. The Pubols did not find that hand and face had shifted partly into the Sylvian fissure.

In 1972, Paul *et al.*, using microelectrodes, defined for the first time separate representations of slowly and rapidly adapting mechanoreceptors of the hand in Brodmann's areas 3 and 1 of *Macaca mulatta*. Earlier Powell and Mountcastle (1959) had reported that area 3 receives input principally from cutaneous receptors, area 2 principally from deep-lying receptors and that area 1 receives input from both deep and cutaneous receptors. Further, recent microelectrode studies on some primates have led to the description of somatotopic body maps for the separate cytoarchitectonic areas of the postcentral somatosensory cortex in the owl monkey (Merzenich *et al.*, 1978, 1981) and in owl monkey, cebus, squirrel monkey and cynomolgus monkey (Kaas, *et al.*, 1981; Sur *et al.*, 1982; Kaas, this meeting).

The results of our studies on the chimpanzee and gibbon have not been published in full, but maps for both were included in Carol Welt's Ph.D. thesis (1962). Both studies were made with a 0.5 stainless steel pickup electrode. Figure 15, reproduced from Woolsey (1964, Fig. 3.3), illustrates the leg, the trunk and the upper arm areas of the chimpanzee on the medial wall, medial postcentral gyrus andthe corresponding postcentral bank of the central sulcus. This shows that the cortical representation of the leg in chimpanzee is quite different from that in *M. mulatta* and other monkeys. The foot is represented most medially and does not split the leg representation into postaxial and preaxial areas. The same plan of organization was found in gibbon (Welt, 1962) and a similar plan of representation in man has been described by Penfield and Rasmussen (1950), by Penfield and Jasper (1954) and by Woolsey *et al.*, (1957).

Another feature of the chimpanzee's postcentral map, which is shown by Fig. 15, is the representation of the trunk. The back is represented most caudally on the postcentral gyrus, while the ventral aspect of the trunk is represented deep within the central sulcus. The map shows a systematic shift in the surface representation of the trunk from the back to the belly. This region of the cortex includes areas 1 and 3b of Brodmann. It appears unlikely that this region of the cortex will be found to contain two representations of this part of the body surface in this animal.

It has been suggested (Woolsey et al., 1960) that the arrangement of the leg area in tailless primates may be related to the absence of a tail area in the cortex, which permits enlargement of the cortical leg area toward the cingulate sulcus unhindered by an already established tail area.

Comparative studies on many different species of mammals have revealed a basic pattern of organization of the "postcentral" sensory cortex from platypus to prosimian primates. Of particular interest is the expansion of various parts of this area according to the development of sensory functions for the parts. Thus, the face area is enlarged in all animals with well developed receptive fields for the face. The basic plan of the face area remains fundamentally the same throughout the animal series. The pig (Woolsey and Fairman, 1946) has a greatly enlarged snout area. The raccoon has a greatly enlarged hand area, associated with the development of the use of the hand. In the primates all areas of the postcentral gyrus have enlarged, as these animals have developed greater use of all parts of the body. Particularly striking is the greatly enlarged tail area of the spider monkey.

Recent studies with microelectrodes have demonstrated that each of Brodmann's areas of the postcentral gyrus of some of the primates has a separate representation for body surface or deep structures. In this, these animals differ from lower forms, which, as yet, have been found to have only one body representation. Further detailed studies may be able toreveal the phylogenetic development of these multiple representations.

ACKNOWLEDGMENTS

I wish to express my appreciation to Shirley Hunsaker and Terrill Stewart for their photographic work and to Evadine Olson for preparation of the typed copy with a Victor 9000 Word Processor and a Diablo 630 Printer.

I also wish to thank the related publishers for permission to reproduce the following figures:

Fig. 1. The Johns Hopkins University Press, Baltimore, MD
Fig. 4. Harper and Rowe, NY (for P. B. Hoeber).
Figs. 5 & 15. The University of Wisconsin Press, Madison, WI.

Figs. 9 & 10. The American Physiological Society, Bethesda, MD.
Fig. 12. S. Karger AG., Basel, Switzerland.
Fig. 14. Craftsman Press, Singapore.

REFERENCES

Adrian, E.D. (1940). Double representation of the feet in the sensory cortex of the cat. J. Physiol., Lond., 98, 16P.

Adrian, E.D. (1941). Afferent discharges to the cerebral cortex from peripheral sense organs. J. Physiol., Lond., 100, 159-191.

Bard, P. (1938). Studies on the cortical representation of somatic sensibility. Harvey Lecture. Bull. N.Y. Acad. Med., 14, 585-607.

Bard, P. (1938). Macleod's Physiology in Modern Medicine, 8th Edition. Mosby, St. Louis, Mo., pp. 168-173.

Bard, P. (1956). Medical Physiology, 10th Edition. Mosby, St. Louis, MO. (Fig. 398, p. 1165).

Benjamin, R.M., and Welker, W.I. (1957). Somatic receiving areas of cerebral cortex of squirrel monkey (Saimiri sciureus). J. Neurophysiol., 20, 286-299.

Blomquist, A.J., and Ambrogi-Lorenzini, C. (1965). Projection of dorsal roots and sensory nerves to cortical sensory motor regions of squirrel monkey. J. Neurophysiol., 28, 1193-1205. (Fig. 6)

Boisacq-Schepens, N., Gerebtzoff, M.A., and Goffart, M. (1977). Sensory projections to the cerebral cortex in Perodicticus potto edwarsi (Bouvier). Primates, 18, 401-416.

Burton, H., Mitchell, G., and Brent, D. (1982). Second somatic sensory area in the cerebral cortex of cats: Somatotopic organization and cytoarchitecture. J. comp. Neurol., 210, 109-135.

Burton, H., and Robinson, C.J. (1981). Organization of S II parietal cortex. Multiple somatosensory representations within and near the second somatic sensory area of cynomolgus monkeys. Pp. 67-119, in Woolsey, C.N. (Ed.), Cortical Sensory Organization, Vol. 1, Multiple Somatic Areas. Humana Press, Clifton, N.J.

Carlson, M., and Welt, C. (1980). Somatic sensory (SmI) of the prosimian primate, Galago crassicaudatus: Organization of mechanoreceptive input from the hand in relation to cytoarchitecture. J. comp. Neurol., 189, 249-271.

Carlson, M., and Welt, C. (1981). The somatic sensory cortex SmI in prosimian primates. Pp. 1-27, in Woolsey, C.N., (Ed.), Cortical Sensory Organization, Vol. 1, Multiple Somatic Areas. Humana Press, Clifton,N.J.

Celesia, G.G. (1963). Segmental organization of cortical afferent areas in the cat. J. Neurophysiol., 26, 193-206.

Chang, H.-T., Woolsey, C.N., Jarcho, L.W., and Henneman, E. (1947). Representation of cutaneous tactile sensibility in the cerebral cortex of the spider monkey. Fedn. Proc. Fedn. Am. Socs. exp. Biol, 6, 89.

Clemo, H.R., and Stein, B.E. (1982). Somatosensory cortex: a "new" somatotopic representation. Brain Res., 235, 162-168.

Darian-Smith, I., Ibister,J., Mok, H., and Yokota, T. (1966). Somatic sensory cortical projection areas excited by tactile stimulation of the cat: A triple representation. J. Physiol., Lond., 182, 671-689.

Friedman, D.P. (1981). Body topography in the second somatic sensory area: Monkey S II somatotopy. Pp. 121-165, in: Woolsey, C.N., (Ed.) Cortical Sensory Organization, Vol. 1, Multiple Somatic Areas. Humana Press, Clifton, N.J.

Haight, J.R. (1972). The general organization of somatotopic projections to S II cerebral neocortex in the cat. Brain Res., 44, 483-502.

Hayes, G.J., and Woolsey, C.N. (1944). The pattern of organization within the primary tactile area of the cerebral cortex of the cat. Fedn. Proc. Fed. Am. Socs. exp. Biol., 3, 18.

Hirsch, J.F., and Coxe, W.S. (1958). Representation of cutaneous tactile sensibility in cerebral cortex of Cebus. J. Neurophysiol., 21, 481-498.

Jameson, H.D., Kanagasuntheram, R., DeCoursey, G.E., Murray, M., O'Brien, G.S., and Woolsey, C.N. (1963). Cortical localization in Galago senegalensis. Neurology, Minneap. 13, 352.

Kaas, J.H., Sur, M., Nelson, R.J., and Merzenich, M.M. (1981). The postcentral somatosensory cortex: Multiple representations of the body in primates. Pp. 29-45, in Woolsey, C.N. (Ed.), Cortical Sensory Organization, Vol. 1, Multiple Somatic Areas. Humana Press, Clifton, N.J.

Kanagasuntheram, R. (1963). Some sensory areas of the brain. Inaugeral lecture, University of Singapore, Craftsman Press, Singapore, pp. 1-11; Fig. 6.

Krishnamurti, A., Sanides, F., and Welker, W.I. (1976). Microelectrode mapping of modality specific somatic sensory cerebral neocortex in slow loris. Brain, Behav. Evol., 13, 267-283.

Landgren, S., and Olsson, K.Å. (1980). Low threshold afferent projections from the oral cavity and the face to the cerebral cortex of the cat. Exper. Brain Res., 39, 133-147.

Lende, R.A. (1970). Cortical localization in the tree shrew (Tupaia). Brain Res., 18, 61-75.

Lende, R.A., and Woolsey, C.N. (1956). Sensory and motor localization in cerebral cortex of porcupine (Erthizon dorsatum). J. Neurophysiol., 19, 544-563.

Marshall, W.H. (1949). Presence of multiple cortical foci in the primary projection systems of the cat. Fedn. Proc. Am. Socs. exp. Biol., 8, 107.

Marshall, W.H., Woolsey, C.N., and Bard, P. (1937a). Cortical representation of tactile sensibility as indicated by cortical potentials. Science, N.Y., 85, 388-390.

Marshall, W.H., Woolsey, C.N., and Bard, P. (1937b). Representation of tactile sensibility in the monkey's cortex as indicated by cortical potentials. Am. J. Physiol., 119, 372-373.

Marshall, W.H., Woolsey, C.N., and Bard, P. (1941). Observations on cortical somatic sensory mechanisms of cat and monkey. J. Neurophysiol., 4, 1-24.

Merzenich, M.M., Kaas, J.H., Sur, M., and Lin, C.-S. (1978). Double representation of the body surface within cytoarchitectonic areas 3b and 1 in "S I" in the owl monkey (Aotus trivirgatus). J. comp. Neurol., 181, 41-73.

Merzenich, M.M., Sur, M., Nelson, R.J., and Kaas, J.H. (1981). Organization of S I cortex: Multiple cutaneous representations in 3b and 1 of the owl monkey. Pp. 47-66, in Woolsey, C.N. (Ed.), Cortical Sensory Organization. Vol. 1, Multiple Somatic Areas. Humana Press, Clifton, N.J.

Murray, E.A., and Coulter, J.D. (1981). The medial parietal cortex. Pp. 167-195 in Woolsey, C. N. (Ed.), Cortical Sensory Organization, Vol. 1, Multiple Somatic Areas. Humana Press, Clifton, N.J.

Nelson, R.J., Sur, M., Felleman, D.J., and Kaas, J.H. (1980). Representation of the body surface in postcentral cortex of Macaca fascicularis. J. comp. Neurol., 192, 511-544.

Paul, R.L., Merzenich, M., and Goodman, H. (1972). Representation of slowly and rapidly adapting cutaneous mechanoreceptors of the hand in Brodmann's areas 3 and 1 of Macaca mulatta. Brain Res., 36, 229-249.

Penfield, W., and Jasper, H. (1954). Epilepsy and the Functional Anatomy of the Human Brain. Little, Brown, Boston, MA.

Penfield, W., and Rasmussen, T. (1950). The Cerebral Cortex of Man. Macmillan Press, NY.

Pinto Hamuy, T., Bromiley, R.B., and Woolsey, C.N. (1956). Somatic afferent areas I and II of dog's cerebral cortex. J. Neurophysiol., 19, 485-499.

Powell, T.P.S., and Mountcastle, V.B. (1959). Some aspects of the functional organization of the cortex of the postcentral gyrus of the monkey: A correlation of findings obtained in a single unit analysis with cytoarchitecture. Bull. Johns Hopkins Hosp., 105, 108-131.

Pubols, B.H., and Pubols, L.M. (1971). Somatic organization of spider monkey somatic sensory cerebral cortex. J. comp. Neurol., 141, 63-76.

Robinson, C.J., and Burton, H. (1980). The organization in the second somatosensory area of M. fascicularis. J. comp. Neurol., 192, 69-92.

Sur, M., Nelson, R.J., and Kaas, J.H. (1980a). The representation of the body surface in somatic koniocortrex in the prosimian, Galago. J. comp. Neurol., 189, 381-402.

Sur, M., Nelson, R.J., and Kaas, J.H. (1982). Representations of the body surface in cortical areas 3b and 1 of squirrel monkeys: comparisons with other primates. J. comp. Neurol., 211, 177-192.

Sur, M., Weller, R.E, and Kaas, J.H. (1980b). The representation of the body surface in the first somatosensory area of the tree shrew, Tupaia glis. J. comp. Neurol., 194, 71-95.

Sur, M., Weller, R.E., and Kaas, J.H. (1981a). The organization of somatosensory area II in tree shrews. J. comp. Neurol., 201, 121-133.

Sur, M., Weller, R.E., and Kaas, J.H. (1981b). Physiological and anatomical evidence for a discontinuous representation of the trunk in S-I of tree shrews. J. comp. Neurol., 201, 135-147.

Tanji, D.G., Wise, S.P., Dykes, R.W., and Jones, E.G. (1978). Cytoarchitecture and thalamic connectivity of third somatosensory area of cat cerebral cortex. J. Neurophysiol., 41, 268-284.

Tasker, R.R. (1960). A third somatic tactile sensory area. Physiologist, Wash., 3, 162.

Welker, W.I., and Seidenstein, S. (1959). Somatic sensory representation in the cerebral cortex of the raccoon (Procyon lotor). J. comp. Neurol., 111, 469-502.

Welt, C. (1962). Topographical Organiation of Somatic Sensory and Motor Areas in the Cerebral Cortex of the Gibbon (Hylobates) and Chimpanzee (Pan). Unpublished doctoral dissertation. Univ. Chicago.

Whitsel, B.L., Petrucelli, L.M., and Werner, G. (1969). Symmetry and connectivity in the map of the body surface in somatosensory area II of primates. J. Neurophysiol., 32, 170-183.

Woolsey, C.N. (1952). Patterns of localization in sensory and motor areas of the cerebral cortex. In: The Biology of Mental Health and Disease. P. B. Hoeber, Inc., New York, pp. 193-206. (Fig. 59)

Woolsey, C.N. (1954). Localization patterns in a lissencephalic primate (Hapale jacchus). Am. J. Physiol., 179, 686.

Woolsey, C.N. (1958). Organization of somatic sensory and motor areas of the cerebral cortex. Pp. 63-81 in: Harlow, H.F., and Woolsey, C.N. (Eds.), Biological and Biochemical Bases of Behavior. University of Wisconsin Press, Madison, WI,

Woolsey, C.N. (1964). Cortical localization as defined by evoked potentials and electrical stimulation. Pp. 17-26, in Schaltenbrand, G., and Woolsey, C.N. (Eds.), Cerebral Localization and Organization. University of Wisconsin Press, Madison, WI.

Woolsey, C.N., Erickson, T.C., and Gilson, W.E. (1979). Localization in somatic sensory and motor areas of human cerebral cortex as determined by direct recording of evoked potentials and electrical stimulation. J. Neurosurg., 51, 476-506.

Woolsey, C.N., and Fairman, D. (1946). Contralateral, ipsilateral and bilateral representation of cutaneous receptors in somatic areas I and II of the cerebral cortex of pig, sheep and other mammals. Surgery, 19, 684-702.

Woolsey, C.N., and LeMessurier, D.H. (1948). The pattern of cutaneous representation in the rat's cerebral cortex. Fedn. Proc. Am. Soc. exp. Biol., 7, 137-138.

Woolsey, C.N., Marshall, W.H., and Bard, P. (1942). Representation of cutaneous tactile sensibility in the cerebral cortex of the monkey as indicated by evoked potentials. Bull. Johns Hopkins Hosp., 70, 399-441.

Woolsey, C.N., Tasker, R., Welt, C., Ladpli, R., Campos, G., Potter, H.D., Emmers, R., and Schwassman, H. (1960). Organization of pre- and postcentral leg areas in chimpanzee and gibbon. Trans. Am. neurol. Ass., 85, 144-146.

Woolsey, C.N., and Wang, G.-H. (1945). Somatic sensory areas I and II of the cerebral cortex of the rabbit. Fedn. Proc. Am. Soc. exp. Biol., 4, 79.

ADDENDA

Adrian, E.D. (1943). Afferent areas in the brain of ungulates. Brain, 66, 89-104.

Adrian, E.D. (1946). The somatic receiving area in the brain of the Shetland pony. Brain, 69, 1-5.

Allison, T., and Goff, W.R. (1972). Electrophysiological studies of the echidna, Tachyglossus aculeatus. III. Sensory and interhemispheric responses. Arch. ital Biol., 110, 195-216.

Bodemer, C.W., and Towe, A.L. (1963). Cortical localization patterns in the somatic sensory cortex of the opossum. Expl. Neurol., 8, 380-394.

Bohringer, R.C., and Rowe, M.J. (1977). The organization of of sensory and motor areas of the cerebral cortex in the platypus (Ornithorynchus anatinus). J. comp. Neurol., 174, 1-14.

Campos, C.B., and Welker, W.I. (1976). Comparisons between brains of a large and a small hystrichomorph rodent: capybara, Hydrochoerus and guinea pig, Cavia; neocortical projection regions and measurements of brain subdivisions. Brain, Evol. Behav., 13, 243-266.

Carlson, M., and Welker, W.I. (1976). Some morphological, physiological and behavioral specializations in North American beavers (Castor canadensis). Brain, Behav. Evol., 13, 302-326.

Haight, J.R., Weller, W.L. (1973). Neocortical topography in the bush-tailed possum: Variability and functional signifance of sulci. J. Anat., 116, 473-474.

Johnson, Jr., J.I. (1980). Morphological correlates of specialized elaborations in somatic sensory cerebral neocortex. Chap. 14 in: Ebbesson, S.O.E., ed., Comparative Neurology of the Telencephalon. Plenum Press, N.Y. and London.

Johnson, J.I., Haight, J.R., and Megirian, D. (1973). Convolutions related to sensory projections in cerebral neocortex of marsupial wombats. J. Anat., 114, 153.

Johnson, Jr., J.I., Rubel, E.W., and Hatten, G.I. (1974). Mechanosensory projections to cerebral cortex of sheep. J. comp. Neurol., 158, 81-107.

Johnson, J.I., Switzer, R.C., and Kirsch, J.A.W. (1982). Phylogeny through brain traits: The distribution of categorizing characters in contemporary mammals. Brain, Behav. Evol., 20, 97-117.

Lende, R.A. (1963). Sensory representation in the cerebral cortex of the opossum (Didelphis virginiana). J. comp. Neurol., 121, 395-403.

Lende, R.A. (1964). Representation in the cerebral cortex of a primitive mammal. Sensorimotor, visual and auditory areas in the echidna (Tachyglossus aculeotus). J. Neurophysiol., 27, 34-48.

Lende, R.A. (1969). A comparative approach to the neocortex: localization in monotremes, marsupials and insectivores. Ann. N.Y. Acad. Sci., 167, 262-274.

Lende, R.A., and Sadler, K.M. (1967). Sensory and motor areas in neocortex of hedgehog (Erinaceus). Brain Res., 5, 390-405.

Lende, R.A., and Welker, W.I. (1974). An unusual sensory area in the cerebral cortex of the bottle nose dolphin, Tursiops truncatus. Brain Res., 45, 555-560.

Magalhães-Castro, B., and Saraiva, P.E.S. (1971). Sensory and motor representation in cerebral cortex of the marsupial (Didelphis azarae azarae). Brain Res., 34, 291-299.

Nelson, R.J., Sur, M., and Kaas, J.H. (1979). The organization of the second somotosensory area (Sm II) of the grey squirrel. J. comp. Neurol., 184, 473-489.

Pimental-Souza, F., Cosenza, R.M., Campos, G.B., and Johnson, Jr., J.I., (1980). Somatic sensory cortical regions of the agouti (Dasypracta aguti). Brain, Behav.Evol., 17, 218-240.
Pubols, Jr., P.H. (1977). The second somatic sensory area (Sm II) of opossum neocortex. J. comp. Neurol., 174, 71-78.
Pubols, Jr., B.H., Pubols, L.M., DiPette, D.J., and Sheely, J.C. (1976). Opossum somatic sensory cortex: a microelectrode mapping study. J. comp. Neurol., 165, 229-245.
Royce, G.J., Martin, G.F., and Dom, R.M. (1975). Functional localization and cortical architecture in nine-banded armadillo (Dasypus novemcinctus mexicanus). J. comp. Neurol., 164, 495-522.
Saraiva, P.E.S., and Magalhães-Castro, B. (1975). Sensory and motor representation in the cerebral cortex of the three-toed sloth (Bradypus tridactylus). Brain Res., 90, 181-193.
Sur, M., Nelson, R.J., and Kaas, J.H. (1977). The representation of the body surface in S II of the grey squirrel. Neurosc. Abs., 3, 492.
Sur, M., Nelson, R.J., and Kaas, J.H. (1978). The representation of the body surface in somatosensory area I of the grey squirrel. J. comp. Neurol., 179, 425-449.
Welker, C. (1971). Microelectrode delimitation of fine grain somatotopic organization of SmI cerebral neocortex in albino rat. Brain Res., 26, 259-275.
Welker, W.I., and Campos, G.B. (1963). Physiological significance of sulci in somatic sensory cerebral cortex in mammals of the family Procyonidae. J. comp. Neurol., 120, 19-36.
Welker, W.I., and Carlson, M. (1976). Somatic sensory cortex of hyrax (Procavia). Brain, Behav. Evol., 13, 294-301.
Welker,W.I., Adrian, H.O., Lifschitz, W., Kaulen, R., Caviedes, E. and Gutman, W. (1976). Somatic sensory cortex of llama (Lama glama). Brain, Behav. Evol.,13, 284-293.
Welker, W.I., and Lende, R.A. (1980). Thalamocortical relationships in echidna (Tachyglossus aculeatus). Chap. 15 in: Ebbesson, S.O.E., ed., Comparative Neurology of the Telencepalon. Plenum Press, N.Y. and London.
Weller, W.L., Haight, J.R., Neylon, L., and Johnson, J.I. (1976). Single representation of mystacial vibrissae in S I neocortex of rufous wallaby, Thylogale billardierii. Neurosc. Abst., 2, 926.
Woolsey, C.N., and Fairman, D. (1946). Contralateral, ipsilateral and bilateral representation of cutaneous receptors in somatic areas I and II of pig, sheep and other mammals. Surgery, 19, 684-692.
Woolsey, T.A. (1967). Somotosensory, auditory and visual cortical areas in the mouse. Johns Hopkins Med. J., 121, 91-112.

Woolsey, T.A., and Van der Loos, H. (1970). The structural organization of layer IV in the somatosensory region (S I) of mouse cerebral cortex: the description of a cortical field composed of discrete cytoarchitectonic units. Brain Res., 17, 205-240.

Zeigler, H.P. (1964). Cortical sensory and motor areas of the guinea pig (Cavia porcellus). Arch. ital. Biol., 102, 587-598.

THE ORGANIZATION OF SOMATOSENSORY CORTEX IN PRIMATES AND OTHER MAMMALS

JON H. KAAS

Department of Psychology, Vanderbilt University, Nashville, TN 37240, USA

INTRODUCTION

In pioneering studies on macaque monkeys, Marshall, Woolsey and Bard (1937) defined a region of postcentral parietal cortex that was responsive to stimulation of the body. The responsive region, which became known as the primary or first somatosensory region, S-I, contains four generally recognized architectonic fields, Areas 3a, 3b, 1, and 2. Since the original report of Marshall et al. (1937), there have been a number of interpretations of the significance of these four fields, and of the organization of the S-I region of primates (for recent examples, see Carlson et al., 1982; McKenna et al., 1982; Jones et al., 1978; Paul et al., 1972). In a series of papers, stemming from our first report on the organization of the S-I region in owl monkeys (Merzenich et al., 1978), my colleagues and I have argued that each of the four architectonic fields constitutes a functionally distinct area of somatosensory cortex. In the same sense that visual areas I and II (Areas 17 and 18) are functionally distinct and yet obviously interrelated, we postulated that each of Areas 3a, 3b, 1 and 2 contains a separate and parallel representation of the body; that Areas 3b, 1, and probably 2 are further distinguished by mirror reversals of somatopic order at their common borders, and that each Area has its own distinctive pattern of anatomical connections, neuron response types, and intrinsic organization. In contrast to monkeys (and other higher primates), multiple representations have not been found in the S-I region of prosimians and non-primates. The many similarities in S-I organization and other features in these mammals with the Area 3b representation in monkeys suggests that the S-I representation, as commonly described in most mammals, is clearly a homologue of the Area 3b representation in monkeys, and comparisons across species should be made on that basis. Finally, the organization of cortical somatosensory maps is not static, but subject to modification even in adults, perhaps in a manner related to patterns of neuronal activity. Evidence for these points of view has been presented elsewhere, most recently in Kaas (1983) and Kaas et al. (1983). In this report, some of this evidence is briefly outlined, and several recent advances are emphasized.

MICROELECTRODE MAPPING IN MONKEYS

Most of our evidence for the distinctiveness of Areas 3a, 3b, 1 and 2 comes from multineuron microelectrode mapping experiments in which minimal excitatory receptive fields were determined for extensive arrays of recording sites in individual anesthetized monkeys. In our initial experiments with owl monkeys (Merzenich et al., 1978), we discovered that Areas 3b and 1 were activated throughout by lightly touching skin or hairs, while Areas 3a and 2 generally required more intense stimuli such as pressure or taps, or the movement of joints and muscle. These observations suggested a basic division into two adjoining fields related to cutaneous receptors, and two flanking fields largely related to deep noncutaneous receptors. Detailed microelectrode maps revealed that Areas 3b and 1 contained separate representations of the body surface that proceeded from tail to tongue in two parallel mediolateral cortical sequences. The two representations were somatotopically matched, or congruent, along their common border, and opposite or mirror reversal sequences of representation were noted in rostral and caudal cortical progressions from the 3b/1 border. Thus, for example, the representations of the glabrous digits of the hand pointed rostrally in Area 3b, and caudally in Area 1, and were joined along the palm or the bases of the digits at the 3b/1 border. Both Area 3b and Area 1 had populations of neurons that could be distinguished by their slowly or rapidly adapting response characteristics to maintained skin displacement, but only Area 1 had neurons with large receptive fields that were sensitive to low threshold and high frequency vibration, perhaps indicating inputs from Pacinian receptors. Because the deep receptors related to cortical neurons in Areas 3a and 2 were difficult to localize precisely, the detailed organizations of these fields were not determined. However, the limited mapping results did indicate that the somatotopic organization in each of these fields proceeded in parallel to those observed in Areas 3b and 1.

Subsequent mapping experiments strongly supported the contention that Areas 3a, 3b, 1 and 2 constitute functional subdivisions of somatosensory cortex in monkeys. The basic parallel and mirror image somatotopic organizations of Areas 3b and 1 were found in other species of monkeys, including New World squirrel (Sur et al., 1982) and cebus (Fellman et all., 1983) monkeys, and Old World cynomolgus macaque monkeys (Nelson et al., 1980). The major species difference was that parts of the trunk and limbs appeared to be reversed in somatotopic sequence in squirrel and cebus monkeys as compared with owl and macaque monkeys. However, in all species, Areas 3b and 1 represented the same body parts in parallel mediolateral cortical sequences, and in mirror image or reversed sequences in rostrocaudal cortical sequences.

More recently, we have obtained similar results for the hand representations in Areas 3b and 1 of rhesus macaque monkeys (Pons et al., 1983). In each field, the glabrous hand surfaces were represented separately with digits pointing in opposite directions. The two representations were not serial, as suggested by Paul et al. (1972), and separate representations of hairy and glabrous skin surfaces did not distinguish the two fields as suggested by Carlson et al. (1983). Instead, the two representations of the hand were not notably different from those found in cynomolgus macaque monkeys, and we conclude that there are no major differences in hand representations between the two species of macaque monkeys.

For Area 3a, results from other monkeys were also similar to those obtained for owl monkeys. Area 3a was largely, but not completely, unresponsive to cutaneous stimuli, and there is considerable evidence from other experimentors that the major driving influence is from muscle spindle receptors (See Jones and Porter, 1980). Our experiments to date consistently indicate that body parts are represented in Area 3a in a pattern that parallels those in the cutaneous representations in Areas 3b and 1. It is not yet certain if the 3a representation mirrors that of 3b.

Information on the organization of Area 2 is more complete. Area 2 is clearly activated by inputs from both cutaneous and noncutaneous receptors. The cutaneous inputs seem less secure and susceptible to anesthetics and other disruptions, but under many recording conditions the cutaneous inputs predominate. Under these conditions, we have been able to determine some of the organization of the cutaneous inputs in Area 2, and limited observations suggest that the somatotopic order of the noncutaneous activation matches that of the cutaneous activation.

Early explorations of the hand representation in Area 2 of macaque monkeys (Kaas et al., 1979) and squirrel monkeys (Sur et al., 1982) indicated that the cutaneous input is roughly organized as a mirror reversal of that in Area 1. Thus, somatotopic organization proceeds from the base to the tip of the digits in rostrocaudal progressions across Area 1, and from the tip to the base of the digits for rostrocaudal progressions across Area 2. More recent experiments in rhesus macaques (Pons et al., 1983), confirm that the representation of the hand in Area 2 mirrors that in Area 1, and show that a mirror reversal exists for the representations of other body parts as well. Overall, the Area 2 representation closely parallels that of Area 1 in somatotopic order, the two representations have a somatopically congruent border, and they are roughly mirror reversals of each other in progressions away from that border.

Our experiments on rhesus monkeys also suggest the existence of one or more additional somatotopic representations starting near the caudal border of Area 2. The caudal border of Area 2 is sometimes difficult to distinguish architectonically, and opinions vary on how to define the location of this border. In the somatosensory cortex devoted to the hand, the pattern of organization changed in caudal Area 2 near the Area 5 border. The change in organization was characterized in part by a reversal so that digit tips were found again in Area 5. The point of change in the somatotopic pattern appeared to start in what some investigators would call Area 2. Thus, there may be reason to reevaluate the cytoarchitecture of the region, since there is evidence for a systematic cutaneous representation, in addition to that found in rostral Area 2, that may include parts of Areas 2 and 5.

PATTERNS OF THALAMIC AND CORTICAL CONNECTIONS

The connection patterns of Areas 3a, 3b, 1 and 2 with each other and the thalamus are consistent with the contention that they are four distinct representations. Both Areas 3b and 1 receive input from a single thalamic nucleus that we prefer to term the ventroposterior nucleus (VP), although it is smaller than the VP of some investigators. The nucleus contains a single systematic map of the body surface (Nelson et al., 1982); however, there may be separation of rapidly and slowly adapting inputs at the modular level (Dykes et al., 1981). Because VP contains a single map, and Areas 3b and 1 contain two mirror image maps, a given restricted located in VP might project to two quite separate locations, one at the rostral border of Area 3b and the other at the caudal border of Area 1, to joined regions at the 3b/1 border, or to locations in between (Lin et al., 1979; Nelson et al., 1981;

Kaas, 1983). The thalamic region immediately dorsorostral to VP is activated by noncutaneous stimuli (see Dykes et al., 1981), and there is evidence for dividing this region into a more rostral nucleus relaying information from muscle spindles to Area 3a, and a more caudal nucleus relaying joint and other deep receptor information to Area 2 (see Kaas, 1983).

At the cortical level, our experiments have been more limited and they are in general agreement with previous findings (see Kaas, 1983 for review of previous reports by other investigators). Area 3b projects to the granular or input layers of Area 1, and therefore, it is likely that this input has a major role in determining the response characteristics of Area 1 neurons (Sur, 1980). Area 1 in turn relays to Area 2 and the caudal portion of Area 5, presumably providing the source of the cutaneous information in this cortex. The majority of projecting neurons in any Area are in somatotopic register with those of the termination site.

Recently, we have studied the relation of corpus callosum connections to the representation in Areas 3a, 3b, 1 and 2 (Killackey et al., 1983). In all four fields, callosal connections are unevenly distributed so that the representations of the hands and feet have the fewest callosally projecting cells and terminations. However, the four representations also differ in the overall density of connections, so that Area 3b has the fewest, Areas 3a and 1 have intermediate amounts, and Area 2 has the most connections. In Area 2, even the representations of the hand and foot have many callosaly projecting neurons and terminations. The stepwise increase in density of callosal connections across fields 3b, 1, and 2 suggests, as does the pattern of ipsilateral connections, that these fields form a serial processing sequence. At the first stage in Area 3b, little information is sent to the opposite hemisphere; the next stage sends more, and the third stage sends the most. Information for the hand and foot is relayed in significant amount only after relatively processed cutaneous information from Area 1 is combined in Area 2 with deep receptor information from the thalamus.

THE DYNAMIC NATURE OF MAP STRUCTURE

A number of observations suggest that the observed patterns of somatotopic organization in Areas 3b and 1, as determined by minimal receptive fields for given recording sites, does not reflect the full anatomical diversity of connections. First, the spread of the terminal arbors of incoming thalamic axons (Pons et al., 1982) is greater than that expected from the receptive field sizes and the precision of the somatotopic order. Second, there is seldom any evidence in the contralateral excitatory receptive fields of cortical neurons for any effects of the callosal connections. Third, there is some scatter in the ipsilateral relay of information. An injection of a tracer confined to the representation of the distal phalanx of a digit in Area 1, for example, will label a number of cells in the representation of the distal phalanx in Area 3b, but there will be some scatter of labeled cells in the representations of adjoining phalanges and digits. Fourth, and more directly, if the patent input to a limited part of the maps in these fields is removed, for instance by cutting the median nerve to the radial half of the glabrous hand, then the deprived cortex rapidly becomes responsive to other inputs (see Kaas et al., 1983; Merzenich et al., 1983a & b). Not only can the organization of these cortical maps change rapidly (hours to weeks) but after such changes the original organization can be restored. Crushing as well as cutting the median nerve deactivates it, and the deprived region of cortex becomes responsive to inputs from other parts of the hand. However, the crushed nerve is capable of regenerating with great fidelity, and normal organization of the hand representations can be restored (Wall et al., 1983). Thus, the somatotopic organization seen at any one time does not reflect all of the existing connections, and sensitivity to different inputs is subject to change, perhaps in an activity dependent manner (see Merzenich et al., 1983a).

S-I IN OTHER MAMMALS

The first somatosensory area as commonly described in cats, rats, squirrels, tree shrews, opossums, and even prosimian primates, consists of a single systematic representation of the body surface (see Kaas, 1983 & 1982 for references and review). The overall organization of this representation is similar across species, and it is similar to the Area 3b, but not the Area 1, cutaneous representation in monkeys. Like Area 3b, the cortex of this representation has a well developed granular cell layer, and like Areas 3b and 1, there is input from the ventroposterior nucleus. These similarities suggest that S-I, as commonly described in most mammals, and Area 3b in monkeys are homologous, i.e., that the Area 3b representation in monkeys and the "S-I" representations in other mammals evolved from the same representation in a common ancestor. Thus, comparisons across species for the same cortical area should be between S-I of most mammals and Area 3b of monkeys or higher primates.

Clear evidence for an Area 3a, as a muscle receptor field on the rostral border of a cutaneous field, exists only for carnivores and primates. Since this field appears to exist in these two groups, and they are not closely related, a homologous 3a field may be widespread in mammals. There are no obvious homologues of Areas 1 and 2 of monkeys in nonprimates, or even in prosimians.

REFERENCES

Carlson, M., Dianhua, S., Bingxuan, X. (1982), Significance of topography and submodality in the organization of Brodmann's Area 1 of Macaca. Soc. Neurosci Abstr. 8, 36.

Dykes, R. W., Sur, M., Merzenich, M. M., Kaas, J. H., and Nelson, R. J. (1981). Regional segregation of neurons responding to quickly adapting, slowly adapting, deep and pacinian receptors within thalamic ventroposterior lateral and ventroposterior inferior nuclei in the squirrel monkey. Neurosci., 6, 1687-1692.

Felleman, D. J., Nelson, R. J., Sur, M., and Kaas, J. H. (1983). Representations of the body surface in Areas 3b and 1 of postcentral parietal cortex of cebus monkeys. Brain Res., in press.

Jones, E. G., Coulter, J. D., and Hendry, S. H. C. (1978). Intracortical connectivity of architectonic fields in the somatic sensory, motor and parietal cortex of monkeys. J. Comp. Neur. 181, 291-348.

Jones, E. G. and Porter, R. (1980). What is Area 3a? Brain Res. Rev. 2, 1-43.

Kaas, J. H. (1983). What, if anything, is S-I? The organization of the "first somatosensory area" of cortex. Physiological Reviews, 63, 206-231.

Kaas, J. H. (1982). The somatosensory cortex and thalamus in galago, In: E. E. Haines, (Ed.), The lesser bush baby (galago) as an animal model: Selected topics, CRC Press, Inc., pp. 169-181.

Kaas, J. H., Merzenich, M. M., and Killackey, H. P. (1983). Changes in the organization of somatosensory cortex following peripheral nerve damage in adult and developing mammals. Ann. Rev. Neurosci., 6, 325-356.

Kaas, J. H., Nelson, R. J., Sur, M., Lin, C.-S., and Merzenich, M. M. (1979). Multiple representations of the body within "SI" of primates. Science, 204, 521-523.

Killackey, H. P., Gould, H. J., III, Cusick, C. G., Pons, T. P., and Kaas, J. H. (1983). The relation of corpus callosum connections to architectonic fields and body surface maps in sensorimotor cortex of New and Old World monkeys. J. Comp. Neuro. In press.

Lin, C.-S., Merzenich, M. M., Sur, M., and Kaas, J. H. (1979). Connections of Areas 3b and 1 of the parietal somatosensory strip with the ventroposterior nucleus in the owl monkey (Aotus trivirgatus). J. Comp. Neurol., 185, 355-372.

Marshall, W. H., Woolsey, C. N., and Bard, R. (1937). Cortical representation of tactile sensibility as indicated by cortical potentials. Science 85, 388-390.

Merzenich, M. M., Kaas, J. H., Wall, J. T., Nelson, R. J., Sur, M., and Felleman, D. J. (1983a). Topographic reorganization of somatosensory cortical Areas 3b and 1 in adult monkeys following restricted deafferentation. Neuroscience, 8, 33-35.

Merzenich, M. M., Kaas, J. H., Wall, J. T., Sur, M., Nelson, R. J., and Felleman, D. J. (1983b). Progression of change following median nerve section in the cortical representation of the hand in Areas 3b and 1 in adult owl and squirrel monkeys. Neuroscience, in press.

Merzenich, M. M., Kaas, J. H., Sur, M., Lin, C.-S. (1978). Double representation of the body surface within cytoarchitectonic Areas 3b and 1 in "SI" in the owl monkey (Aotus trivirgatus). J. Comp. Neurol., 181, 41-74.

Nelson, R. J., and Kaas, J. H. (1981). Connections of the ventroposterior nucleus of the thalamus with the body surface representations cortical Areas 3b and 1 of the cynomolgus macaque (Macaca fascicularis). J. Comp. Neurol., 199, 29-64.

Nelson, R. J., Kaas, J. H., Merzenich, M. M., Dykes, R. W., and Sur, M. (1982). Somatotopic organization of the ventroposterior nucleus of the squirrel monkey (Saimiri sciureus). Soc. Neurosi. Abstracts, 8, 38.

Nelson, R. J., Sur, M., Felleman, D. J., and Kaas, J. H. (1980). The representations of the body surface in postcentral somatosensory cortex of (Macaca fascicularis). J. Comp. Neurol., 192, 611-643.

Paul, R. L., Merzenich, M. M., and Goodman, H. (1972). Representation of slowly and rapidly adapting cutaneous mechanoreceptors of the hand in Brodmann's Areas 3 and 1 of Macaca mulatta. Brain Res. 36, 229-249.

Pons, T. P., Cusick, C. G., Garraghty, P. E., and Kaas, J. H. (1983). The representation of the body in parietal cortex of macaque monkeys: The organization of Area 2. Anat. Rec., 105, 226A.

Pons, T. P., Sur, M., and Kaas, J. H. (1982). Axonal arborizations in area 3b of somatosensory cortex in the owl monkey, Aotus trivirgatus. Anat. Rec., 202, 151A.

Sur, M. (1980). Receptive fields of neurons in areas 3b and 1 of somatosensory cortex in monkeys. Brain Res. 198, 465-471.

Sur, M., Nelson, R. J., and Kaas, J. H. (1982). Representations of the body surface in cortical Areas 3b and 1 of squirrel monkeys: Comparisons with other primates. J. Comp. Neurol., 211, 177-192.

Sur, M., Wall, J. T., and Kaas, J. H. (1981). Modular segregation of functional cell classes within the postcentral somatosensory cortex of monkeys. Science, 212, 1059-1061.

Sur, M., Nelson, R. J., and Kaas, J. H.. (1980). The representation of the body surface in somatic koniocortex in the prosimian (Galago senegalensis). J. Comp. Neurol., 180, 381-402.

Wall, J. T., Felleman, D. J. and Kaas, J. H. (1983). Recovery of normal topography in the somatosensory cortex of monkeys after nerve crush and regeneration. Science, In press.

IMAGING THE RESPONDING NEURONAL POPULATION WITH 14C-2-DEOXYGLUCOSE: THE SOMATOSENSORY CEREBRAL CORTICAL SIGNATURE OF A TACTILE STIMULUS

B.L. WHITSEL and S.L. JULIANO

Department of Physiology, and Dental Research Center, University of North Carolina at Chapel Hill, NC 27514, USA

INTRODUCTION

The single unit method has been used to great advantage for several decades and, as a result, there exists a wealth of information about the properties of individual somatosensory neurons and their capacity to signal specific stimulus events. While single unit studies have advanced our understanding of individual somatosensory neuron behavior (e.g., response properties, input-output relations, receptive field organization, stimulus feature extraction, etc.), they have provided limited information about the spatial distribution of the population of neurons activated at a given level of the somatosensory projection pathway by a given stimulus. As a consequence, there is currently little evidence which bears directly on the possibility that population response patterns might encode important information about somatsensory stimulus conditions, or the possibility that the information encoded at the level of somatosensory neuron populations might differ from that encoded by single neurons. A related possibility for which there also is little evidence is that the spatiotemporal population response patterns set up at cerebral cortical levels of the projection pathway (e.g., the SI or SII cortex) by somatic stimuli are decoded by higher levels (e.g., posterior parietal cortex) which process somatosensory input.

Topographic maps derived by reconstructing the observations gathered in neurophysiological experiments are available for each level of the somatosensory projection pathway, and these maps frequently are used to derive inferences about the spatial distribution of neural activity set up at a given level by a single presentation of a somatic stimulus. While topographic maps certainly provide valuable information about the responding population (i.e., they define the boundaries of the cortical sector in which activity is set up by a stimulus), one should not be surprised if they serve as poor predictors of the detailed spatial pattern of neural activity set up by a given somatic stimulus. More specifically, even if the inaccuracies introduced by the pooling of observations from different animals and the reconstruction process could be neglected, the approach used to define single neuron functional properties in typical neurophysiological mapping experiments would not be expected to generate maps clearly relatable to the detailed spatial pattern of activity set up at a given level by a somatic stimulus. The reasons for this conclusion are: (i) each observation obtained in a neurophysiological mapping experiment is obtained at a different time during the experiment, but the population response to a stimulus consists of the activity occurring within a restricted and precisely defined time period (usually corresponding to the time during and immediately after stimulus application) and (ii) each neurophysiological observation used for the construction of a topographic map is obtained using optimal stimulus conditions, whereas a given somatic stimulus may be optimal for some members of the responding population, it is surely less than optimal (perhaps even inhibitory) for many others.

The suggestion that observations of single neuron activity or of small groups of neurons obtained at different times might not disclose important aspects of the neural population response set up by a sensory stimulus is supported by results obtained using mathematical models of nervous tissue. As an example, Wilson and Cowan (1973) showed that a locally redundant neural network comprised of neurons having simple properties and interconnections (this particular model was designed to approximate the recurrent lateral connections characteristic of neocortex) exhibits complex dynamic modes of response which mimic certain basic attributes of sensation (e.g., threshold, spatial and temporal summation, and contrast enhancement). One particularly interesting aspect of the behavior of the Wilson and Cowan network is that "sensory" inputs lead to the establishment of spatially inhomogeneous and maintained changes in network responsitivity that are dependent on the characteristics of the triggering stimulus (i.e., the changes in network status are "stimulus-specific"). While the network model of Wilson and Cowan (1973) is especially attractive because it exhibits behaviors comparable to well known and quantifiable sensory phenomena, model neural networks differing greatly in design and level of complexity can exhibit a class of properties (termed "self-organizing properties") which transcend the relatively simple properties of their constituent neurons (Wilson and Cowan, 1973; for recent reviews see Amari and Arbib, 1982, and Szentagothai and Arbib, 1974). Given these higher-order properties of relatively simple networks, it seems reasonable to expect that the complex neural networks which exist at different levels of the somatosensory nervous system may also exhibit properties not predictable from either the "static" or "dynamic" properties of single neurons.

To date, there have been no studies specifically addressed to the possibility that spatial patterns of somatosensory neuron activity encode useful stimulus information. This is undoubtedly due to technical limitations, since current neurophysiological methods do not enable one to obtain simultaneous recordings from populations of neurons as large and widely distributed as those present at thalamic and cortical levels of the somatosensory nervous system. Since neurophysiological methods appeared inappropriate for the task, we sought to measure the spatial distribution of neuronal activity evoked by precisely defined somatic stimuli in unanesthetized animals by employing the ^{14}C-2-deoxyglucose method of metabolic mapping (the 2DG method; Sokoloff, 1977). This work, begun in 1979 with the collaboration of Dr. Peter Hand of the University of Pennsylvania and more recently with Dr. Surindar Cheema and Mr. Oleg Favorov in the Physiology Department of the University of North Carolina, has now progressed to the point where we feel that, whenever appropriate controls are employed, the method provides a highly reproducible image of stimulus related somatosensory neural metabolic activity in unanesthetized animals. In addition, we have developed methods enabling the features of the stimulus-evoked labeling pattern to be correlated with the neural activity recorded with microelectrodes under the same experimental and stimulus conditions. In this paper we shall (i) briefly describe representative data obtained at thalamic and cortical levels of the somatosensory projection pathway, and (ii) outline a view of somatosensory cortical stimulus representation consistent with both the 2DG and neurophysiological mapping data obtained from unanesthetized animals and with current knowledge about the anatomical connections linking the thalamic and cortical levels of the somatosensory system.

METHODS

The 2DG procedure we use closely follows that described by Sokoloff, et al. (1977); see also Juliano, et al., 1981. The surgical procedures were performed on monkeys or cats maintained under halothane anesthesia; they consisted of the insertion of a tracheal cannula and a venous catheter. If extracellular recordings were to be carried out, a lucite chamber was attached with dental acrylic over a small opening in the skull, and all the recordings were obtained prior to the conduct of the 2DG labeling experiment. For details of the recording procedure see McKenna, et al., 1982. Following the surgery, the animal was removed from the anesthetic and neuromuscular blockade was achieved with gallamine triethiodide; artificial respiration was initiated. End-tidal CO2 and body temperature were monitored and maintained within normal limits throughout the experiments.

Each cat or monkey rested on a soft, molded cushion throughout the experiment and received gentle, but precisely controlled somatic stimulation. The stimulus commenced 10-15 min prior to the single pulse, intravenous injection of (14C)2-deoxy-D-gluocse (10um Ci/100g). After 45 min of continuous stimulation, an overdose of general anesthetic was administered and the animal was perfused with 3.3% buffered formalin. The brain was quickly removed and prepared for autoradiography.

RESULTS

<u>2DG labeling in the thalamus</u>

Metabolic labeling was consistently found in the ventrobasal complex of the thalamus (VB) following delivery of controlled somatic stimulation to a specific body region (Figures 1 and 2). Different modes of somatic stimuli were applied: intermittent vertical displacement of a localized region of skin (flutter, at 15 hz and 1mm skin indentation), unidimensional, constant velocity brush strokes (15-27 cm/sec) and constant velocity joint rotations (22°/sec). In the series of animals stimulated in this way, the distribution of 2DG labeling shifted systematically within the mediolateral dimension of VB with changes in the locus of stimulation. The general topographic relationships between the VB territories labeled by the different somatic stimuli were consistent with

Figure 1. A tactile stimulus produces 2DG labeling that occupies a lamella-shaped territory in the contralateral VB. A: Direct print of film autoradiograph prepared from coronal section through monkey VB. White areas indicate regions of elevated metabolic activity. B: Section which produced the autoradiograph shown in A after Nissl staining. The stimulus was intermittent vertical displacement of the volar skin of the distal tips of the first and second toes. Note that many of the irregular boundaries of the labeled VB territory correspond precisely to the boundaries between regions comprised primarily of fiber fascicles and others of neuropil. LP, lateralis posterior; CM, centre median; VPI, ventralis posterior inferior; VPM, ventralis posterior medialis; VPLc, ventralis posterior lateralis, pars caudalis.

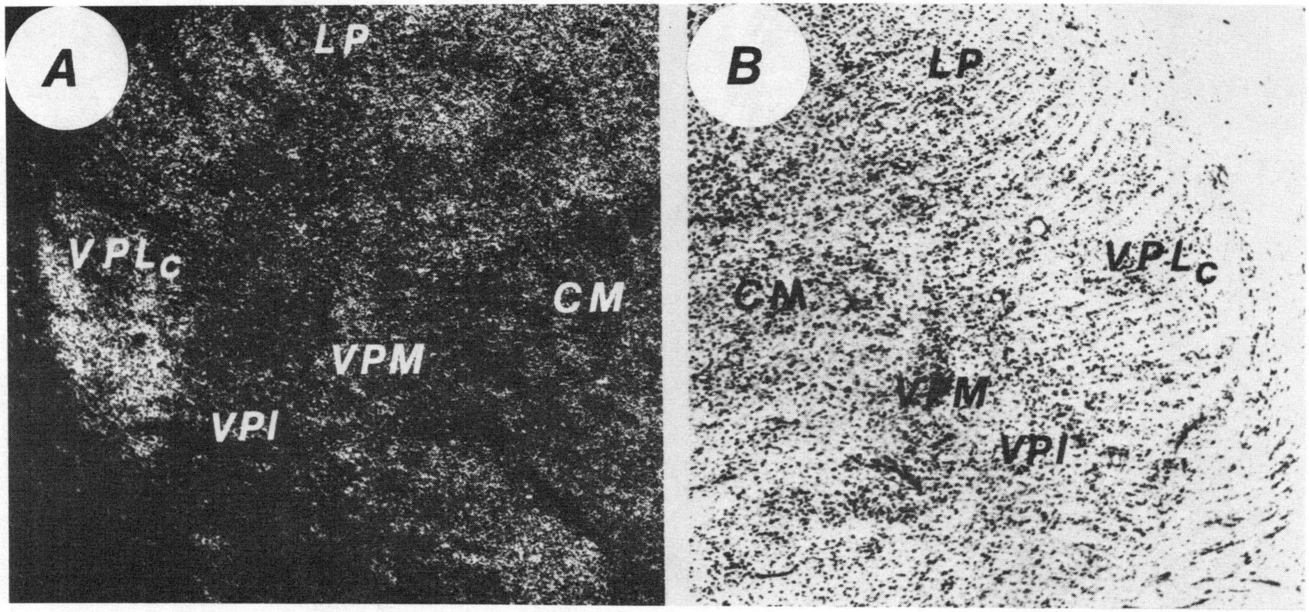

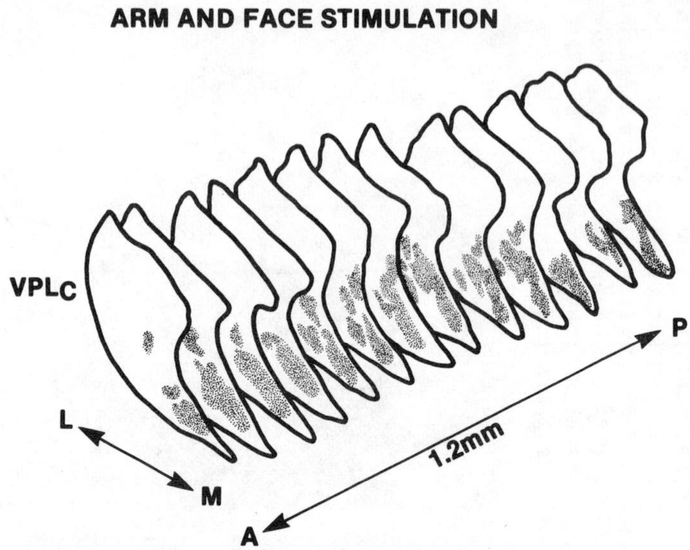

Figure 2. 3-dimensional view of stimulus-related 2DG labeling in VPLc. Labeling is shown by stippling. The contralateral stimuli were unidirectional brush strokes applied sequentially to the dorsal upper arm and to the skin overlying the lower jaw. The labeling is lamellar in configuration at each level and extends a large anteroposterior distance. The reconstruction method of Johnson and Copowski (1983) was used. A, anterior; P, posterior; L, lateral; M, medial.

those predicted by the available neurophysiological maps (Henneman and Mountcastle, 1952; Loe, et al, 1977; Jones, et al., 1982). These indicate that the face is represented most medially, the lower extremity most laterally, and the hand in a central position surrounded both medially and laterally by regions receiving input from the arm. Additionally, the stimulus-evoked VB labeling in all animals exhibited certain common attributes: (i) it was dorsoventrally extensive and occupied a cresecent-shaped territory with its concave surface directed medially, (ii) within the crescent of label, fluctuations in optical density exist, revealing regions of dense metabolic activity embedded in the larger territories of more diffuse labeling; (iii) the stimulus-related labeling extended across long antero-posterior distances (see figures 1 and 2); this was true even when topographically restricted stimuli were delivered (e.g., flutter to the skin of the distal tip of the first and second toes).

All except one of the above attributes of VB 2DG labeling fit well with the lamellar concept of VB topographical organization derived from neurophysiological experiments (Mountcastle and Henneman, 1952; Loe, et al., 1977; Jones, et al., 1982). Although the fluctuations in optical density within a labeled VB lamella clearly exist they are not adequately accounted for by the lamellar concept of VB organization. In some instances, the fluctuations in stimulus-related label within VB appear to be the result of the inhomogeneities in VB cyto- and myeloarchitecture (see figure 1). It is well known, for example, that neurons in VB aggregate into clusters, and that groups of neuronal clusters tend to be separated from their neighbors by mediolaterally oriented fiber bundles (c.f., Olszewski, 1952; Boivie, 1978); it thus seems reasonable that these morphological inhomogeneities should be paralled by inhomogeneities in DG labeling. However, in some instances, the fluctuations in VB labeling density are not clearly attributable to such inhomogeneities in VB architecture, for example see Figure 4 of Juliano, et al., 1983b.

One model of thalamic organization which may help to explain some of the fluctuations in the density of stimulus-evoked metabolic label within a given VB lamella is that proposed by Jones and co-workers (1982). They describe an anatomical and functional organization for VB which consists of groups of neurons which are organized into units called "rods". Each rod extends for a substantial antero-posterior distance, but has a limited dorso-ventral dimension. A rod is definable functionally since it receives input from a specific peripheral locus and a restricted class of mechanoreceptors; it is also definable anatomically since it is comprised of one or several clusters of VB neurons and receives its input from a bundle of afferents projecting from the dorsal column nuclei. Thus, a rod is defined as both a functional and anatomical unit, and may correspond to some of the focal accumulations of 2DG label observed within a VB lamella following delivery of a restricted tactile stimulus.

Figure 3. A tactile stimulus produces a spatially non-homogeneous pattern of 2DG labeling within SI. A: Low power view of pericentral cortex of <u>Macaca fascicularis</u> monkey at the level of the hand representation; Nissl stained section, postcentral cortex appears at left of central sulcus (CS), precentral cortex appears at right of CS; oblique plane of section perpendicular to CS. The cytoarchitectural fields 3a, 3b, 1 and 2 are identified. B: Direct print of film autoradiograph prepared from section shown in A. C: Computerized image of autoradiograph shown in B. Optical scanning was carried out at 50 um intervals using a 100 um aperture (see Goochee, et al., 1980 for details). Light areas indicate cortical regions of high optical density. It should be noted that within the cortical mantle there are no notable morphological inhomogeneities like those present in VB, yet the 2DG labeling pattern is strikingly inhomogeneous.

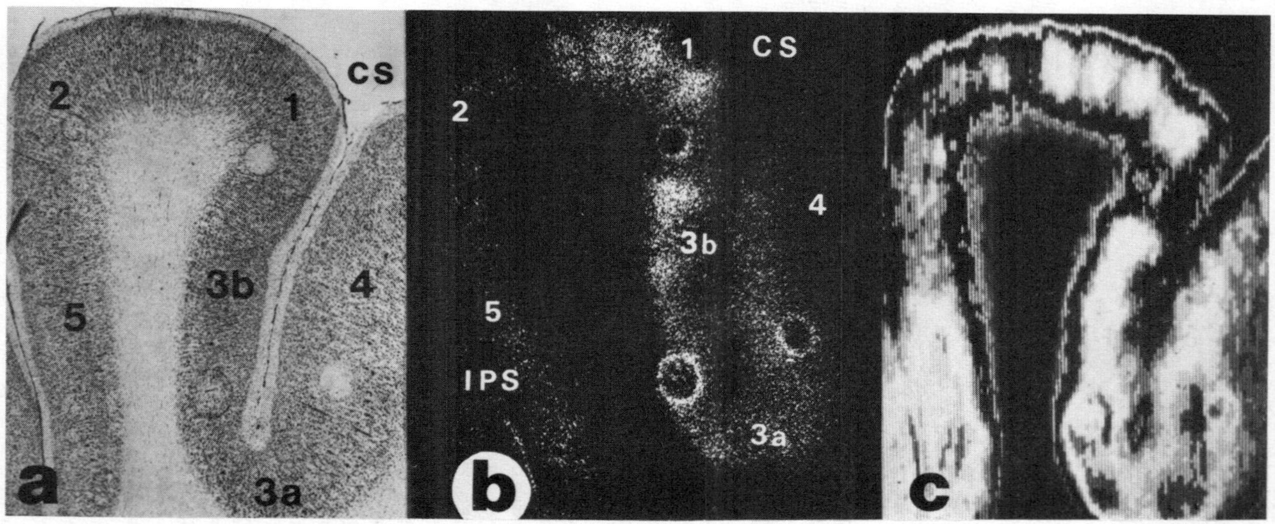

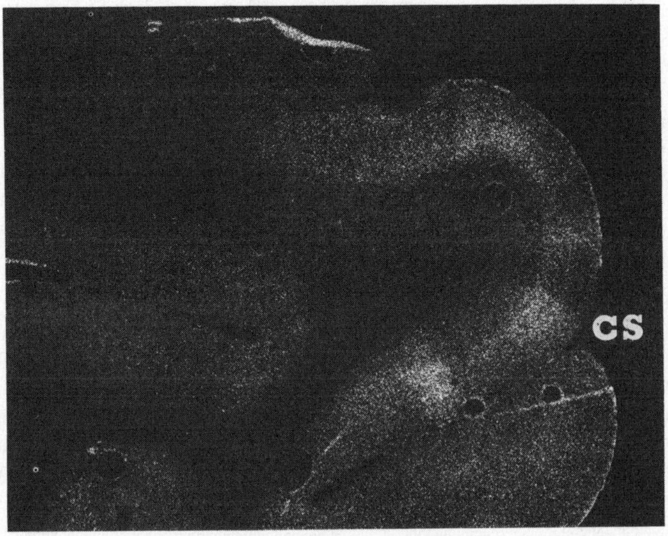

Figure 4. A spatially non-homogeneous pattern of activity evoked in SI by electrocutaneous stimulation. Autoradiograph of coronal section through areas 3b and 1 of the SI cortex of a cat. Constant current (1 milliamp) trains of square wave electrical stimuli were delivered to the central pad of the contralateral forepaw. Stimulus parameters were: train duration 100 msec, train frequency 0.5/sec, 5 pulses/train, pulse duration 10msec. Two S44 Grass solid state stimulators and a CCU-1 constant current unit were configured to give stimulus trains having the above parameters. When such stimuli were applied to human subjects they produced a sensation of skin touch referred to the region contacted by the electrodes.

2DG labeling in the cortex.

Metabolic labeling routinely occurs in the cortex (both SI and SII) in animals subjected to somatic stimulation. The 2DG labeling evoked in either SI or SII by the natural stimuli used in this study exhibited common characteristics. Perhaps the most striking was the observation that in all animals the SI and SII labeling occupied column-like zones which extended through cortical laminae II-V, and were of limited tangential width (Figure 3). These column-like zones of heavy labeling were typically located adjacent to regions of less dense 2DG uptake, thus giving the impression of an intermittent pattern of columnar aggregates of activity (see figures 3 and 4). Within the column-like extensions of label, the optical density was greatest in the cortical layers which receive primary thalamic terminations, i.e. laminae IIIb and IV. In certain cortical areas (particularly striking in area 3b of SI, but also evident in area 1), the radial zones of heavy labeling were superimposed on a continuous band of activity which occupied layers IIIb and IV. The metabolic pattern was thus relatively continuous at the level of the central cortical layers, but discontinuous or intermittent at the level of the supra- and infragranular laminae. However, within this tangential band of label in the central layers are periodic fluctuations in optical density, with the most intense labeling occurring at locations where the labeling extends into the supra- and infragranular layers. Alignment of adjacent cortical sections revealed that the column-like aggregates of label form strips of activity extending for varying lengths through the cortex. The orientation of these strips were different for SI and SII: for both cortical areas they paralled the long axis of the gyrus (i.e., roughly a mediolateral orientation in SI, and a posteromedial to anterolateral orientation in SII). It is of interest that these orientations bear a consistent relation to the body representation in both areas (McKenna, et al., 1982, Nelson, et al., 1980, Robinson and Burton, 1980).

While the periodic fluctuations of labeling density in layers IIIb and IV in both SI and SII appear similar to the rod-like aggregations of label within a VB lamella, the discontinuities between adjacent column-like aggregates at the level of the supra- and infrgranular cortical laminae are not predicted by the thalamic labeling, nor are they predicted by any neurophysiological maps of SI or SII (McKenna, et al., 1982, Nelson, et al., 1980, Robinson and Burton, 1980). One possiblity underlying such an intermittent pattern of activity is that it may be set up by the activity in a certain class of mechanoreceptors which project preferentially to multiple and disjoint columnar groupings of neurons in the cortex. Although such a pattern may exist (c.f., Sur, et al., 1981) it seems unlikely that such a mechanism caused the intermittent 2DG labeling pattern in SI or SII. For example, the cutaneous stimuli we used (flutter, and brush strokes) are known to activate both rapidly and slowly adapting cutaneous mechanoreceptors, and the joint rotation stimulus (which presumably activated rapidly adapting and slowly adapting deep receptors) all produced an intermittent pattern. To obtain evidence bearing directly on the possibility that the intermittency might reflect the pattern of mechanoreceptor representation in the cortex, a series of experiments was carried out in which square-wave electrical stimuli were delivered to the skin of cats. Since this type of stimulation is known to bypass the receptor organs and to directly activate the axons of fibers supplying the skin, such stimuli should effectively activate all the large diameter mechanoreceptive afferents in the skin field stimulated. This would lead to non-selective activation of the mechanoreceptor afferents types (all supplied by Aβ afferents) which innervate the skin. Figure 4 clearly demonstrates that even such presumably non-selective mechanoreceptor afferent activation produces an intermittent pattern of activity in the cortex and thus the data do not support the idea that such patterns reflect selective mechanoreceptor input to the cortex.

In another series of studies it was demonstrated that the labeling obtained from a given somatic stimulus was highly reproducible (Juliano, et al., 1983a). This was accomplished by subjecting a series of animals to the same stimulus (e.g., flutter to the tip of the index finger). Although the pattern of metabolic activity produced in areas 3b and 1 was in each case complex, it was virtually identical from one animal to the next.

In a final series of experiments we sought to determine the relationship between cortical metabolic activity and neural activity recorded from the same cortical field in the same animal. Figure 5 summarizes the results from a representative experiment. The data indicate that the relationship between stimulus-evoked metabolic activity and single neuron properties is not simple. Inspection of the data shown in Figure 5 reveals that metabolic label only occurs at cortical locations where the neurons possessed functional properties consistent with their activation by the stimulus used during the 2DG experiment. However, the single neuron observations obtained at loci between the column-like aggregates of 2DG label indicate that the properties of neurons in sites lacking such aggregates were, in a number of instances, quite similar to those in the neighboring regions containing metabolic label (see legend to Figure 5 for details). These findings suggest that the highly reproducible pattern of metabolic activity produced in SI by a somatic stimulus is determined by multiple factors. Since labeling only occurs in regions where the single neurons possess receptive fields and submodality

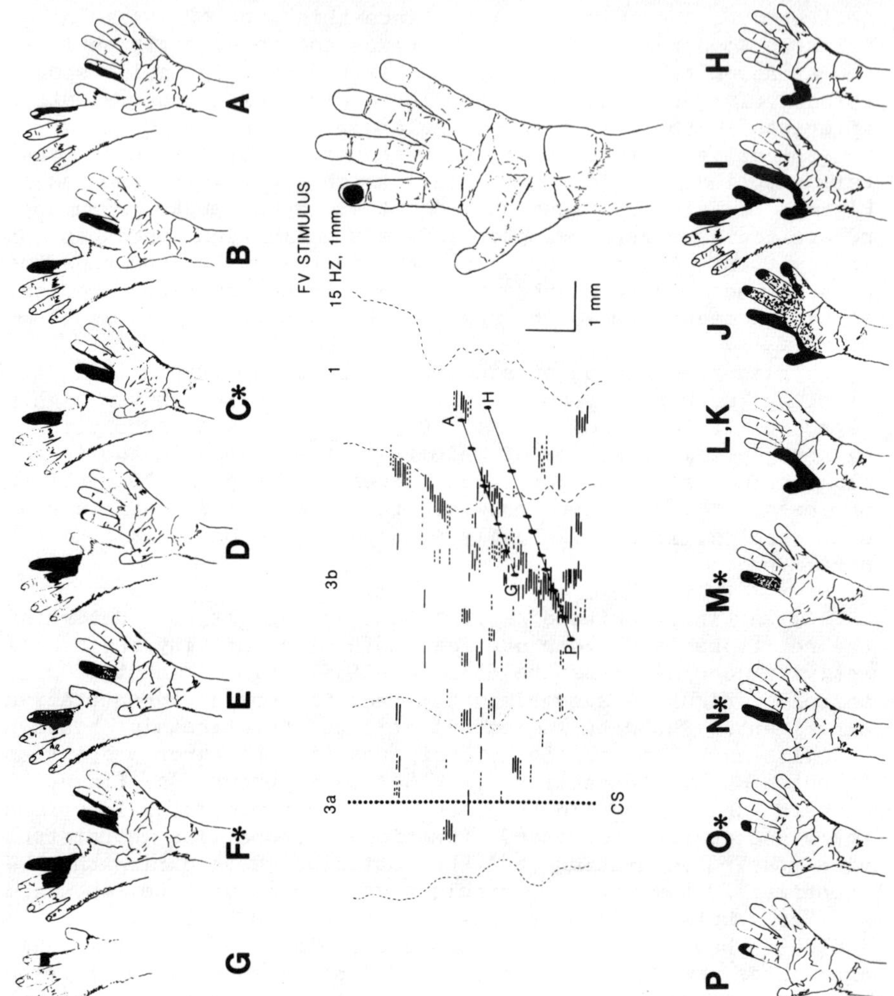
A
B
C*
D
E
F*
G
FV STIMULUS
15 HZ, 1mm
1 mm
1
3b
3a
CS
A
H
G
P
H
I
J
L,K
M*
N*
O*
P

Figure 5. Comparison of single unit and 2DG data obtained from the SI cortex of the same animal. Shown in the center is an "unfolded projection map" showing the distribution of 2DG label in the anterior parietal fields of a Macaca fascicularis monkey which was subjected to a flutter stimulus (FV) on the tip of the contralateral index finger. Such a map summarizes the labeling data observed in a large number of autoradiographs prepared from sections cut serially from a block of cortex extending between the lateral end of the intraparietal sulcus and the lateral end of the postcentral sulcus. For each autoradiograph the locations of labeled column-like aggregates were projected onto layer IV and the distances (along layer IV) of all such aggregates from a landmark present in each section (the fundus of the CS) was measured and plotted. The borders between cytoarchitectural fields are indicated by broken lines. The locations of the columnar 2DG aggregates are indicated by horizontal line segments; solid line segments indicate aggregates of high mean optical density, broken segments indicate aggregates of relatively weak labeling. Superimposed on the map of 2DG labeling are 2 tracks of microelectrode penetrations performed in the same brain. Points at which single unit receptive field (RF) and response properties were determined are indicated by vertical ticks; the sequence of observations collected in each penetration is indicated by letters: observations A-G in the more medial penetration, H-P in the more lateral penetration. Figurine drawings of the cutaneous receptive fields are shown at the top and bottom of the figure. Figurines identified with an asterisk indicate RFs isolated within 2DG labeled aggregates. It should be noted that other units had virtually identical RFs and response characteristics (e.g., units B,E,I,J, and P), but were recorded at loci outside the labeled aggregates.

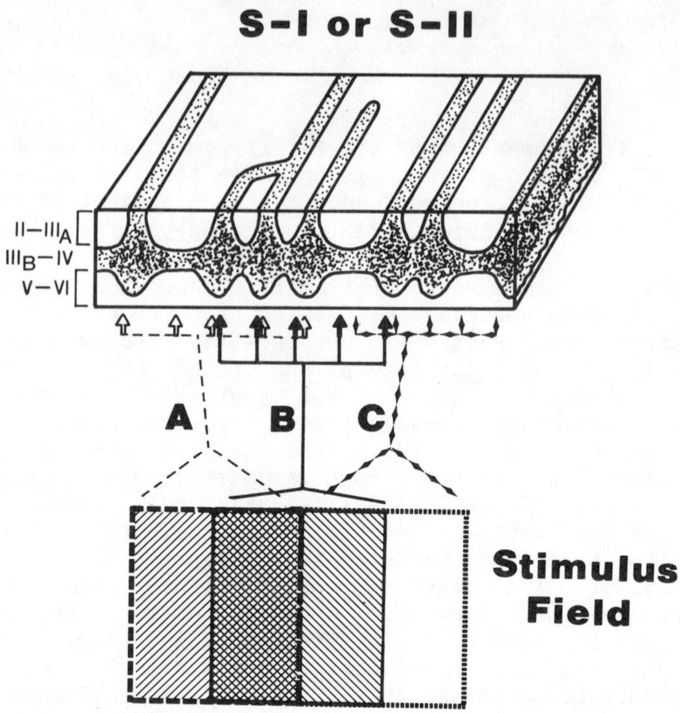

Figure 6. A schematic summary of the 2DG labeling pattern produced within the SI or SII region that represents a stimulated field. At the bottom: 3 subdivisions (indicated by shading) of the peripheral field engaged by a somatic stimulus. At the center: the orderly, but partially shifted and overlapping central projections of the mechanoreceptors within the stimulated field. At the top: the region of primary somatosensory cortex that receives the ascending projections from the stimulated mechanoreceptors, and the spatial and intensive pattern of metabolic activity (presumed to be tightly coupled to cortical neuroelectric activity) evoked by stimulation of the peripheral field. The density of the stippling indicates the density of 2DG labeling. Note that (i) although layer IIIb-IV is labeled continuously throughout the cortex that receives input from the stimulated field, the labeling density fluctuates periodically in this layer, and (ii) the columnar aggregates of label extend for appreciable distances forming strips of label. The overall pattern of labeling is perhaps most easily appreciated from the point of view of the cortical surface, and the pattern changes as stimulus parameters are varied.

properties matching the stimulus, these properties are clearly a major determinant of the pattern. RF properties are not the sole factor at work, however, since many neurons in the regions separating the strips possess RF and response properties consistent with activation by the 2DG mapping stimulus.

DISCUSSION

General properties of 2DG labeling in the somatosensory cerebral cortex.

Figure 6 schematically summarizes 4 fundamental properties of the 2DG labeling pattern observed in SI and SII of all animals studied to date. The first property is that stimulus-evoked 2DG labeling is confined to those sectors of SI and SII which receive afferent input of the appropriate submodality class from the body region stimulated. Most frequently a skin field was stimulated, but in several experiments joint rotation was employed. It should be noted that the field of SI or SII cortex containing stimulus-related labeling can be extensive. For example, even a spatially restricted stimulus (such as a flutter stimulus to a circular field 0.5 cm in diameter on the volar distal phalange of the index finger in monkeys) led to metabolic labeling within a SI territory of at least 4-6 mm^2. In every case, however, the distribution of label was confined to cortical regions in which previous neurophysiological studies had detected units whose receptive fields included the body region stimulated in the 2DG experiment (McKenna, et al., 1982). The second property of the 2DG labeling pattern is that wherever labeling occurs, it exists as intermittent, strip-like aggregates; the aggregates typically extend across layers II-V and, at the level of layer IV, possess an average width of 500 um. In areas 3b and 1, these radial aggregates are frequently superimposed on a continuous, tangential band of label limited to the vicinity of the principle thalamocortical afferent terminal plexus (layers IIIb and IV; see figure 6). Unlike the radial extensions of label which involve the supra-and infragranular layers, the tangential band of label in layers IIIb and IV extends across the entire cortical region representing the body part stimulated. It does, however, exhibit periodic fluctuations in intensity with the most intense labeling occurring at the loci at which labeling extends into the supra- and infragranular layers. The third fundamental property of the spatial labeling pattern is that a given stimulus evokes a pattern that is highly reproducible from one animal to the next (Juliano, et al., 1983a). The fourth property is that the pattern is "stimulus-specific": i.e., the details of the labeling pattern set up within the field representing the body region stimulated vary systematically with variations in the parameters of the somatic stimulus applied to the same body region. One example of the sensitivity of the pattern to stimulus parameters which has been described elsewhere in greater detail (Juliano, et al.,1981) is that a gentle brushing stimulus applied to a defined region on the distal volar pad of the index finger of a monkey leads to a pattern of labeling in SI different from that produced by the application of a flutter-vibration stimulus to the same body region. Another example is that variations in stimulus intensity are accompanied by changes in the patterns of labeling (unpublished observations, Juliano and Whitsel, 1983). Because of its dependency on the parameters of the somatic stimulus, the "stimulus-specific" pattern of labeling associated with a given stimulus has been referred to as the cortical "signature" of that stimulus (Juliano, et al., 1983c).

Functional implications and directions for future research.

The stimulus-evoked 2DG labeling pattern summarized above has been described in detail in previous publications (Juliano, et al.,1981,1983a,1983b). For no stimulus used to date was the pattern of labeling obtained for cerebral cortical areas SI and SII predictable in all details by any neurophysiological map. Moreover, the distribution of labeling is not predicted by single unit data obtained from the same cortical field of the same animal at a time just prior to carrying out the 2DG experiment (see Figure 5). The demonstration that labeling (i) is confined to the cortical field receiving afferent input from the stimulated body region, (ii) is organized into strip-like aggregates of constant dimensions superimposed on a tengential band limited to the region of thalamic afferent termination, (iii) is spatially complex, yet the pattern for a given stimulus is highly reproducible from one animal to the next, (iv) varies systematically with variations in stimulus parameters, and (v) is minimal or absent in the absence of intentional somatic stimulation (Juliano, et al., 1981, 1983a) are interpreted as evidence that patterns of labeling encode information about both the locus and intensity of stimulus evoked neuronal activity.

While additional data are required to quantify and extend the observations obtained to date, the available data appear sufficient to propose a working hypothesis having 3 ingredients: (i) the spatial distribution pattern of somatosensory cortical neural activity encodes information about stimulus conditions, (ii) the same stimulus evokes essentially the same spatial activity pattern in different animals, and (iii) because the pattern is stimulus-specific and reproducible, the levels of the nervous system efferent to the supra- and infragranular layers of SI and SII receive different spatially patterned inputs when different somatic stimuli are delivered. Quantification of the stimulus-evoked patterns of cerebral cortical activity is a requirement if we are to assess the capacity of SI and SII population response patterns to convey stimulus information. Until such quantitative analyses are carried out, relative comparisons of the information encoded by the responding population and its constituent neurons will not be possible.

Mechanisms of cortical signature generation

While the mechanisms of cerebral cortical signature generation must depend on the structure of the underlying network (i.e., the extrinsic and intrinsic connections of the cortex), they appear to transcend the underlying structure in much the same way as the stimulus-evoked, spatially non-homogeneous changes in network status transcend the simple rules of connectivity of the model network studied by Wilson and Cowan (1973). If this is the case, it would be most appropriate if the somatosensory cortical "signature" of a stimulus were regarded as the outcome of a dynamic process involving lateral interactions between the many elements exhibiting common functional properties (RF, submodality and response properties) within the cortical field of interest. Moreover, since the outcome of dynamic processing depends as much on lateral interactions as on differences in single unit functional properties, the boundaries of the strips of active neural aggregates (modules) demonstrated in the somatosensory cortex by the 2DG method should not be expected to exhibit any rigid correspondence to boundaries identified in single unit recording experiments or by procedures for tracing extrinsic connections. Rather, modules would be "formed" at spatially distributed loci within the field receiving input from the body region stimulated, the precise boundaries of the labeled loci being determined by the self-organizing, higher-order properties of the cortical network. An extension of this view of dynamic processing is that cortical signatures will exhibit exquisite sensitivity to manipulations of the intrinsic cortical circuitry such as alterations of the efficacy of inhibitory networks using GABA as the neurotransmitter substance. For the foreseeable future, the concept of stimulus-specific spatial patterns of neural activity (the cortical signatures) and their role in somatosensory stimulus coding appear to be topics most ammenable to experimental investigation with the metabolic mapping approach.

ACKNOWLEDGEMENTS

Supported in part by NIH Grant NS-10865 and by ONR Contract N00014-83-K-0387. The authors express their thanks to Dr. S. Cheema and Mr. O. Favorov for their permission to describe unpublished experimental materials.

BIBLIOGRAPHY

Amari, S. and Arbib, M.A. (1982) Competition and cooperation in neural nets. In: Lecture Notes in Biomathematics, ed. S. Levin, vol. 45, Springer-Verlag, New York.

Boivie, J. (1978) Anatomical observations on the dorsal column nuclei, their thalamic projection and the cytoarchitecture of some somatosensory thalamic nuclei in the monkey. J. Comp. Neurol. 178: 17-48.

Goochee, C., Rasband, W. and Sokoloff, L. (1980) Computerized densitometry and color coding of (14C) deoxyglucose autoradiographs. Ann. Neurol. 7: 359-370.

Johnson, E.M. and Copowski, J.J. (1983) A system for the three-dimensional reconstruction of biological structures. Computers and Biomedical Res. 16: 79-87.

Jones, E.G., Friedman, D.P. and Hendry, S.H.C. (1982) Thalamic basis of place-and- modality-specific columns in monkey somatic sensory cortex: a correlative anatomical and physiological study. J. Neurophysiol. 48: 521-544.

Juliano, S.L., Favorov, O., and Whitsel, B.L. (1983a) The reproducibility of 2DG patterns in monkey SI and their relationship to single unit mapping data. Soc. Neurosci. Abst. 9: 922.

Juliano, S.L., Hand, P. and Whitsel, B.L. (1981) Patterns of increased metabolic activity in the somatosensory cortex of monkeys subjected to controlled cutaneous stimulation: a 2-deoxyglucose study. J. Neurophysiol. 46: 1260-1284.

Juliano, S.L., Hand, P. and Whitsel, B.L. (1983b) Patterns of metabolic activity in cytoarchitectural area S-II and surrounding cortical fields of the monkey. J. Neurophysiol. 50: 961-980.

Juliano, S.L., Whitsel, B.L. and Hand, P.J. (1983c) Patterns of ventrobasal thalamic activity evoked by controlled somatic stimuli: a preliminary analysis. In: Somatosensory Integration in the Thalamus, eds. G. Macchi, A. Rustioni and R. Spreafico, pp. 107-124.

Kaas, J.H. (1983) What, if anything, is S-I? Organization of First Somatosensory Area of Cortex. Physiological Rev. 63: 206-231983.111983.

Loe, P.R., Whitsel, B.L., Dreyer, D.A., and Metz, C.B. (1977) Body representation in ventrobasal thalamus of macaque: a single unit analysis. J. Neurophysiol. 40: 589-607.

Mountcastle, V.B. (1978) An organizing principle for cerebral function: the unit module and the distributed system. In: The Mindful Brain, ed. O. Creutzfeldt, p. 7-50, MIT Press, Cambridge, MA.

Mountcastle, V.B. and Henneman, E. (1952) The representation of tactile sensitivity in the thalamus of the monkey. J. Comp. Neurol. 97: 400-440.

McKenna, T.M., Whitsel, B.L. and Dreyer, D.A. (1982) Anterior parietal cortical topographic organization in macaque monkey: a reevaluation. J. Neurophysiol 48: 289-317.

McKenna, T.M., Whitsel, B.L., Dreyer, D.A. and Metz, C.B. (1981) Organization of cat anterior parietal cortex: Relations among cytoarchitecture, single neuron functional properties, and interhemispheric connectivity. J. Neurophysiol. 45: 667-697.

Nelson, R.J., Sur, M., Felleman, D.J. and Kaas, J.H. (1980) Representations of the body surface in postcentral parietal cortex of Macaca fascicularis. J. Comp. Neurol. 192: 611-643.

Olszewski, J. (1952) The thalamus of the Macaca mulatta. Karger.

Robinson, C.J. and Burton, H. (1980) Somatotopic organization in the second somatosensory area of Macaca fascicularis. J. Comp. Neurol 192: 43-68.

Sokoloff, L. (1981) The deoxyglucose method for the measurement of local glucose utilization and the mapping of local functional activity in the central nervous system. Int. Rev. Neurobiol. 22: 287-333.

Sokoloff, L., Reivich, M., Kennedy, C., Des Rosiens, M.H., Patlak, C.S., Pettigrew, K.D., Sakurada, O. and Shinghana, M. (1977) The (14C) deoxyglucose method for the measurement of local cerebral glucose utilization: Theory, procedure, and normal values in the conscious and anesthetized albino rat. J. Neurochem. 28: 897-916.

Sur, M., Wall, J.T., and Kaas, J.H. (1981) Modular segregation of functional cell classes within postcentral cortex of monkeys. Science. 212: 1059-1061.

Szentagothai, J. and Arbib, M.A. (1974) Conceptual Models of Neural Organization, NRP Bulletin 12: 310-479, (also: The MIT Press, Cambridge, MA.)

Wilson, H.R. and Cowan, J.D. (1973) A mathematical theory of the functional dynamics of cortical and thalamic nervous tissue. Kybernetik. 13: 55-80.

ORGANIZATION OF NOXIOUS AND NON-NOXIOUS INPUTS IN SmI CORTEX: COMPARISON IN NORMAL AND IN ARTHRITIC RATS

G. GUILBAUD, Y. LAMOUR, and J.C. WILLER

Laboratoire de Neurophysiologie Pharmacologique INSERM U. 161, 2, rue d'Alésia 75014, Paris, France

Experimental evidence for a noxious input to the cerebral somatosensory neocortex is so far very limited. Observations suggesting that the cerebral cortex is actually not directly involved in pain mechanisms have been reported (Penfield and Rasmussen 1950). However, preliminary observations of somatosensory cortical neurones driven by noxious stimuli have been reported in cat and monkey (see references in Lamour et al., 1983 a et b).

In the ventro-basal (VB) complex of the rat's thalamus, many neurones receive noxious inputs (Guilbaud et al., 1980). Since in this species most of VB neurones seem to be thalamo-cortical relay cells (Saporta and Kruger 1977) it was of interest to know wether the noxious information is transmitted to the cerebral cortex, and wether it converges or not with non-noxious inputs on single cortical neurones. Moreover it was of interest to investigate the neuronal responses to the different submodalities and the columnar and laminar organization of these inputs in this species. Indeed the columnar organization of the cerebral cortex has been studied almost exclusively in cat and monkey since the pioneering studies of Mountcastle (1957), and has not been so far systematically investigated in the SmI cortex of the rat, although they have been suggested by several authors(Armstrong-James 1975 ; Simons 1978 ; Chapin et al., 1980).

Important questions are raised by the process of cortical reorganization following peripheral injuries (see ref. in Merzenich and Kaas 1982) and it appeared attractive to determine if the long-lasting peripheral modifications induced by arthritis are also accompanied by modifications of the functional properties of SmI cortical neurones. Therefore a comparative study of SmI cortical neurones responses to peripheral stimulation was performed in normal and in arthritic rats.

Method. Procedures for animals preparation and electrophysiological recordings have been extensively described elsewhere (Lamour et al., 1983 a et b). Rats (male albino Sprague-Dawley) were anesthetized with a gazeous mixture (1/3 O_2 - 2/3 N_2O + halothane) immobilized with Flaxedil and artificially ventilated. The anesthesia, deep during surgical procedure (2.5% halothane) moderate during recordings (0.5%) was continously checked by an electrocorticographic recording. Unitary recordings were performed during penetrations perpendicular to the cortical surface with glass micropipettes, filled with a mixture of 2% Pontamine sky blue in 1M NaCl. A dye deposit was made at the final recording site of each penetration which was subsequently reconstructed on camera lucida drawings of sections of the whole brain stained with cresyl violet.

Results

A) Normal rats

Forty penetrations were made in the hind limb, four in the tail and three in the forelimb representation.

1) There are in SmI cortex of normal rats neurones which are driven by noxious stimulations. Among 694 neurones recorded from 26 animals, 292 had a receptive field. a) 201 were driven by non-noxious mechanical stimuli : 161 by superficial inputs (fig. 1A)(most of them by brushing), 40 by deep inputs (joint movement, deep pressure on muscle and fascia). The majority of these neurones responded phasically, adapted very rapidly. Their receptive fields were contralateral, usually of small size.

b) 91 neurones were driven by noxious mechanical stimulations (fig. 1B,C) : 56 responded to noxious stimuli only (nociceptive specific) 35 to noxious as well as non-noxious stimuli (nociceptive non-specific or convergent neurones) ; these convergent neurones were frequently inhibited (13/35) by the stimulation of parts of their receptive field (RF). The nociceptive specific neurones displayed more frequently tonic than phasic responses which often outlasted the stimulation period. The receptive fields of the nociceptive specific neurones were usually large (Fig. 2A). RFs restricted to a part of the contralateral limb or of the tail were observed in 20% of the nociceptive specific neurones only, but in 40% of the convergent neurones. Both types of neurones could also be excited by noxious heat (tested for 24 neurones specific nociceptive and 16 convergent neurones) (Fig. 1C). The threshold of activation was between 46°C and 50°C . Relation between neuronal discharge and temperature was tested in 21 cases (hot water bath from 46° to 60 °, 15 sec duration, every 3 minutes) : in 17 cases there was an increase of the discharge in parallel with temperature and in some cases a linear relationship was found. A good correlation was also observed by increasing the surface of stimulation for some neurones (see however fig. 1C). Phenomena

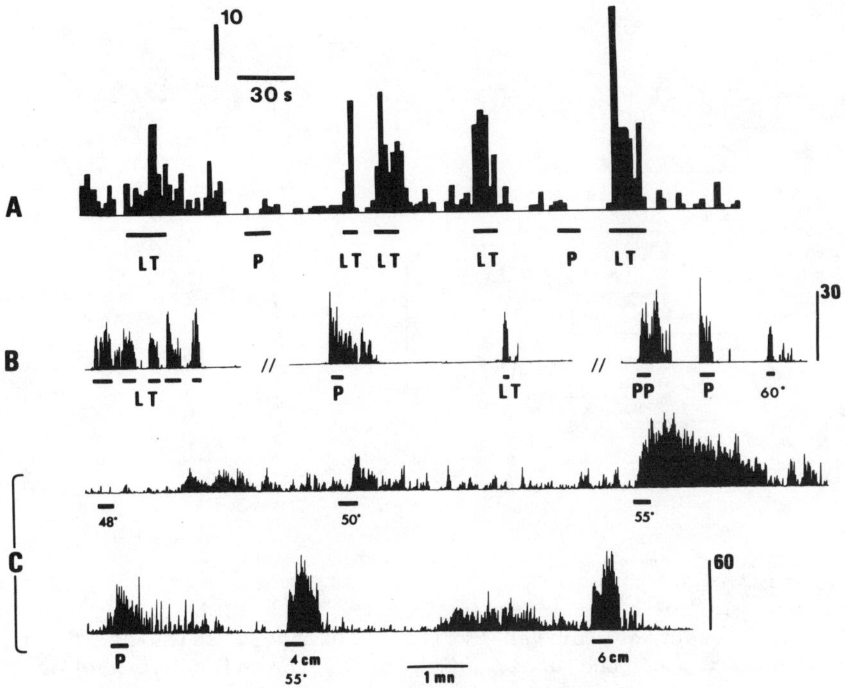

Figure 1

Responses of three cortical neurones (located respectively in layers Vb, VIa and VIb), recorded in different normal rats, to the mechanical or thermal stimulation of their receptive fields (RFs).

A. Example of a non-noxious neurone. This unit was a pyramidal tract neurone (latency of the antidromic response : 2.2 msec). Its RF was restricted to a punctate area of the fifth digit, around the nail (contralateral hind limb). The neurone was easily driven by repetitive light tactile stimulation (LT) but not by pinching (P).

B. Example of a nociceptive non-specific neurone. This unit was driven by repetitive light touch (LT), pinch (P), pin prick (PP) or radiant heat (60° C). The RF was located on the fifth digit of the contralateral hindlimb.

C. Example of a nociceptive specific neurone. The RF was located on the tail. This neurone was driven by noxious heat application (hot water bath) with a threshold between 48° Cand 50°C. Noxious heat applied to a segment of 4 cm of tail was as effective as a stimulation of 6 cm. This neurone was also excited by pinching the tail (P). Vertical bar : number of impulses per second.

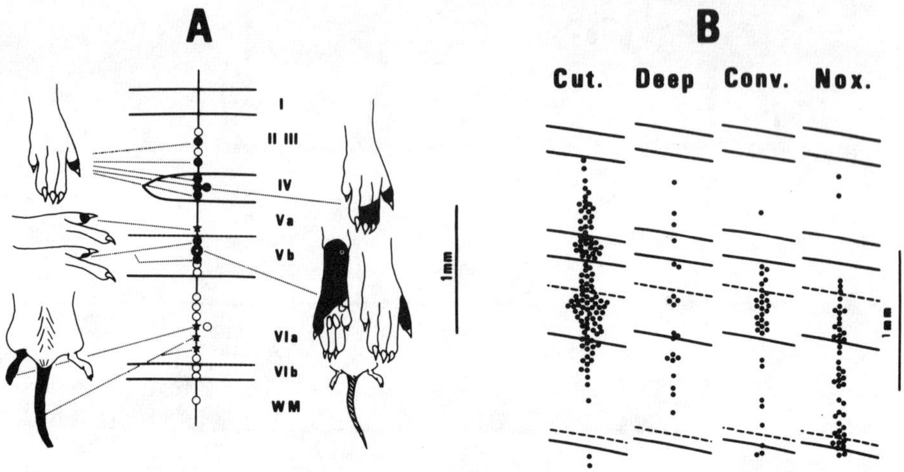

Figure 2
A. Example of a penetration in the hind limb representation including neurones driven by light cutaneous stimulation only, and neurones of the nociceptive specific and convergent categories. Filled circles: cutaneous non-noxious neurones ; open circles: neurones with no RF ; filled stars: nociceptive specific neurones ; open stars: convergent neurones. Shaded areas: excitatory RFs ; hatched area: inhibitory RF.
B. Laminar distribution of the neurones in the different categories (cut: neurones driven by non-noxious cutaneous stimulation, deep: neurones driven by non-noxious deep stimulation, conv: neurones driven by non-noxious as well as noxious stimulation, nox: neurones driven by noxious stimulation only). Each dot represents one cell. Notice the contrast between the repartition of the nonnoxious cutaneous units (most of them being in layers II to V), the convergent (layer V) and the nociceptive specific units (layers Vb and VI). Adapted from Lamour et al. (1983b).

of sentization were noticed when 2 series of stimulations could be performed in the same neurone.

2) There is a laminar organization of noxious and non-noxious inputs but little submodality segregation in the rat SmI cortex. Neurones driven by cutaneous non-noxious stimulation were mainly recorded from layers II to V. Their distribu-

tion, quite similar to that of neurones driven by deep inputs, contrasts sharply with that of neurones driven by noxious stimuli, which were found mainly in layers Vb to VIb, (fig. 2B).

The nociceptive specific neurones were more frequent in layer VI whereas convergent neurones were concentrated in layer Vb. There was not a single noxious neurone in lamina IV. As mentionned above the size of the RF of noxious neurones was often large, specially for the nociceptive specific neurones.

The RF size of the non-noxious neurones seems to vary as a function of their laminar position. Small RFs (below 0.8cm^2) predominated in layer IV, whereas neurone with large RFs (above 3cm^2) were more numerous in layer VI (43%) than in layers II to IV (about 10%).

In the majority of the cases it was possible to record in the same penetration different neurones, which were activated by different modalities i.e. superficial and deep. The "modality pure" columns represented only 48% of the penetrations (41 for the superficial 7 for the deep inputs). Moreover there was almost no segregation among columns between non-noxious and noxious inputs in contrast with their clear laminar segregation. The overlap between successive RFs observed for non-noxious cutaneous neurones, was not necessarily observed between RF of noxious and non-noxious neurones.

B) Arthritic rats. Arthritic rats were injected with Freund's adjuvant into the tail at the breading center (Charles River France). The experiments were performed between the 3rd and 4th weeks following the injection, when the arthritic disease reaches its maximum intensity (Gouret et al., 1976 ; Pearson et al., 1959). Anesthetic conditions were as described above.

1) The functional properties of cortical SmI neurones are dramatically changed. Among 380 neurones, 131 had a receptive field. We used a classification of neurones different from the one used in normal rats. It was indeed difficult to assume that a given, even moderate stimulation was not noxious in the arthritic rat in view of the important inflammatory lesions. Six neurones only were driven by brushing applied on a restricted contralateral RF (type 1) and only 4 neurones were exclusively excited by intense mechanical stimuli (i.e. pinch), (type 3).

Most of the neurones (121) were driven by mild mechanical stimulation applied to joints and/or on cutaneous oedematous areas (type 2) : 50% were driven by joints movement only, 25% by light pressure, and another 25% by both. Fifty one only had

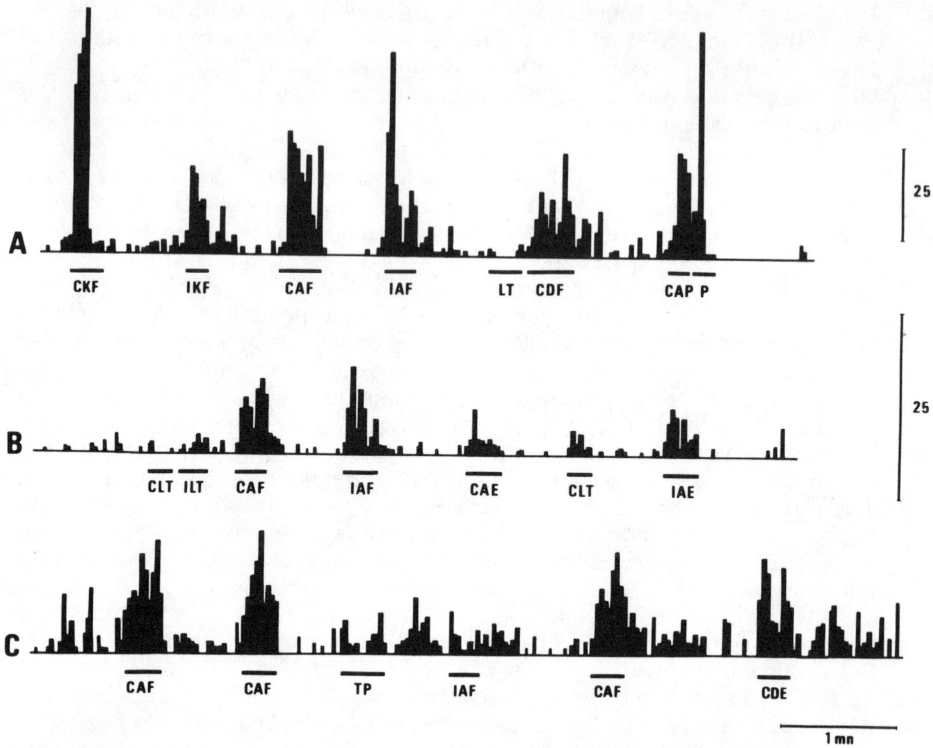

Figure 3

Responses of three cortical neurones (located respectively in layers IV, Va and VIa) recorded in an arthritic rat, to the mechanical stimulation of their peripheral receptive fields.

A. This unit was driven by flexion of the contralateral knee (CKF) and ankle (CAF), the ipsilateral knee (IKF) and ankle (IAF) as well as by contralateral digit flexion (CDF). Contralateral ankle pressure was also effective (CAP). In contrast light touch (i.e. brushing, LT) or pinch on the ankle (P) did not result in a sustained discharge of this neurone.

B. This unit was driven by contralateral and ipsilateral ankle flexion (CAF and IAF) and to a lesser extent by contralateral and ipsilateral ankle extenxion (CAE and IAE). Contra or ipsilateral light touch was only marginaly efficient.(CLT and ILT).

C. Unit driven by contra but not by ipsilateral ankle flexion (CAF, IAF) and contralateral digit extension (CDE). A pinch applied on the tail was also ineffective (TP). Vertical bar: number of impulses per second.

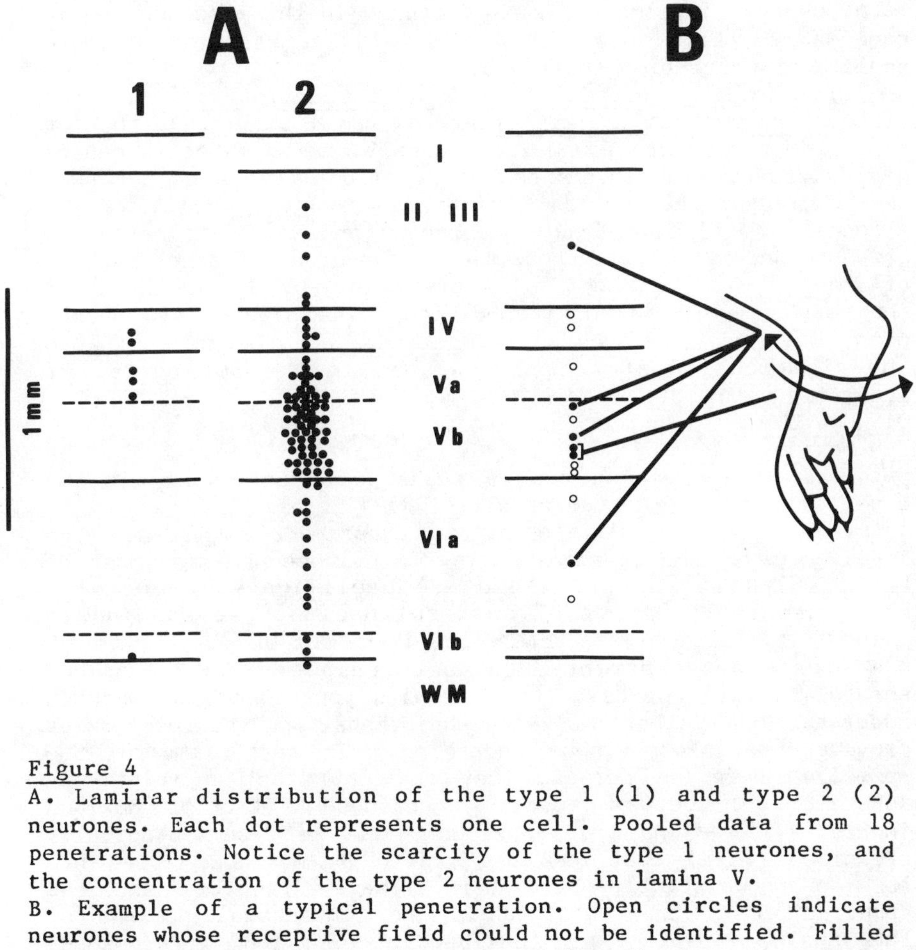

Figure 4
A. Laminar distribution of the type 1 (1) and type 2 (2) neurones. Each dot represents one cell. Pooled data from 18 penetrations. Notice the scarcity of the type 1 neurones, and the concentration of the type 2 neurones in lamina V.
B. Example of a typical penetration. Open circles indicate neurones whose receptive field could not be identified. Filled circles indicate neurones driven by either flexion or extension of the ankle (arrows). (From Lamour et al., Brain Research, in press).

a strictly contralateral RF, whereas 60 had a bilateral, and 10 an ipsilateral RF. The vast majority of the neurones were driven by stimulation applied to the ankle and less often to the digits or both. The effective movement was either a flexion or an extension of the ankle or sometimes both (fig. 3).

The responses had a rapid onset and were usually sustained, lasting at least as long as the stimulation was maintained, with little or no adaptation. Long after-discharges outlasting the period of stimulation were rare (see however Fig. 3A). Examples of rapidly adapting responses or "on-off" responses were observed in some cases. A decrease in the response amplitude was sometimes observed when repetitive stimulations were applied at short time intervals.

2) There is a profound change of the laminar organization of somatic neurones in SmI cortex. As shown in Fig. 4, the few type 1 neurones were found in layers IV and Va, whereas neurones driven by limb movements and/or mild pressure on the skin were found in all cortical layers, specially in layer V. The trend toward finding type 2 neurones in layer V is certainly strengthened by the electrode bias. However a different pattern of laminar distribution was observed in normal animals under the very same experimental conditions. The bias toward recording from large neurones in layer V does not therefore explain the distribution of type 2 neurones in the SmI cortex of arthritic rats.

DISCUSSION

Noxious inputs. The role of the cerebral cortex in the perception of pain has been a subject of controversy in the past. (see ref. in Lamour et al., 1983 a et b).

There is, very little direct experimental evidence for involvement of the cerebral cortex in pain mechanisms. Most of the results so far published are preliminary, or briefly mentioned the presence of cortical neurones responding to painful stimulation (see references in Lamour et al., 1983 a et b). Others are restricted to tooth pulp projection to the somato sensory cortex. Our results obtained in animals moderately anaesthetized with halothane and nitrous oxide provide new informations concerning pain mechanisms in the mammalian cerebral cortex. They show (i) that a relatively substantial number of cortical neurones do receive noxious inputs, (ii) that this noxious input is somatotopically organized, (iii) that neurones driven by noxious inputs are able to encode stimulus parameters. These results can be compared with those obtained in the ventrobasal nucleus of the thalamus in the rat (Guilbaud et al., 1980 ; Peschanski et al., 1980). However, the overall percentage of SmI neurones driven by noxious stimulation was much lower in our study than that observed in the VB thalamus and the ratio of convergent neurones to nociceptive specific neurones higher (Guilbaud et al., 1980). Moreover the threshold of the responses to noxious heat application was slightly higher in the cerebral cortex. It appears that, despite the massive projection from the VB thalamus to the SmI cortex (and even if the general characteristics of the responses are unchanged), some features of the neuronal responses to noxious stimulation differ in these structures.

Columnar and laminar organizations.Our results concerning responses to non-noxious stimulation are consistent with those reported in the vibrissa representation. Neurones responded preferentially to stimulus transients and a columnar pattern of organization was observed during radial penetrations. The somatotopic organization of cutaneous responses we observed is in general agreement with that described by Welker (1976), if one considers the RFs of neurones located in the superficial layers. Since neurones located in infragranular layers tended to have larger RFs, there was more overlap between the RFs of these neurones.

Mouncastle (1957) reported that in the cat SmI cortex 84% of the penetrations were place and modality specific. We observed only 48% of place and modality specific penetrations in the hind limb representation of the rat SmI cortex. We conclude that this type of organization which also exist in the monkey (Powell and Mouncastle 1959) is not as well developed in the rat, at least in the hind limb representation.

However our results support the concept of a type of cortical organization based on differences in the properties of neurones located in different layers. Indeed there is an obvious trend for the noxious inputs to drive neurones is layers V and VI. This trend is even more pronounced for the neurones of the nociceptive specific type which were numerous in layers VIa and VIb, whereas neurones of the convergent type were more often found in layer V. There is also a clear segregation between neurones driven by noxious and neurones driven by non-noxious stimulations. This is evidence for the specificity of the noxious input to the cerebral cortex. This information might not travel along the same pathways as non-noxious information. The complete lack of responses to noxious stimulation in layer IV is yet more evidence in favor of this hypothesis. In contrast there is apparently little difference between the laminar distribution of neurones driven by non-noxious cutaneous and deep stimulations.

Layer IV appears very specific among other cortical layers in terms of the type of input it receives. This is consistent with autoradiographic investigations (Herkenham 1980) showing a heavy termination of thalamo-cortical axons in layer IV of the cortical zone containing aggregates. However, one might wonder why layer IV neurones never responded to noxious stimulation. Indeed, it was shown that many neurones in the rat VB respond to noxious stimulation (Guilbaud et al., 1980). Since most of VB neurones are likely to be thalamo-cortical relay cells (Saporta and Kruger 1977) these noxious VB neurones might also project to layer IV. Do thalamo-cortical terminals relaying noxious inputs terminate specifically on dendrites of deeper neurones but not on layer IV neurones themselves ? or does the specificity rely upon intrinsic cortical organization ? As yet, we do not have evidence favoring either one of these possibilities. It seems likely that neurones in layer V and VI receive direct noxious inputs since they cannot be relayed by layer IV neurones and since thalamo-cortical axons also terminate in layer VI (Herkenham 1980 ; Landry and Deschênes 1981). However cortical projections of the intralaminar thalamic nuclei could also participate in the transfer of noxious information from the thalamus to the cerebral cortex. These nuclei send terminals in layers V and VI (Herkenham 1980), and receive noxious information (Peschanski et al., 1981).

In any case this laminar organization supports the hypothesis of White (1978) : there is in the SI cortex, beside a sequential processing of information (whose first step would be in layer IV), a parallel processing of thalamic input by different neurones located in different layers. Our results also agree with recent anatomical data (Penny et al., 1982) and with the possibility, suggested by Jones et al. (1982) that spino-thalamic recipient cells projecting to the cortex may form rods independent of those relaying lemniscal inputs only.

Arthritic rats. The present results obtained in arthritic rats reveal a very profound change in the properties of cortical SmI neurones in rats suffering from a chronic arthritis, as compared with normal animals. The functional categories of neurones, the properties of their responses to peripheral mechanical stimulation as well as their respective laminar distribution were dramatically modified.

The possibility that these changes are due to a direct effect on the nervous system of the arthritic-inducing substance seems to be quite remote (Pearson et al., 1961). Whether the changes are due to a peripheral or a central process is a complex issue.

The changes in the responsiveness of articular nerve fibers (in arthritic rats by comparison with normal) (Iggo and coll., 1983) could be at the origin of the increased responsiveness of CNS neurones to moderate stimulations of inflammed joints in the spinal dorsal horn (Menetrey and Besson 1982), the thalamic nuclei (Gautron and Guilbaud 1982 ; Kayser and Guilbaud in preparation) and the SmI cortex. However they cannot explain all the features observed at the cortical level and central mechanisms have to be involved.

Although similar changes were observed in the VB nucleus of the arthritic rat, (lack of neurones driven by intense mechanical stimuli whereas most of them are driven by mild joint stimuli), some differences appear in the detail of the results (long after-discharges outlasting the period of stimulation not so frequent, and lack of type 1 neurones in SmI cortex). In any case the comparison between the laminar distribution of cortical neurones driven by intense mechanical stimuli in normal rats and by mild joint stimulation in arthritic rats, is an indirect evidence that VB neurones considered as "noxious" in the first group (Guilbaud and al., 1980) and those driven by inflamed joint stimuli in the second are not of the same type as it was previously suggested (Gautron and Guilbaud 1982).

These results, obtained in arthritic rats which are a useful model of experimental pain, do suggest a considerable degree of functionnal plasticity in the somato-sensory cortex.

REFERENCES

Armstrong-James M. (1975) The functional status and columnar organization of single cells responding to cutaneous stimulation in neonatal rat somatosensory cortex SI. J. Physiol (Lond) 246, 501-538.

Chapin J.K., Lin C.S., Woodward D.J. (1980) Laminar differences in the size and shape of receptive fields in rat somatosensory (SI) cortex. 10th Annual Meeting, Cincinnati. Neuroscience Abstr. 6, 62.

Gautron M., Guilbaud G. (1982) Somatic responses of ventrobasal thalamic neurones in polyarthritic rats. Brain Res. 237, 459-471.

Gouret C., Mocquet G., Raynaud G. (1976) Use of Freund's adjuvant arthritis test in anti-inflammatory drug screening in the rat : value of animal selection and preparation at the breeding center. Lab-Anim-Sci 26, 281-287.

Guilbaud G., Peschanski M., Gautron M., Binder D. (1980) Neurones responding to noxious stimulation in VB complex and caudal adjacent regions in the thalamus of the rat. Pain 8, 303-318.

Herkenham M. (1980) Laminar organization of thalamic projections to the rat neocortex. Science 207, 532-535.

Iggo A. Guilbaud G., Tegner R. (1983) Sensory mechanisms in arthritic rats . In press.

Jones E.G., Friedman D.P., Hendry S.H.C. (1982) Thalamic basis of place-and modality-specific columns in monkey somatosensory cortex: a correlative anatomical and physiological study. J. Comp. Neurol. 48, 545-568.

Lamour Y., Willer J.C., Guilbaud G. (1983) Rat somatosensory SmI cortex: I characteristics of neuronal responses to noxious stimulation and comparison with responses to non-non noxious stimulation. Exp. Brain Res. 49, 39-45.

Lamour Y., Guilbaud G., Willer J.C. (1983) Rat somatosensory SmI cortex: II laminar and columnar organization of noxious and non-noxious inputs. Exp. Brain Res. 49, 46-54.
Landry P., Deschenes M. (1981) Intracortical arborizations and receptive fields of identified ventrobasal thalamo-cortical afferents to the primary somatic sensory cortex in the cat. J. Comp. Neurol. 199, 345-371.
Ménétrey D., Besson J.M. (1982) Electrophysiologcial characteristics of dorsal horn cells in rats with cutaneous inflammation resulting from chronic arthritis. Pain 13, 343-364.
Merzenich M.M., Kaas J.H. (1982) Reorganization of mammalian somatosensory cortex following peripheral nerve injury. Trends in Neuroscience. dec, 434-436.
Mountcastle V.B. (1957) Modality and topographic properties of single neurons of cat's somatic sensory cortex. J. Neurophysiol. 20, 408-434.
Pearson C.M., Wood F.D. (1959) Studies of polyarthritis and other lesions induced in rats by injection of mycrobacterial adjuvant. I: General clinical and pathological characteristics and some modifying factors. Arthr Rheum 2, 440-459.
Penfield W., Rasmussen T. (1950) The cerebral cortex of man. A clinical study of localization of function. Hafner, New York.
Penny G.R., Itoh K., Diamond I.T. (1982) Cells of different sizes in the ventral nuclei project to different layers of the somatic cortex in the cat. Brain Res. 242, 55-65.
Peschanski M., Guilbaud G., Gautron M., Besson J.M. (1980) Encoding of noxious heat messages in neurons of the ventrobasal thalamic complex of the rat. Brain Res. 197, 401-413.
Peschanski M;, Guilbaud G., Gautron M. (1981) Posterior intralaminar region in rat: Neuronal responses to noxious and non-noxious cutaneous stimuli. Exp. Neurol. 72, 226-238.
Powell T.P.S., Mountcastle V.B. (1959) Some aspects of the functional organization of the cortex of the post-central gyrus of the monkey: A correlation of findings obtained in a single unit analysis with cytoarchitecture. Bull Johns Hopkins Hosp. 105, 133-162.
Saporta S., Kruger L. (1977) The organization of thalamo-cortical relay neurons in the rat ventrobasal complex studied by the retrograde transport of horseradish peroxidase. J. Comp. Neurol. 174, 187-208.
Simons D.J. (1978) Response properties of vibrissa units in rat SI somatosensory neocortex. J. Neurophysiol. 41, 798-820.
Welker C. (1976) Receptive fields of barrels in the somatosensory neocortex of the rat. J. Comp. neurol. 166, 173-190.
White E.L. (1978) Identified neurons in mouse SmI cortex which are postsynaptic to thalamocortical axon terminals: A combined Golgi-electron microscopic and degeneration study. J. Comp. neurol. 181, 627-662.

CORTICAL NEURONAL MECHANISMS UNDERLYING THE PERCEPTION OF MOTION ACROSS THE SKIN

ESTHER P. GARDNER

Department of Physiology and Biophysics, New York University School of Medicine, 550 First Avenue, New York, NY 10016, USA

The tactile sense is normally used to perceive contact of the body with external objects, and to measure their size, geometric form, texture, density and weight. This information is used not only for conscious awareness of the external environment, but also to guide motor activity. Most of these sensory funtions involve motion of the object into and across the skin, as motion enhances one's sensory discriminative abilities, particularly for size, shape and texture recognition (Gibson, 1962; Lederman, 1974; Morley, et al., 1983). Motion of the skin can be active, as when exploring one's environment, or passive, as when contacting another living animal, or a moving object. Psychophysical studies have shown that active and passive modes are equally effective in providing information for texture perception (Lederman, 1974), but active motion is superior for shape recognition (Gibson, 1962).

For the last several years, my colleagues and I have studied the responses of neurons in the cerebral cortex of alert monkeys to passively applied moving stimuli. The data suggest that individual cortical neurons, or even groups of neurons of a given type, do not encode all of the properties of moving objects. Rather, different pieces of information are processed in parallel by specialized groups of neurons, which detect particular aspects of the stimulus.

There are at least two different populations of neurons in the parietal lobe providing information about stimulus motion on the skin. One population consists of the simple tactile neurons in areas 3b and 1 whose response properties were first described by Mountcastle and Powell (1959). We find that activity of these cells provides information about instantaneous position of a moving stimulus.

The second population of cortical neurons is more specialized, responding preferentially to moving stimuli (Whitsel, et al., 1972; Hyvarinen and Poranen, 1978; Costanzo and Gardner, 1980). Some of these motion sensitive neurons have the further requirement that the stimuli move in particular directions across the skin, and are called direction sensitive neurons. I propose that motion sensitive and direction sensitive neurons serve to abstract the concept of motion across the skin, rather than encode all of the perceptual details of the moving stimulus.

To characterize the responses of simple tactile neurons, we initially mapped the spatial organization of their receptive fields using airpuffs. Airpuffs provide a highly localized, reproducible tactile stimulus which is relatively selective for rapidly-adapting mechanoreceptors. Simple tactile neurons respond to such brief stimuli with a high frequency burst of impulses lasting 15-25 ms. Responses are usually graded in intensity as a function of stimulus location on the skin, and are largest at the center of the receptive field (Gardner and Costanzo, 1980a).

The excitatory receptive fields of such cortical neurons are quite large, and often span entire functional areas of the hand or arm such as the palm, volar surface of the forearm, or an entire digit (Fig. 1). However, the spatial acuity of these cells is much greater than the receptive field area would suggest because of the gradation of sensitivity within the field. Most SI neurons show a distinct region of maximum sensitivity at the center of the receptive field, and a relatively smooth decrease in responsiveness as stimuli move away from the field center. Some cortical neurons, particularly those with receptive fields near the wrist or on the digits, display a distal skewing of the point of maximum sensitivity (Fig. 1A, B left), reflecting in part the gradient in innervation density of these skin areas (Darian-Smith and Kenins, 1980; Johansson and Vallbo, 1979). Occasionally we find neurons which lack a clear central focus in their fields, and instead show fairly uniform sensitivity to punctate stimuli (Fig. 1B, right).

These response profiles indicate that point localization on the skin involves a population coding mechanism, as originally proposed by Mountcastle and Powell (1959). For neurons with large receptive fields, as in the forearm area, it is necessary for the nervous system to evaluate the relative distribution of activity within the population to determine where the stimulus is situated on the skin. On average, the most active cells in the population are those whose receptive field centers are contacted by the stimulus, while the other neurons show relatively weaker responses.

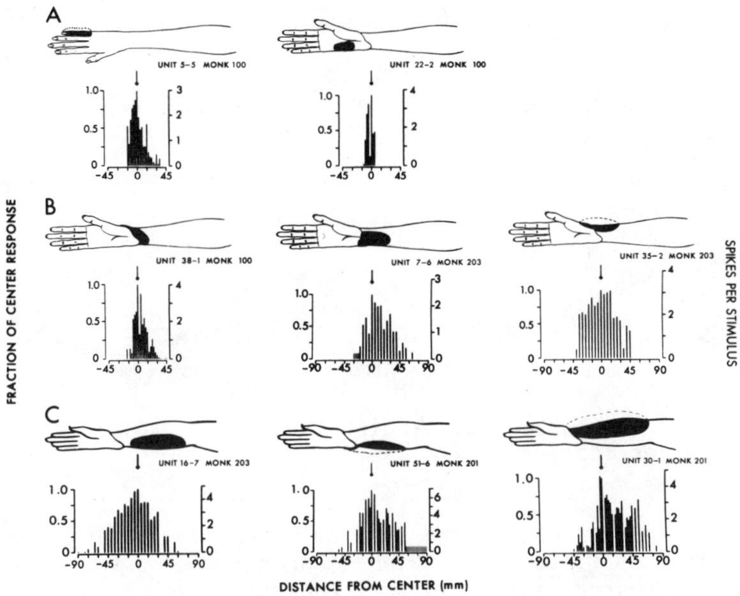

Fig. 1. Excitatory response profiles of 8 cortical neurons showing mean number of impulses elicited at points along the longitudinal axis of the receptive field (RF). RF locations: A. Hand. B. Wrist. C. Forearm. Note increase in RF size with more proximal locations.

In addition to these excitatory receptive fields, the simple tactile neurons also have inhibitory receptive fields, which when stimulated depress the firing rate of the neuron. The spatial organization of inhibitory receptive fields has been described by two different models which are outlined in Fig. 2. In the surround inhibition model (Mountcastle and Powell, 1959), the receptive field consists of an excitatory field in the center, which is surrounded by another region of skin from which inhibition is evoked (the "inhibitory surround"). In this scheme, the excitatory and inhibitory receptive fields are spatially distinct from one another, with the inhibitory field forming an annulus around the excitatory field. The strongest excitation is produced in the middle of the receptive field, while the most intense inhibition is evoked from just outside of the excitatory field in the inhibitory surround. However, neurons with inhibitory surrounds are not very common in SI cortex; only 10-15% of these cortical neurons have actually been shown to have an inhibitory surround (Mountcastle and Powell, 1959; Baker, et al., 1971).

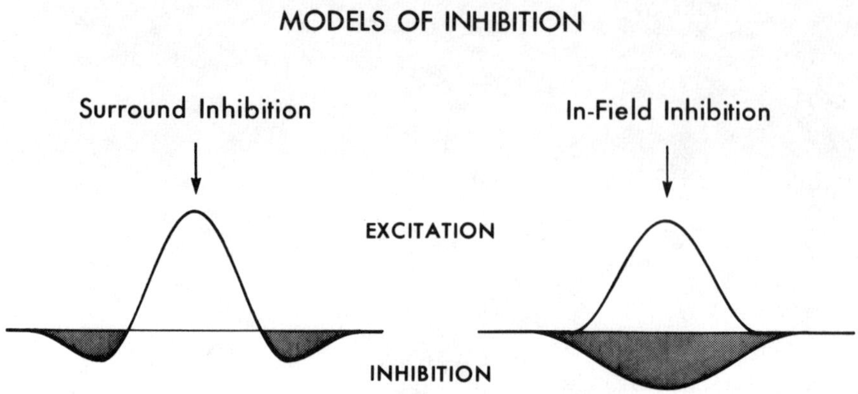

Fig. 2. Models of the spatial distribution of excitation and inhibition in cortical receptive fields.

We have proposed a second type of receptive field organization called in-field inhibition (Gardner and Costanzo, 1980b). Here, the excitatory and inhibitory receptive fields are superimposed on each other. In this model, inhibition is generated from within the excitatory receptive field, and the point on the skin evoking the most excitation also produces the most inhibition. The reason this is meaningful, and that inhibition does not cancel out excitation, is that inhibition takes longer to develop than excitation, and lasts for a longer period of time. Therefore, in the in-field inhibition model, excitation and inhibition are separated not in space, but in time.

For some neurons, the inhibitory receptive field may be slightly larger than the excitatory field, and extend beyond the border to form an inhibitory surround. However, inhibition generated from the periphery or surround of the receptive field is trivial compared to the inhibition generated from within the receptive field, at least for rapidly-adapting cortical neurons.

There are two lines of evidence which support the in-field inhibition model. First, intracellular recordings from cortical neurons show that stimulation of the skin produces first an EPSP and spiking for 15-20 ms, followed by an IPSP (Innocenti and Manzoni, 1971; Whitehorn and Towe, 1968). The IPSP reaches maximum negativity 5-15 ms after the EPSP, and lasts 50-150 ms. Similar EPSP-IPSP sequences have also been demonstrated in the dorsal column nuclei and ventrobasal nucleus of the thalamus. Thus excitation, followed by inhibition, is characteristic of the response patterns of the entire lemniscal pathway.

Secondly, the effects of the IPSP can be demonstrated extracellularly, as a period of decreased excitability immediately following a tactile stimulus. In Fig. 3, we stimulated the skin sequentially with two airpuffs (C and T), each of which excites the cortical neuron. When the stimuli are paired (C-T), the neuron's response is reduced. For example, if the test airpuff follows the conditioning airpuff by 20 ms, the response to the test stimulus is practically abolished. Instead of responding with three spikes to the test stimulus, the neuron now fires only intermittently. In the period 20-40 ms following the conditioning stimulus, the response to the test stimulus remains small, returning to control levels only at C-T intervals greater than 100 ms. The time course of test response suppression thus closely parallels that of IPSPs recorded intracellularly from cortical neurons.

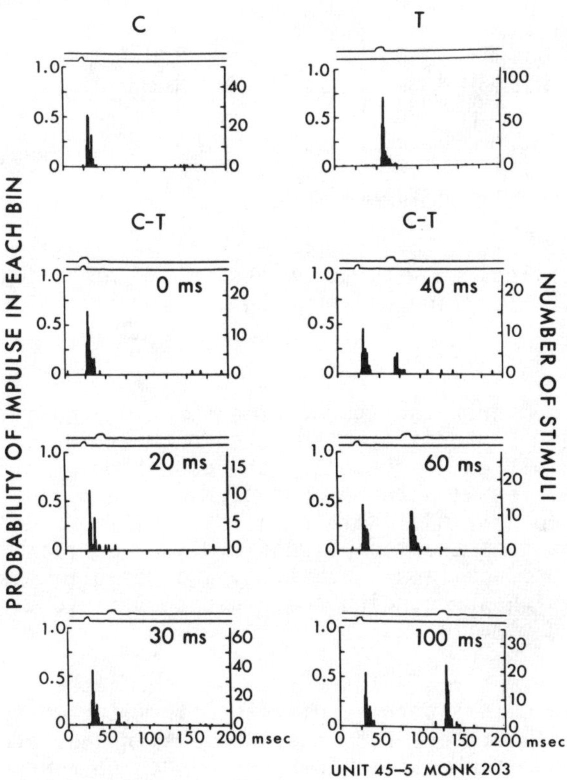

Fig. 3. PSTHs showing the time course of in-field inhibition produced by airpuffs to the field center. (Reproduced from Gardner and Costanzo, 1980b).

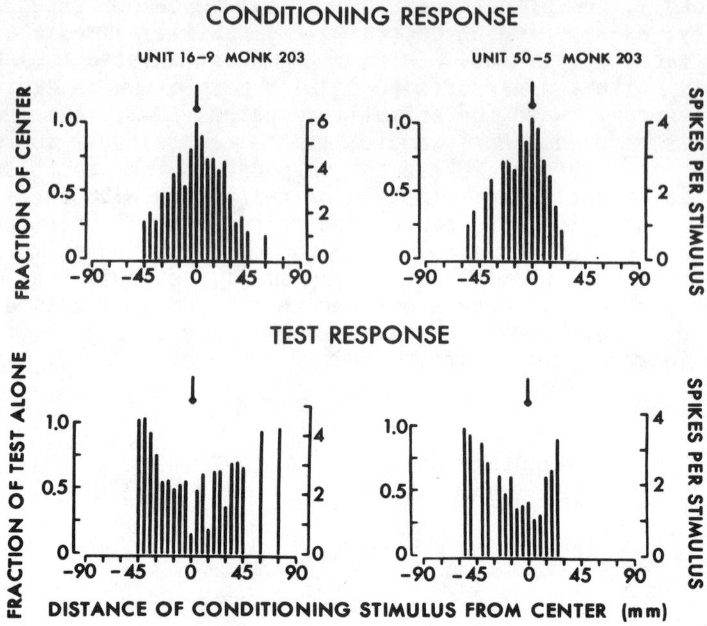

Fig. 4. Top: Excitatory response profiles. Bottom: Inhibitory response profiles showing suppression of the test response as a function of the position of the conditioning stimulus in the receptive field.

Fig. 4 compares the amount of excitation and inhibition generated by stimuli at different points in the receptive field. The amount of inhibition generated by the conditioning stimulus seems to be correlated with the amount of excitation that it produces. Like excitation, maximum inhibition is usually produced by stimuli near the center of the receptive field. Stimuli in the periphery of the field produce less inhibition, and stimuli near the edge and in the region surrounding the excitatory field evoke the weakest inhibition of all.

In some receptive fields (Fig. 4 right), the center of the inhibitory field is laterally displaced from the center of the excitatory field. Different points on the skin thus evoke the most excitation and the most inhibition in such neurons. These cells may provide the inputs to direction sensitive neurons.

These data show that the excitatory and inhibitory receptive fields are superimposed in space, and that the in-field inhibition model best describes their spatial geometry. Over 80% of the rapidly-adapting simple cortical neurons tested showed in-field inhibition, while very few showed inhibitory effects from stimuli outside of their excitatory receptive fields (Gardner and Costanzo, 1980b). Similar results have been obtained in the cat (Janig, et al., 1977, 1979; Laskin and Spencer, 1979). We have not tested slowly-adapting neurons with this paradigm, because they do not respond to airpuffs. It is still possible that surround inhibition may correctly characterize their receptive fields.

These studies with airpuffs were done with stimuli applied to individual points on the skin, rather than with moving stimuli. Nevertheless, they may shed light on how simple tactile cells respond when a probe or edge is moved across the skin. Fig. 5 left summarizes the events occurring in SI cortex when the skin is briefly touched. 10-20 ms after the stimulus is presented to the skin, a broad population of cortical cells is excited. The degree of excitation of each neuron depends upon the distance of the stimulus from the center of that neuron's receptive field. Following the excitatory response, this same population of neurons becomes inhibited. The neurons which were initially the most excited, subsequently become the most inhibited, and those that received the weakest excitation will be the least inhibited.

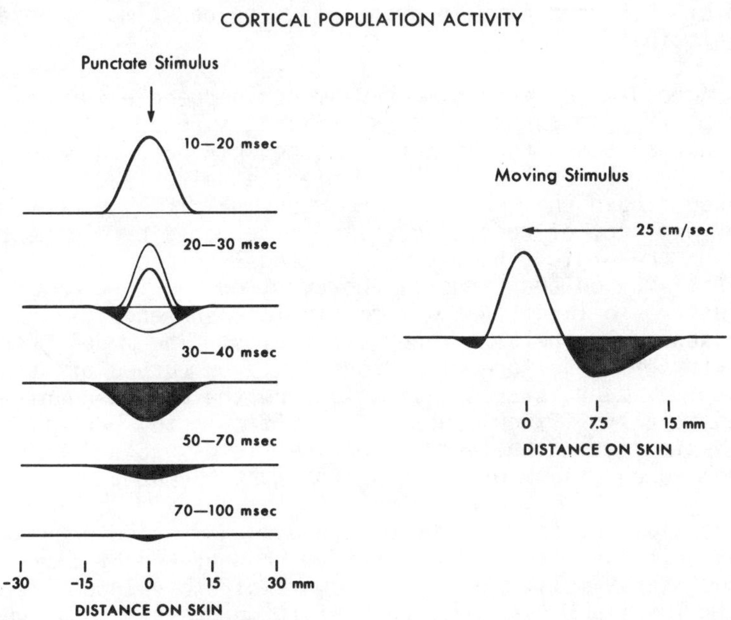

Fig. 5. Left: Reconstruction of the time course of population responses to a brief tactile stimulus. Right: Theoretical distribution of population responses to a 25 cm/s moving stimulus.

Fig. 5, right, diagrams the theoretical distribution of activity in the population of simple tactile neurons when a stimulus is moved across the skin. As the stimulus moves, successive populations of neurons are engaged by the stimulus. Therefore, at any moment in time, there should be a population of cortical neurons in each of the states of activity shown on the left. Each population should be initially excited, and then inhibited by the stimulus. Etched in the population activity profile appears a picture of where the stimulus is at any moment in time, as indicated by the locus of peak excitation. The population profile should also indicate where the stimulus has been in the previous 100 ms, by the inhibited neurons. Therefore, a moving stimulus is postulated to leave behind a trail of inhibition in the wake of peak excitation representing the instantaneous location of the stimulus on the skin.

This population profile contains within it all of the perceptual details necessary for specifying an individual event on the skin. The population profile depicts the direction of motion by the trail of inhibition, and the velocity of motion, by both the firing rates of neurons within the active population, and the spatial distance in the cortex between columns containing the most active and the most inhibited neurons. Thus, a population coding mechanism of this sort can, in principle, depict stimulus movement across the skin.

Such a population coding mechanism could operate for neurons with receptive fields on the hand, which are relatively small in area. As normal scanning movements of the skin occur at speeds of 10-20 cm/s (Dreyer, et al., 1978; Morley, et al., 1983), a moving stimulus would have shifted only 4 mm when inhibition becomes maximum. A new set of receptive fields is engaged by movements of this size on the digits, permitting the model to account for motion sensitivity on the fingers. However, on the forearm, receptive fields are so large that a 4 mm displacement engages neurons having essentially the same receptive fields. The model predicts that activity of these forearm neurons would be turned off shortly after the movement starts. If this were the only mechanism for motion sensitivity, it would be rather difficult to perceive motion on skin areas with large receptive fields. Since this perception does occur, some other mechanism must operate.

One possibility is that the continuous excitation provided by a moving stimulus is different from that evoked by temporally spaced punctate stimuli. The continuous excitatory input from a moving stimulus might overcome the inhibition one sees when a tactile stimulus ends, and the excitatory activity alone would be used to locate the moving stimulus. Support for this idea is indicated in Fig. 6, which shows a slowly adapting cortical neuron with a receptive field on the glabrous skin of digit 3. The cell was most sensitive to stimuli on the lower portion of the distal phalange. When the skin was stroked with a cotton swab, or a toothed wheel was rotated over the surface, the firing rate increased to a maximum when the receptive field center was crossed. However, although the cell's firing rate diminished, it did not turn off after the field center was crossed, as would be predicted by the in-field inhibition model.

In-field inhibition is either inoperative for such slowly adapting neurons, or else is overcome by the continuous excitatory barrage elicited by the moving stimulus. This is true even when the receptive field center is continuously stimulated. For example, when the monkey actively grasped a toothed wheel (Fig. 6C), the cell fired continuously, at a reasonably steady rate, for the entire period of several seconds until he released his grasp. Similarly, when a large surface was pulled across the skin, contacting the entire receptive field, the neuron fired throughout the period of motion (Fig. 6E, F).

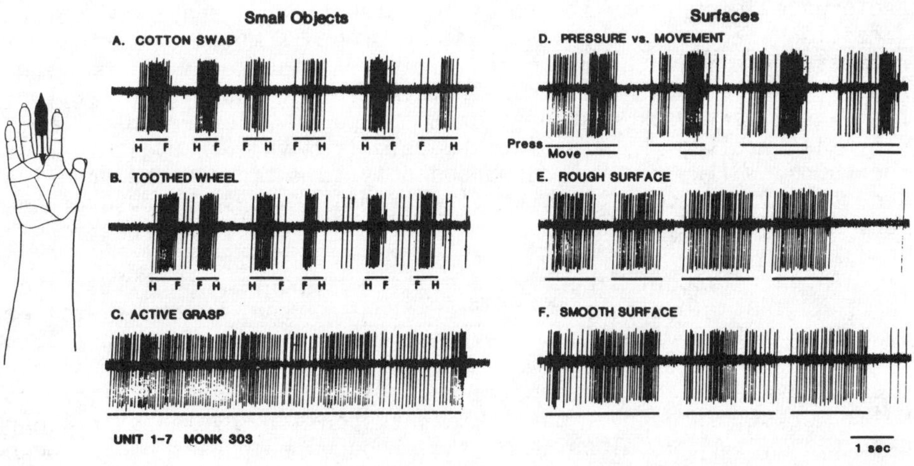

Fig. 6. Simple cortical neuron responding to motion of small objects (A, B), active grasp (C), and pressure and motion of large textured surfaces (D-F).

Nevertheless, the peak firing during motion of the surface was lower than that achieved by a small object crossing the receptive field center. The neuron fired in a graded fashion when small objects sequentially traversed different portions of the receptive field, but discharged at a steady, lower rate when a large surface continuously covered the entire area of the receptive field. Thus, as the area of contact between the moving object and the skin increases, we find a lowering of the peak firing rate, rather than a summated response. This change in firing patterns may result from complex interactions of excitation and inhibition within the receptive field.

In addition to the simple tactile neurons we studied in areas 3b and 1, we found neurons in the more posterior portions of SI which respond preferentially to moving stimuli, or to pressure from large surfaces, and are poorly excited by punctate stimuli such as airpuffs or Von Frey hairs (Costanzo and Gardner, 1980). Motion sensitive neurons are usually located in the posterior portion of area 1, and in areas 2 and 5, arrayed in columns crossing several cortical layers. Neurons within a column usually display similar response properties, such as similar direction or orientation preferences.

Examples of the most common type of motion sensitive neuron we observed are shown in Fig. 7. These cells showed a clear preference for motion in specific directions, and are called direction sensitive neurons. When a probe is moved through their receptive field in the preferred direction, the cells fire vigorously, while when the probe is moved in the opposite direction, there is a very weak response, or no response. For example, cell 9-6 was excited by brushing the palm from the 5th digit towards the thumb, while cell 9-7 responded only to motion in the opposite direction. We call these unidirectional neurons, because of the very weak activity elicited by motion in at least one direction.

Another type of direction sensitive response is illustrated in Fig. 8. This is a multidirectional neuron, which is excited by motion in any direction. Nevertheless, this neuron showed a direction preference for movements in the volar to dorsal direction. Movements in the opposite direction, from dorsal elbow to volar wrist, produced the weakest response. Movements in other directions, such as elbow to wrist, produced responses that were intermediate in intensity.

Unlike the simple cells in the anterior portion of SI, motion sensitive neurons show uniform rather than graded responses to stimuli moving across their receptive fields, and therefore lack a point of maximum sensitivity. Since their receptive fields tend to be larger than those of simple tactile neurons, motion sensitive neurons seem poorly suited for localizing the exact position of a moving stimulus.

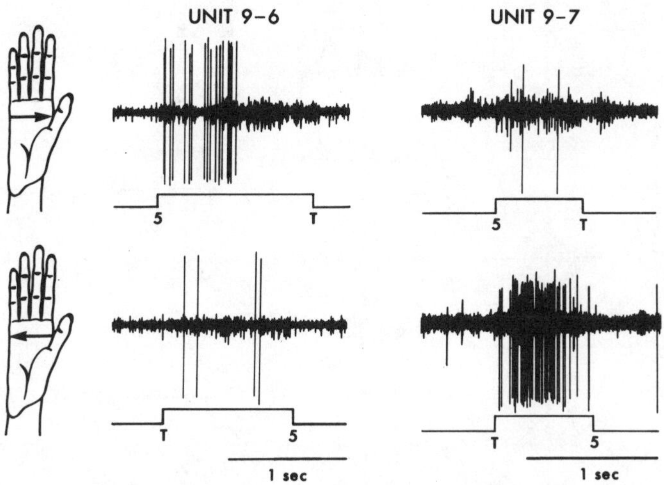

Fig. 7. Unidirectional neurons with receptive fields on the hand. (Reproduced from Costanzo and Gardner, 1980).

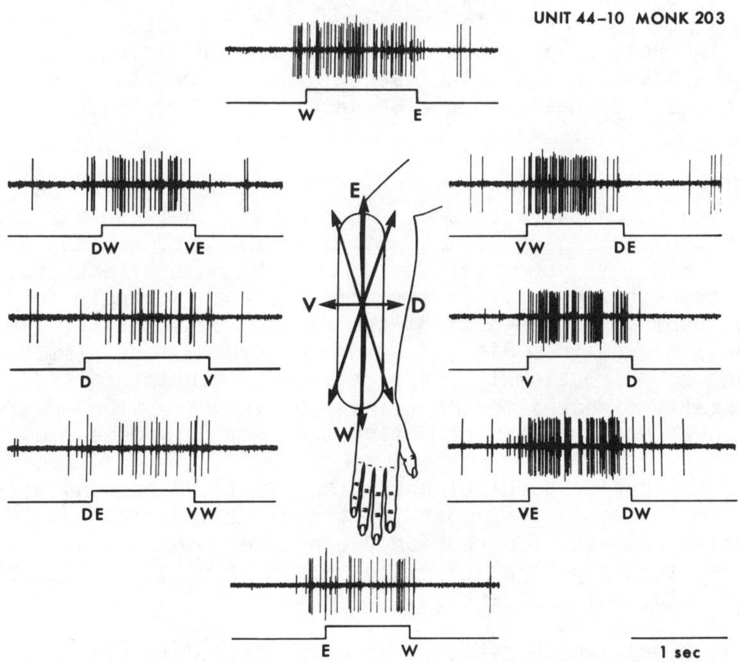

Fig. 8. Multidirectional neuron with a forearm receptive field.

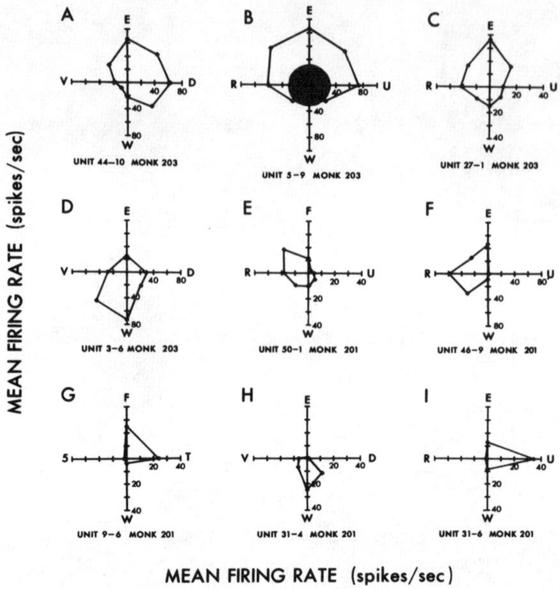

Fig. 9. Degree of tuning of 9 cortical neurons to different directions of motion. The radial distance of each point from the origin in these polar plots measures the mean firing rate, and the angular displacement indicates the direction of motion. (Reproduced from Costanzo and Gardner, 1980).

Direction sensitive neurons vary considerably in the range of motion to which they respond. Multidirectional neurons are tuned to much broader angles of motion than unidirectional neurons. For example, Fig. 9A summarizes data from the multidirectional neuron illustrated in Fig. 8. This neuron responds strongly to stimuli moving over more than half of the range. By contrast, the unidirectional neuron from Fig. 7 responded over a more limited arc, and showed directional preferences for both laterally directed, and distally directed motion (Fig. 9G). Direction preferences along more than one axis of motion were commonly observed.

In the entire population studied, we found neurons with response preferences for each of the directions tested. However, more cells had distal direction preferences than proximal ones, and the radial to ulnar direction was more often represented than the ulnar to radial direction.

The neuron whose responses are summarized in Fig. 9B was the least common type of direction sensitive neuron. It had a relatively high tonic discharge which persisted over a 2 hour recording period. This cell was strongly excited by stimuli moving proximally (Fig. 10B) and transversely, but was inhibited by distal movements (Fig. 10C). This apparent inhibition of activity suggested to us that direction sensitivity might be conferred by in-field inhibition distributed asymmetrically within the receptive field. To test the idea that the off-direction response involved an inhibitory process, we paired on- and off-direction stimuli. Thus at the same time that we moved one edge from wrist to elbow, another edge was moved from the elbow to the wrist. Pairing the stimuli produced a marked decrease in the excitatory response (Fig. 10D). These data suggest that the off-direction stimulus activates an inhibitory mechanism which suppresses not only the spontaneous activity of the cell, but also its responses to excitatory stimuli.

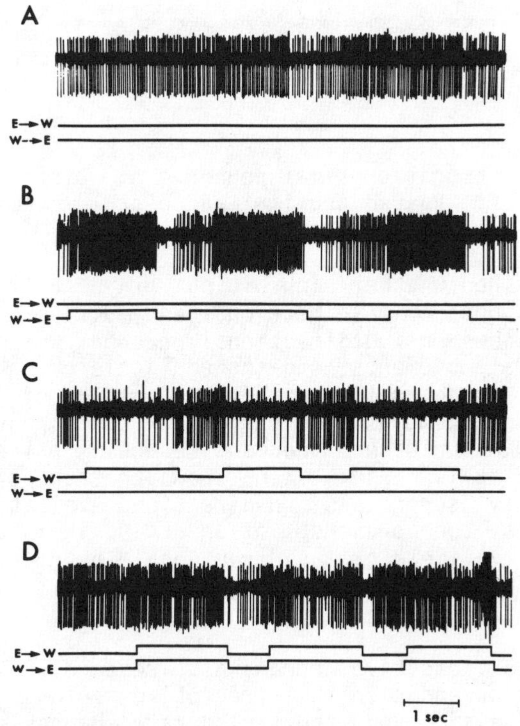

Fig. 10. Responses of an opponent-direction neuron to moving edges. A. Spontaneous discharge, Mean instantaneous firing rate (IFR), 30 spikes/s. B. Proximal motion, IFR, 90 spikes/s. C. Distal motion, IFR, 10 spikes/s. D. Paired proximal and distal movements, IFR, 46 spikes/s.

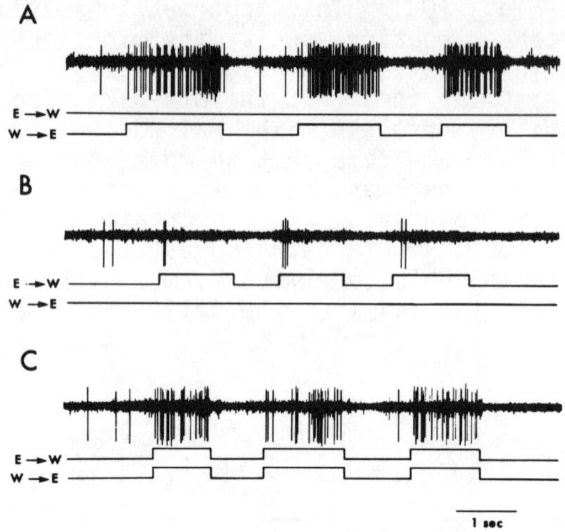

Fig. 11. Unidirectional cortical neuron. A. Distal motion. B. Proximal motion. C. Paired proximal and distal movements.

We then examined whether inhibition plays a role in the off-direction response of unidirectional and multidirectional neurons. Fig. 11 shows a unidirectional neuron which responded strongly to distal movements, and showed little or no response to proximal motion. To determine whether the absence of activity to the off-direction stimulus involves a lack of responsiveness, or is the result of an inhibitory process, we again paired on- and off-direction stimuli. Fig. 11C shows that the off-direction stimulus partially suppresses on-direction excitation. This strongly suggests the presence of an active inhibitory process within the receptive field of unidirectional neurons. However, multidirectional neurons, which are excited by all directions of motion, showed little suppression when paired stimuli were tested.

Many direction sensitive neurons also respond to steady pressure of surfaces or edges in their receptive fields. However, if the surface or edge is moved, the firing rate changes significantly depending upon the direction of motion. Fig. 12 shows a neuron which responded to pressure of a 10 mm edge on the volar forearm, with the greatest sensitivity near the elbow. When the edge was moved from the elbow to the wrist, the cell stopped firing, even though the edge crossed skin areas that were excited by static pressure. This suggests that a major effect of motion in specific directions is the inhibition of responses to steady pressure.

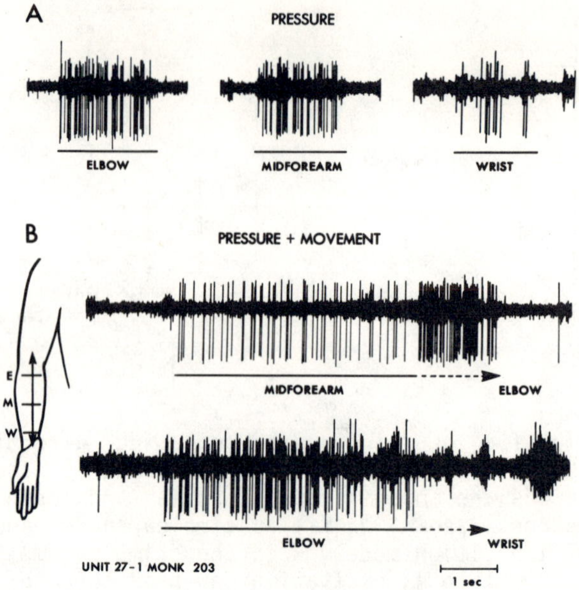

Fig. 12. A. Responses of a direction sensitive neuron to steady pressure from edges on the forearm. B. Responses to pressure (solid line) followed by motion (dashed line). Note the sharp increase in firing when movement is directed towards the elbow, and the cessation of activity when directed towards the wrist. (Reproduced from Gardner and Costanzo, 1980c).

Richard Costanzo and I proposed a simple model, which is outlined in Fig. 13, to explain these findings (Gardner and Costanzo, 1980c). Our underlying assumption is that direction sensitive neurons receive inputs from simple tactile neurons showing in-field inhibition. We further assume that the point of maximum inhibition is laterally displaced from the excitatory center, so that when the stimulus moves across the field in the preferred direction (from 1 to 4), it first crosses the excitatory center before reaching the point of maximum inhibition. When the stimulus moves in the least preferred direction (from 4 to 1), it crosses the inhibitory center before the excitatory one. Direction sensitive neurons are further presumed to receive inputs from a group of simple tactile neurons, all of which have the same directional asymmetries between excitation and inhibition, but each of which has a slightly different receptive field location.

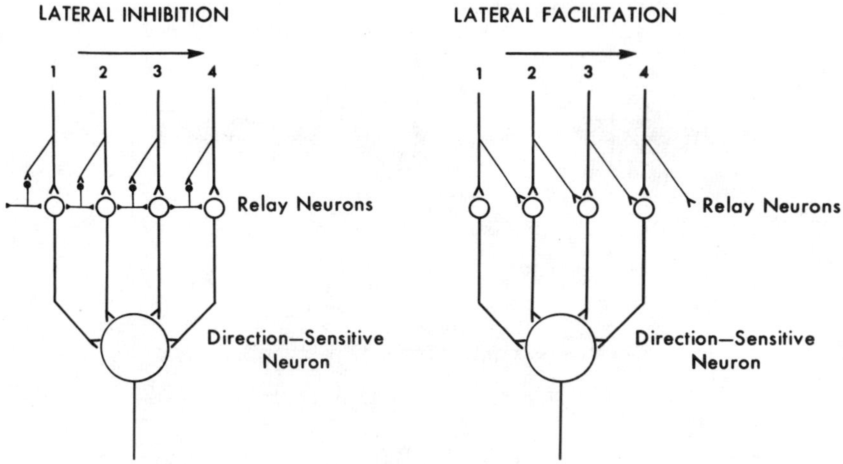

Fig. 13. Simplified neural circuits producing direction sensitivity through lateral inhibition or lateral facilitation. Excitatory neurons are shown as open circles and open synapses. Inhibitory neurons shown as filled circles and closed synapses. In the lateral inhibition model, when the stimulus moves in the on-direction, from 1 to 4, excitation can pass through each of the relay cells to the direction-sensitive neuron before in-field inhibition is sufficiently strong to terminate the relay cell discharge. When the stimulus moves in the opposite direction, inhibition is fed forward to relay cells representing the adjacent skin area in the off-direction, thus decreasing the input to the direction sensitive neuron. In the lateral facilitation model, excitation is preferentially distributed in the on-direction.

The lateral inhibition model predicts that a direction sensitive neuron will respond not only to stimuli moving along a continuous path on the skin, but also to a series of spaced punctate stimuli if they are presented in a linear sequence, in the correct direction, and in rapid temporal order. Furthermore, if we increase the spacing between points stimulated in the off-direction, the cell should show a decreased direction preference, because the off-direction stimulus should become more effective. To test these features of the model, we stimulated the receptive fields of direction sensitive and motion sensitive neurons with toothed wheels rotated across their receptive fields. The wheels were mounted on low-torque potentiometers to accurately measure wheel rotation. The wheel was rolled in a linear path across the skin using the potentiometer shaft an axle, so that the teeth sequentially contacted the skin as it rotated. As the teeth were not dragged across the skin, only selected points in the path were

stimulated, leaving gaps of specified dimensions uncontacted by the moving wheel. When tested with wheels having small gaps between teeth, the cell responded vigorously to distal movements, and fired only 1 or 2 impulses to proximal motion (Fig. 14, top). When the spatial period of the teeth was increased to 3.2 mm (middle trace), there was no effect on responses to movements in the elbow to wrist direction, but the cell became slightly more responsive to wrist to elbow motions. In the bottom trace, the spatial period was doubled to 6.4 mm, while the tooth size was kept constant. With this wide tooth spacing, we see a further increase in the off-direction response, and the neuron responded on every trial to this direction of motion. The on-direction excitation was slightly weaker, but the change was less pronounced than that seen in the off-direction. However, even with this large separation between stimulated points, the cell still manifested a clear direction preference for distally directed motion. These data, while not conclusive, support the lateral inhibition model we proposed to explain direction preferences.

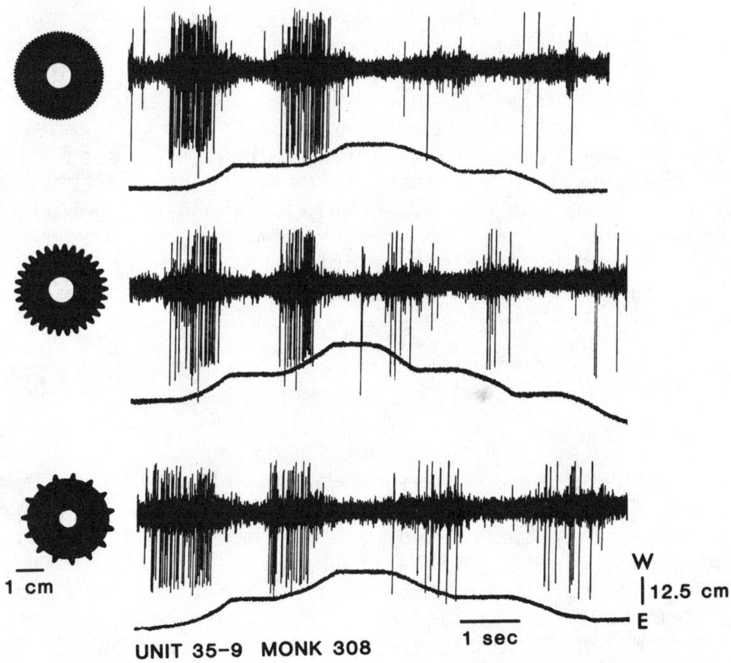

<u>Fig. 14</u>. Responses of a direction-sensitive neuron to rotation of toothed wheels from elbow to wrist (upward deflections) and from wrist to elbow (downward deflections). Spatial period between teeth: <u>Top</u>, 1.2 mm; <u>Middle</u>, 3.2 mm; <u>Bottom</u>, 6.4 mm.

Direction sensitivity and texture differentiation seem to be related, for motion sensitive neurons without direction preferences do not show the differential response to textured surfaces described above. We found little difference in the responses of motion sensitive neurons to wheels of spatial periods of 0.8 to 6.4 mm. Thus the inhibitory mechanisms which contribute to direction sensitivity may also play a role in texture discrimination.

The data presented in this report indicate that moving stimuli activate several neuronal populations in the cerebral cortex which function in parallel. Simple tactile neurons appear to encode the quantitative features of stimulus position frozen in time. Their discharges provide a signal of the instantaneous location of the stimulus on the skin, as well as a history of the previously stimulated points in the trail of inhibition. However, individual simple tactile neurons cannot differentiate lateral motion from static position; instead a population analysis is required to derive this information.

Direction sensitive and motion sensitive neurons involve a higher order of information processing, and are not specifically designed to transmit all of the perceptual details of the stimulus. Instead, they seem to abstract the concept of directed motion from a temporally and spatially ordered series of stimulus events. Their activity may denote particular directions of motion, and in some cases, individual pathways across the skin. Some of these cells are further specialized to characterize specific dimensions of the moving stimulus. These neurons thus serve a feature-detecting role, and may be used by the brain to generalize specific sensory information into more abstract categories or concepts, something which we normally think of as a higher cortical function.

Acknowledgments

Richard M. Costanzo, Susan Warren, Heikki A. Hamalainen Daniel R. Kenshalo, Jr., and Joseph M. Phillips participated in various portions of these studies. We thank Donna Cobelli, Janet Tast and Jane Davis for skilled technical assistance. This investigation was supported by United States Public Health Service Research Grant NS 11862, and Research Career Development Grant NS 00142, from NIH-NINCDS, and by a Research Career Award from the Irma T. Hirschl Trust.

REFERENCES

Baker, M.A., Tyner, C.F. and Towe, A.L. (1971). Observations on single neurons in sigmoid gyri of awake, unparalyzed cats. Exp. Neurol. 32: 388-403.

Costanzo, R.M. and Gardner, E.P. (1980). A quantitative analysis of responses of direction-sensitive neurons in somatosensory cortex of alert monkeys. J. Neurophysiol. 43: 1319-1341.

Darian-Smith, I. and Kenins, P. (1980). Innervation density of mechanoreceptive fibres supplying glabrous skin of the monkey's index finger. J. Physiol. (Lond.). 309: 147-155.

Dreyer, D.A., Hollins, M. and Whitsel, B.L. (1978). Factors influencing cutaneous direction sensitivity. Sensory Processes 2: 71-79.

Gardner, E.P. and Costanzo, R.M. (1980a). Spatial integration of multiple-point stimuli in primary somatosensory cortical receptive fields of alert monkeys. J. Neurophysiol. 43: 420-443.

Gardner, E.P. and Costanzo, R.M. (1980b). Temporal integration of multiple-point stimuli in primary somatosensory cortical receptive fields of alert monkeys. J. Neurophysiol. 43: 444-468.

Gardner, E.P. and Costanzo, R.M. (1980c). Neuronal mechanisms underlying direction sensitivity of somatosensory cortical neurons in alert monkeys. J. Neurophysiol. 43: 1342-1354.

Gibson, J.J. Observations on active touch. (1962). Psych. Rev. 69: 477-491.

Hyvarinen, J. and Poranen, A. (1978). Movement-sensitive and direction and orientation-selective cutaneous receptive fields in the hand area of the postcentral gyrus in monkeys. J. Physiol. (Lond.). 283: 523-537.

Innocenti, G.M. and Manzoni, T. (1972). Response patterns of somatosensory cortical neurones to peripheral stimuli. An intracellular study. Arch. Ital. Biol. 110: 322-347.

Janig, W., Shoultz, T., and Spencer, W.A. (1977). Temporal and spatial parameters of excitation and afferent inhibition in cuneothalamic relay neurons. J. Neurophysiol. 40: 822-835.

Janig, W., Spencer, W.A. and Younkin, S.G. (1979). Spatial and temporal features of the afferent inhibition of thalamocortical relay cells. J. Neurophysiol. 42: 1450-1460.

Johansson, R.S. and Vallbo, A.B. (1979). Tactile sensibility in the human hand: relative and absolute densities of the four types of mechanoreceptive units in glabrous skin. J. Physiol. (Lond.). 286: 283-300.

Laskin, S.E. and Spencer, W.A. (1979). Cutaneous masking. II. Geometry of excitatory and inhibitory receptive fields of single units in the somatosensory cortex of the cat. J. Neurophysiol. 42: 1061-1082.

Lederman, S.J. (1974). Tactile roughness of grooved surfaces: the touching process and effects of macro- and microsurface structure. Percept. Psychophys. 24: 154-160.

Morley, J.W., Goodwin, A.W., and Darian-Smith, I. (1983). Tactile discrimination of gratings. Exp. Brain Res. 49: 291-299.

Mountcastle, V.B. and Powell, T.P.S. (1959). Neural mechanisms subserving cutaneous sensibility, with special reference to the role of afferent inhibition in sensory perception and discrimination. Bull. Johns Hopkins Hosp. 105: 201-232.

Whitehorn, D. and Towe, A.L. (1968). Postsynaptic potential patterns evoked upon cells in sensorimotor cortex of cat by stimulation at the periphery. Exp. Neurol. 22: 222-242.

Whitsel, B.L., Roppolo, J.R. and Werner, G. (1972). Cortical information processing of stimulus motion on primate skin. J. Neurophysiol. 35: 691-717.

INTEGRATION OF SOMATOSENSORY EVENTS IN THE POSTERIOR PARIETAL CORTEX OF THE MONKEY

LEA LEINONEN

Department of Physiology, University of Helsinki, Siltavuorenpenger 20 J, 00170 Helsinki 17, Finland

Introduction

Information entering the somesthetic system through various receptors in various locations is kept rather separate on its way to the cortical projection areas. To serve behavioural purposes, e.g. localization and recognition of objects, information from separate parts of the somatosensory projection areas must be integrated and also compared with the information coming from other sensory organs. The posterior parietal cortex is a terminal for somesthetic association fibres originating in the postcentral gyrus and in the anterior wall of the Sylvian fissure (e.g. Pandya and Kuypers, 1969; Stanton et al., 1977; Seltzer and Pandya, 1980; Pandya and Seltzer, 1982). Additional somesthetic input may reach this area through motor (e.g. Pandya and Kuypers, 1969) and prefrontal (e.g. Pandya and Kuypers, 1969; Mesulam et al., 1977) cortices, and limbic areas 23 and 24 (Mesulam et al., 1977; Pandya et al., 1981). Somesthetic input may further be provided by transcallosal (Pandya et al., 1971) and rich intra-areal connections (Seltzer and Pandya, 1980; Pandya and Seltzer, 1982), and by fibres originating in some subcortical structures (e.g. Pearson et al., 1978).

Ablation experiments have shown that the posterior parietal cortex of the monkey participates in several behaviours requiring somesthetic stimulus processing, e.g. in the detection and localization of contralateral tactual stimuli (Peele, 1944; Heilman et al., 1971), discrimination of three-dimensional forms by contralateral hand (Ruch et al., 1938; Peele, 1944; Ettlinger and Kalsbeck, 1962), discrimination of lifted weights (Ruch et al., 1938; Semmes Blum et al., 1950), route finding (Bates and Ettlinger, 1960; Sugishita et al., 1978; Petrides and Iversen, 1979); it also participates in the maintenance of an adequate limb posture

(Denny-Brown and Chambers, 1958; Faugier-Grimaud et al., 1978), in the control of grasping (Munk, 1881; Fleming and Crosby, 1955); Faugier-Grimaud et al., 1978), and in reaching under visual and somatosensory guidance (Ratcliff et al., 1977; Lamotte and Acuña, 1978). Defects have also been observed in oculomotor performance including saccades, slow pursuit eye movements (Lynch and McLaren, 1979), and optokinetic nystagmus (Lynch and McLaren, 1983). It is not yet known how these eye movements depend on somesthetic feedback from the eye.

Neurones of the posterior parietal cortex of conscious macaque monkeys have been recorded in several laboratories. The number of experiments on visual processing in this area exceeds the number of studies dealing with its somatosensory functions. In the following I shall summarize the fragmentary recording data there are on somesthetic integration. Areas investigated are referred to in the following as areas 5, 7 and Tpt. Area 5 (of Brodmann) covers the superior parietal lobule and area 7 the inferior parietal lobule reaching the anterior wall of the superior temporal sulcus and the anterior wall of the Sylvian fissure; both areas also include parts of the cortex on the medial surface of the parietal lobe. Recently, areas 5 and 7 have been histologically devided to several subareas (Seltzer and Pandya, 1978, 1980; Pandya and Seltzer, 1982). Since the recording data have not yet been correlated with these histological distinctions, the new histological terminology is not used here. Area Tpt of Pandya and Sanides (1973) is located in the upper part of the superior temporal gyrus. According to Pandya and Sanides (1973) area Tpt is histologically not distinguishable from area 7. Since it also resembles functionally area 7 of the parietal cortex (Leinonen et al., 1980), some results from this region are reviewed.

Convergence of Somesthetic, Visual, Vestibular and Auditory Information in Areas 5, 7 and Tpt

Convergence of somesthetic, visual, vestibular and auditory systems at the posterior parietal cortex is presented schematically in Fig. 1. In area 5 all, or nearly all neurones respond to somesthetic stimulation (Duffy and Burchfiel, 1971; Sakata et al., 1973; Mountcastle et al., 1975; MacKay et al., 1978; Murray and Coulter, 1981; Seal et al., 1982). Most responses in area 5 are elicited by rotation of joints, some by touching of the hairs or skin, or palpation or squeezing of tissues (Duffy and Burchfiel, 1971; Sakata et al., 1973; Mountcastle et al., 1975). Responses to noxious stimuli have been recorded in area 5 on the medial surface of the parietal lobe (Murray and Coulter, 1981). The results indicate that information from various somesthetic receptors converges at area 5. Convergence can also be demonstrated in the functions of individual neurones: 14 per cent of the neurones studied by Duffy and Burchfiel (1971) and 37 per cent of the neurones reported by Sakata (1975) responded to both cutaneous and joint stimulation. Visually evoked responses have been observed in area 5 only infrequently (Sakata, 1975; Mountcastle et al., 1975); the effects of vestibular stimulation have not been examined.

The data presented in Fig. 1 on the responsiveness of area 7 is mainly based on multicellular recordings performed by Hyvärinen (1981) in the exposed part of area 7 to map the distribution of responses to somesthetic and visual stimulation. Responses to somesthetic stimulation were elicited in practically all penetrations in the lateral part of area 7 and in some penetrations in the medial part of area 7 too, whereas responses to visual stimulation were concentrated in the medial part of area 7. In about 15 per cent of the penetrations responses were elicited by both visual and somesthetic stimulation. Single neurones in the lateral part of area 7 respond to rotation of joints, palpation of muscles, light touching of the hairs or skin, and to compression of the skin (Hyvärinen and Poranen, 1974; Leinonen and Nyman, 1979; Leinonen et al., 1979; Leinonen, 1980; Robinson and Burton, 1980b). Responses elicited by noxious mechanical or thermal stimulation have been recorded in some neurones near the Sylvian fissure (Robinson and Burton, 1980b). Convergence from various somesthetic receptors ("hair", "skin", "deep") has been observed in about 10 per cent of the neurones (Leinonen et al., 1979; Robinson and Burton, 1980b) and convergence

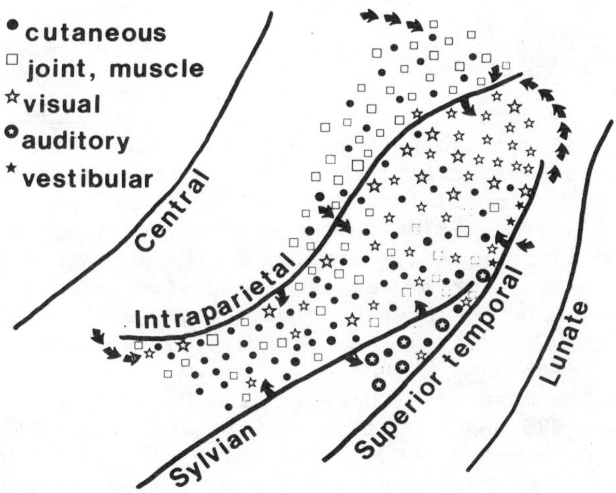

Fig. 1. The distribution of responses to stimulation of various sensory organs in the posterior parietal cortex of the macaque monkey. The arrows indicate the flow of information from the surroundings.

of somatosensory and visual or auditory input has been observed in 5 to 30 per cent of neurones in the lateral part of area 7 and Tpt (Hyvärinen and Poranen, 1974; Robinson and Goldberg, 1978; Leinonen and Nyman, 1979; Leinonen et al., 1979; Leinonen, 1980; Bushnell et al., 1981). Responses to both visual and vestibular stimulation have been documented for neurones in the anterior wall of the superior temporal sulcus (Kawano et al., 1980); responses to sounds have been described only around the posterior end of the Sylvian fissure (Robinson and Burton, 1980b; Leinonen et al., 1980; Bruce et al., 1981).

It has been noticed in several studies with behaving monkeys that some neurones of area 5 (Mountcastle et al., 1975; MacKay et al., 1978; Seal et al., 1982) and of area 7 (e.g. Hyvärinen and Poranen, 1974; Mountcastle et al., 1975; Motter and Mountcastle, 1981) do not respond to sensory stimulation but their discharges are mainly related to active movements. The number of such movement-related neurones varies greatly in different studies, from 0 (Robinson et al., 1978) to 80 (Mountcastle et al., 1975) per cent, probably due to the differences in the type and number of the stimuli used, and in the conceptualization of the results. Many investigators have observed that the terms "motor" and "sensory" were insufficient in the description of all relationships between cellular discharges and the behaviour of the monkey. The terms "directed" (Lynch et al., 1977) and "selective attention" (Bushnell et al. 1981) have been taken into use for the description of some functions.

Size and Stability of Receptive Fields

Somesthetic receptive fields in areas 5 and 7 can often be discovered only when the best stimulation occurs; this is usually a movement in a certain direction. Also, the determination of the size of the receptive fields is greatly affected by the stimulations used. Difficulties in the establishment of the extent of the receptive fields are illustrated by the fact that in area 5 reports on the proportion of neurones receiving input from more than one joint vary from 10 (Mountcastle et al., 1975) to 60 (Sakata et al., 1973) and 90 (Duffy and Burchfiel, 1971) per cent.

The results of most studies in area 5 (Duffy and Burchfiel, 1971; Sakata, 1975; MacKay et al., 1978) and area 7 (Leinonen and Nyman, 1979; Leinonen et al., 1979) indicate that neurones responding to joint stimulation are usually activated through more than one joint. Cutaneous receptive fields, often including both glabrous and hairy skin, extent over entire body parts both in area 5 (Sakata and Iwamura, 1978) and 7 (Leinonen et al., 1979; Robinson and Burton, 1980a). In area 5 cutaneous receptive fields often cover one side of the limb whereas in area 7 they usually cover entire body parts, for instance, hand(s), arm(s), leg(s), or foot (feet). In area 7 convergence from the whole body surface has also been observed (Leinonen et al., 1979; Robinson and Burton, 1980a).

Somatosensory receptive fields are often bilateral and usually symmetrically located: in area 5 bilateral receptive fields have been documented for 20 to 60 per cent of the neurones (Duffy and Burchfiel, 1971; Sakata et al., 1973) and in the lateral part of area 7 and Tpt for 40 to 50 per cent of the neurones (Leinonen and Nyman, 1979; Leinonen et al., 1979, 1980; Leinonen, 1980; Robinson and Burton, 1980a). Most visual receptive fields are bilateral and cover a large part of the visual field (e.g. Motter and Mountcastle, 1981); auditory responses are elicited over a wide range of stimulus frequencies (Leinonen et al., 1980).

In area 5 responses of neurones greatly depend on the alertness of the monkey (Sakata et al., 1973; Mountcastle et al., 1975). In area 7 the strength of somesthetically evoked responses and the size of the receptive field decrease with increasing drowsiness of the animal (Robinson and Burton, 1980a). In area 7 attention without orienting movements towards the stimulus has been observed to strengthen the visual responses (Bushnell et al., 1981). Diverting attention from an effective stimulus by some other stimulation has often a strong unspecific inhibitory effect in both area 7 and Tpt (Leinonen et al., 1979; 1980). It is possible that such inhibitory interactions between stimuli at different locations are generated in the posterior parietal cortex itself.

Effective stimuli

In area 5 most neurones respond to rotation of joints causing a body part to move in a certain direction (Duffy and Burchfiel, 1971; Sakata et al., 1973; Sakata, 1975). Most neurones with cutaneous receptive fields respond to stroking of the skin, usually in a certain direction. Responses elicited by stroking of a cutaneous receptive field on a limb in a certain direction are often influenced by changes in the position of the limb. Many neurones having both cutaneous and joint receptive fields were shown by Sakata (1975) to be such that the best stimulus condition was the movement of a limb that brought it near a certain body part and the simultaneous stimulation of the skin region which was approached by the limb. The studies of both Sakata, and Duffy and Burchfiel suggest that neurones in area 5 respond strongest during certain coordinated, natural movements of one or several limbs.

A great number of neurones in area 7 respond to movements over visual (Yin and Mountcastle, 1977; Robinson et al., 1978; Leinonen et al., 1979; Leinonen, 1980; Motter and Mountcastle, 1981; Sakata et al., 1980), somesthetic (Leinonen and Nyman, 1979; Leinonen et al., 1979) and vestibular (Kawano et al., 1981) receptive fields. Many of them show directional selectivity. Most acoustically drivable neurones in area Tpt prefer sounds with a broad noise spectrum and rapid intensity and frequency modulations. Neurones responding to the stimulation of more than one sensory system resemble neurones in area 5: a single neurone may respond, for instance, to hand movement at the wrist from right to left regardless whether the hand is supinated or maximally pronated, and to cutaneous stimuli moving from right to left across the face, chest, or hand regardless of its position, to blowing into the hairs on the chest from the right but not from the left, and to visual stimuli moving from right to left (Leinonen et al., 1979). In lateral part of area 7 a considerable proportion of neurones (10 to 20 per cent) with cutaneous receptive fields respond to visual stimuli approaching the receptive fields from any direction (Hyvärinen, 1974; Leinonen et al., 1979). In area Tpt neurones have been observed to respond to both touching of a body region and to auditory stimuli near the receptive skin area (Leinonen et al., 1980). All effective stimuli of a neurone activated through more than one sensory system were such that they normally occurred simultaneously during a certain active movement of the monkey (Leinonen and Nyman, 1979; Leinonen et al., 1979; 1980).

Is there Any Somatotopic Organization in the Posterior Parietal Cortex ?

In area 5 receptive fields are spatially arranged in a roughly somatotopic fashion (Sakata, 1975), a "supplementary" somatotopic representation has been discovered on the medial part of the parietal lobe (Murray and Coulter, 1981). Multiunit mappings of Hyvärinen (1981) in area 7 showed that neurones with somesthetic receptive fields on the face are concentrated near the anterior tip of the intraparietal sulcus, receptive fields of the limbs are more posteriorly located; this agrees with the results of Robinson and Burton (1980a), Leinonen and Nyman (1979), and Leinonen et al., (1979). Robinson and Burton (1980a) stated that there is some tendency of neurones with similar receptive fields to be found together but a clear somatotopy cannot be demonstrated: this is in agreement with the findings of other investigators.

In order to see how neurones in nearby locations relate to each other we have compared the results obtained in four arbitrarily chosen neighboring fields in the lateral part of area 7 (Fig. 2, Table 1); the results have been documented in detail elsewhere (Leinonen and Nyman, 1979; Leinonen et al., 1979, 1980; Leinonen,

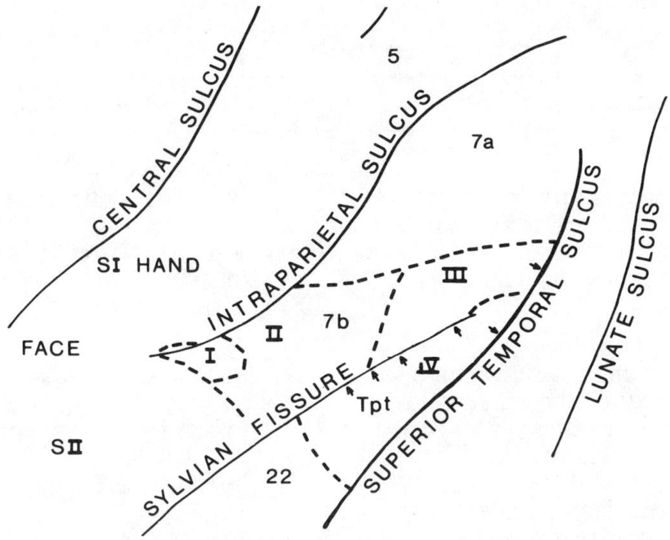

Fig. 2. Location of target areas I, II, III and IV

1980). The comparison of the effective stimuli and the active movements related to cellular activity in these separate parts of area 7 and Tpt, Table 1, showed that in one subarea movement-related neurones discharged during movements resulting in effective stimulation of the stimulus-related neurones; thus the nearby neurones were all related to the same motor behaviours and the neurones further apart to different motor behaviours.

The way in which somatosensory information converges in areas 5 and 7 and the way in which it is integrated in area 7 and Tpt with other sensory modalities suggested us that this integration is generated, guided, and maintained by active movements of the monkey. Integration has its origin in the repetitive simultaneous activation of different receptors at various locations of the sensory spaces determined by the scopes of movements of different body parts. In other words, instead of a somatotopic representation, a sensory map of the behavioural past of the monkey can be demonstrated in the distribution of functions in the posterior parietal cortex.

Development of Functional Associations in Area 7

The study of Hyvärinen et al., (1981a), which was performed to investigate the effect of early visual deprivation on the visual responsiveness of area 7, showed that postnatal deprivation of 7-11 months (with closure of the eye lids) almost completely prevented

Table. 1. Stimuli and behaviours related to the neuronal activity in target areas I, II, III and IV.

STIMULUS-RELATED NEURONES	MOVEMENT-RELATED NEURONES
AREA I	
TOUCHING, PALPATION OF FACE	MANIPULATION WITH LIPS
VISUAL STIMULI MOVING TOWARDS FACE	CHEWING
HAND MOVEMENT TOWARDS FACE	REACHING WITH LIPS
PALPATION OF ARM MUSCLES	BRINGING AN OBJECT WITH HAND TO MOUTH
AREA II	
TOUCHING OF HAND, ARM, CHEST	REACHING WITH ARM
VISUAL STIMULI MOVING TOWARDS HAND, ARM, CHEST	GRASPING FOR AN OBJECT WITH HAND
STROKING OF SKIN IN CERTAIN DIRECTION	MANIPULATION OF AN OBJECT WITH HAND
VISUAL STIMULI MOVING IN CERTAIN TANGENTIAL DIRECTION	
PALPATION OF ARM MUSCLES	
AREA III	
COMPRESSION OF HAND, ARM	GRASPING WITH HAND
PALPATION OF ARM MUSCLES	
FORCED FLEXION OR EXTENSION OF WRIST, FINGERS	DIGGING WITH HAND
VISUAL STIMULI MOVING TOWARDS HAND, ARM	GROOMING
AREA IV	
TOUCHING OF TEMPLE, SHOULDER, BACK	ROTATION OF HEAD
VISUAL STIMULI MOVING TOWARDS TEMPLE, SHOULDER	LIFTING OF SHOULDERS
VISUAL STIMULI MOVING IN PERIPHERAL VISUAL FIELD	
PASSIVE ROTATION OF HEAD	
SOUNDS WITH CERTAIN ANGLE OF INCIDENCE	

the establishment of functional connections from the visual system and caused an overpresentation of the somesthetic modality. Multi-unit recordings in normal monkeys of the same age showed that visual responses could be elicited in 50 per cent of the penetrations but only in 3 per cent of the penetrations in the deprived brain. After the opening of the eyes the use of vision recovers very slowly, and the monkeys remain visually handicapped. Some visual learning occurs, however, and the studies in progess suggest that this learning is accompanied with changes in functions of area 7 (Hyvärinen et al., 1981b).

Behavioural significance of somatosensory analysis in the posterior parietal cortex

The medial part of area 7 has been studied by Mountcastle et al. (e.g. Lynch et al., 1977; Yin and Mountcastle, 1977; Motter and Mountcastle, 1981). Neurones responding to movements towards the center of the visual field or away from it suggested to Motter and Mountcastle (1981) that area 7 "contributes to a continual updating to a central neural image of the spatial frame of the immediate behavioral surround and the perceptual constancy of that space that obtains during body movement". As they did not notice somesthetic, visual and vestibular convergence, they stated that sensory associations do not exist in this area: they concluded that the discharges evoked by stimulation of various sensory organs in the studies of Hyvärinen et al. (rev. Hyvärinen, 1982) form an "abstracted replicate of a complex sensory event". However, the lack of neurones with convergence in their study might derive from methodological reasons which resulted in a great number (30 per cent) of unidentified neurones. Our results suggest that some of their neurones related to the extraction of visual features evoked e.g. by locomotion would also respond to somesthetic stimuli evoked by locomotion. Our results have led us (rev. Hyvärinen, 1982) to conclude, in agreement with Sakata (e.g. Sakata et al., 1980), that the role of the posterior parietal cortex in the control of complex behaviours is to localize stationary or moving objects in reference to various body parts and various sensory systems, and to use this information in the guidance of attention and in the predetermination of the trajectory of a voluntary movement. The results obtained do not necessitate the use of such untestable terms as "perception" or "abstracted replicate of a complex sensory event".

References

Bates, J.A.V., Ettlinger, G. (1960). Posterior biparietal ablations in the monkey. Arch. Neurol., 3, 177-192.

Bruce, C., Desimone, R. and Gross, C.G. (1981). Visual properties of neurons in a polysensory area in superior temporal sulcus of the Macaque. J. Neurophysiol, 46, 369-384.

Bushnell, M., Goldberg, M.E. and Robinson, D.L. (1981). Behavioral enhancement of visual responses in monkey cerebral cortex. I. Modulation in posterior parietal cortex related to selective visual attention. J. Neurophysiol., 46, 755-772.

Denny-Brown, D. and Chambers, R.A. (1958). The parietal lobe and behavior. Res. Publ. Ass. Nerv. Ment. Dis., 36, 35-117.

Duffy, F.H. and Burchfiel, J.L. (1971). Somatosensory system: organized hierarchy from single units in monkey area 5. Science, 172, 273-275.

Ettlinger, G. and Kalsbeck, I.E. (1962). Changes in tactile discrimination and visual reaching after successive and simultaneous bilateral posterior parietal ablations in the monkey. J. Neurol. Neurosurg. Psychiatry, 25, 256-268.

Faugier-Grimaud, S., Frenois, C. and Stein, D.G. (1978). Effects of posterior parietal lesions on visually guided behavior in monkeys. Neuropsychologia 16, 151-168.

Fleming, J.F.R. and Crosby, E.C. (1955). The parietal lobe as an additional motor area. J. Comp. Neurol., 103, 485-512.

Heilman, K.M., Pandya, D.M., Karol, E.A., Geschwind, N. (1971). Auditory inattention. Arch. Neurol., 24, 323-325.

Hyvärinen, J. (1981). Regional distribution of functions in parietal association area 7 of the monkey. Brain Res., 206, 287-303.

Hyvärinen, J. (1982). The Parietal Cortex of Monkey and Man, Springer-Verlag, Berlin.

Hyvärinen, J., Hyvärinen, L. and Linnankoski, I. (1981a). Modification of parietal association cortex and functional blindness after binocular deprivation in young monkeys. Exp. Brain Res., 42, 1-8.

Hyvärinen, J., Hyvärinen, L. and Carlson, S. (1981b). Effect of binocular deprivation on parietal association cortex in young monkeys. Doc. Ophthal. Proc. Series, 30, 177-185.

Hyvärinen, J. and Poranen, A. (1974). Function of the parietal associative area 7 as revealed from cellular discharges in alert monkeys. Brain, 97, 673-692.

Kawano, K., Sasaki, M. and Yamashita, M. (1980). Vestibular input to visual tracking neurons in the posterior parietal association cortex of the monkey. Neurosci. Lett., 17, 55-60.

LaMotte, R.H. and Acuña, C. (1978). Defects in accuracy of reaching after removal of posterior parietal cortex in monkeys. Brain Res., 139, 309-326.

Leinonen, L. and Nyman, G. (1979). Functional properties of cells in anterolateral part of area 7, associative face area, of awake monkeys. Exp. Brain Res., 34, 321-333.

Leinonen, L., Hyvärinen, J., Nyman, G. and Linnankoski. I. (1979). Functional properties of neurons in lateral part of associative area 7 in awake monkeys. Exp. Brain Res., 34, 299-320.

Leinonen. L. (1980). Functional properties of neurones in the posterior part of area 7 in awake monkey. Acta Physiol. Scand., 108, 301-308.

Leinonen, L., Hyvärinen, J. and Sovijärvi, A.R.A. (1980). Functional properties of neurons in the temporo-parietal association cortex of awake monkey. Exp. Brain Res., 39, 203-215.

Lynch, J.C. and McLaren, J.W. (1979). Effects of lesions of parieto-occipital association cortex upon performance of oculomotor and attention tasks in monkeys. Neurosci. Abstr., 5, 794.

Lynch, J.C. and McLaren, J.W. (1983). Optokinetic nystagmus deficits following parieto-occipital cortex lesions in monkeys. Exp. Brain Res., 49,125-130.

Lynch, J.C., Mountcastle, V.B., Talbot, W.H. and Yin, T.C.T. (1977). Parietal lobe mechanisms for directed visual attention. J. Neurophysiol., 40, 362-389.

MacKay, W.A., Kwan, M.C., Murphy, J.T. and Wong, Y.C. (1978). Responses to active and passive wrist rotation in area 5 of awake monkeys. Neurosci. Lett., 10, 235-239.

Mesulam, M.-M., van Hoesen, G.W., Pandya, D.N. and Geschwind, N. (1977). Limbic and sensory connections of the inferior parietal lobule (area PG) in the rhesus monkey: a study with a new method for horseradish peroxidase histochemistry. Brain Res., 136, 393-414.

Motter, B.C. and Mountcastle, V.B. (1981). The functional properties of the light-sensitive neurons of the posterior parietal cortex studied in waking monkeys: foveal sparing and opponent vector organization. J. Neurophysiol., 38, 871-908.

Munk, H. (1881). Ueber die Funktionen der Grosshirnrinde: Gesammelte Mitteilungen aus den Jahren 1877-80. August Hirschwald, Berlin.

Murray, E.A. and Coulter, J.D. (1981). Supplementary sensory area. The medial parietal cortex in the monkey. In Multiple Somatic Areas. (ed. C.N. Woolsey). Humana Press, Clifton. pp. 167-195.

Pandya, D.N., Karol, E.A. and Heilbronn, D. (1971). The topographical distribution of interhemispheric projections in the corpus callosum of the rhesus monkey. Brain Res., 32, 31-43.

Pandya, D.N. and Kuypers, H.G.J.M. (1969). Cortico-cortical connections in the rhesus monkey. Exp. Brain Res., 13, 13-36.

Pandya, D.N. and Sanides, F. (1973). Architectonic parcellation of the temporal operculum in rhesus monkey and its projection pattern. Z. Anat. Entwickl. Gesch. 139, 127-161.

Pandya, D.N. and Seltzer, B. (1982). Intrinsic connections and architectonics of posterior parietal cortex in the rhesus monkey. J. Comp. Neurol. 204, 196-210.

Pandya, D.N., van Hoesen, G.W. and Mesulam, M.-M. (1981). Efferent connections of the cingulate gyrus in the rhesus monkey. Exp. Brain Res., 42, 319-330.

Pearson, R.C.A., Brodal, P. and Powell, T.P.S. (1978). The projection of the thalamus upon the parietal lobe in the monkey. Brain Res., 144, 143-148.

Peele. T.L. (1942). Cytoarchitecture of individual parietal areas in the monkey (Macaca mulatta) and the distribution of the efferent fibers. J. Comp. Neurol., 77, 693-738.

Peele, T.L. (1944). Acute and chronic parietal lobe ablations in monkeys. J. Neurophysiol., 7, 269-286.

Petrides, M. and Iversen, S.D. (1979). Restricted posterior parietal lesions in the rhesus monkey and performance on visuospatial tasks. Brain Res., 161, 63-79.

Ratcliff, G., Ridley, R.M. and Ettlinger, G. (1977). Spatial disorientation in the monkey. Cortex, 13, 62-65.

Robinson, C.J. and Burton, H. (1989b). Organization of somatosensory receptive fields in cortical areas 7b, retroinsula, postauditory and granular insular of M. fascicularis. J. Comp. Neurol., 192, 69-92.

Robinson, C.J. and Burton, H. (1980b). Somatic submodality distribution within the second somatosensory (SII), 7b, retroinsular, postauditory, and granular insular cortical areas of M. fascicularis. J. Comp. Neurol., 192, 93-108.

Robinson, D.L., Goldberg, M.E. and Stanton, G.B. (1978). Parietal association cortex in the primate: sensory mechanisms and behavioral modulations. J. Neurophysiol., 41, 910-932.

Ruch, T.C., Fulton, J.F. and German, W.J. (1938). Sensory discrimination in monkey, chimpanzee and man after lesions of the parietal lobe. Arch. Neurol. Psychiatry 39, 919-938.

Sakata, H. (1975). Somatic sensory responses of neurons in the parietal association area (area 5) of monkeys. In The Somatosensory System. (ed. H.H. Kornhuber). Georg Thieme, Stuttgart. pp. 250-261.

Sakata, H. and Iwamura, Y. (1978). Cortical processing of tactile information in the first somatosensory and parietal association areas in the monkey. In Active Touch. (ed. G. Gordon). Pergamon Press, London. pp. 55-72.

Sakata, H., Takaoka, Y., Kawarasaki, A. and Shibutani, H. (1973). Somatosensory properties of neurons in the superior parietal cortex (area 5) of the rhesus monkey. Brain Res., 64, 85-102.

Sakata, H., Shibutani, H. and Kawano. K. (1980). Spatial properties of visual fixation neurons in posterior parietal association cortex of the monkey. J. Neurophysiol., 43, 1654-1672.

Seal, J., Gross, C. and Bioulac, B. (1982). Activity of neurons in area 5 during a simple arm movement in monkeys before and after deafferentation of the trained limb. Brain Res., 250, 229-243.

Seltzer, B. and Pandya, D.N. (1980). Converging visual and somatic sensory cortical input to the intraparietal sulcus of the rhesus monkey. Brain Res., 192, 339-351.

Seltzer, B. and Pandya, D.N. (1978). Afferent cortical connections and architectonics of the superior temporal sulcus and surrounding cortex in the rhesus monkey. Brain Res., 149, 1-24.

Semmes Blum, J., Chow, K.L. and Pribram, K.H. (1950). A behavioral analysis of the organization of the parieto-temporo-preoccipital cortex. J. Comp. Neurol., 93, 53-100.

Stanton, G.B., Cruce, W.L.R., Goldberg, M.E. and Robinson, D.L. (1977). Some ipsilateral projections to areas PF and PG of the inferior parietal lobule in monkeys. Neurosci. Lett., 6, 243-250.

Sugishita, M., Ettlinger, G. and Ridley, R.M. (1978). Disturbance of cage-finding in the monkey. Cortex, 14, 431-438.

Yin, T.C.T. and Mountcastle, V.B. (1977). Visual input to the visuomotor mechanisms of the monkey's parietal lobe. Science, 197, 1381-1383.

CORTICAL AREAS IN MAN PARTICIPATING IN SOMATOSENSORY DISCRIMINATION OF MICROGEOMETRIC SURFACE DEVIATIONS AND MACROGEOMETRIC OBJECT DIFFERENCES

P.E. ROLAND

Department of Neurology N 2082, Rigshospitalet, 9, Blegdamsvej, DK-2100 Ø, Copenhagen, Denmark

It has been known for 100 years that the ability to reconstruct tactually sensed objects in the mind was often lost after lesions of the cerebral hemispheres in man (Hoffmann, 1884). Since then, to my knowledge, there has been no quantitative measurements published of the somatosensory sensation of microgeometric and macrogeometric object properties after anatomically verified brain lesions in man. Shape and size are macrogeometric objects properties. Microgeometric object properties are the small deviations present in the surface of objects and commonly referred to as roughness or smoothness.

The purpose of the present series of experiments was to measure disturbances of the transmission, demodulation and internal reconstruction of microgeometric and macrogeometric somatosensory signals after anatomically verified lesions to the human brain. An attempt was also made to see whether the noise arising from these lesions was confined to any particular bandwidth or physical aspect of the stimuli. Finally, the purpose was to delimit the cortical areas and subcortical connexions to which lesions elicited disturbances in the transmission, demodulation and reconstruction of microgeometric and macrogeometric information.

For these purposes a series of stimuli, accurately quantified in every physical aspect, were manufactured (Roland, 1975). The subjects and patients had to discriminate these stimuli. They were, however, free to choose their own way of palpation and sampling of information.

Anatomical Measurements

The experiments were carried out with the participation of 93 patients with unilateral circumscribed lesions of the cerebral hemispheres. These patients were selected from 800 patients whose lesions were studied and anatomically measured by the author during craniotomy. The selection criteria have been described in a preliminary report (Roland, 1976).

The size and shape of the lesion was determined from measurements during the operation of 1) the depth and width of the lesion. 2) The distance to identified cortical sulci and other landmarks. 3) The volume of the lesion and the volume of the removed tissue. 4) For later identification silver clips were fixed to the bottom of the operation cavity and the cortical periphery of the lesion. 5) The exposed brain was color photographed and drawn at different stages of the operation. From these operation charts the outlines of the brain lesion was transferred to a proportional stereotaxical system (Talairach et al., 1967).

The lesion of the 93 patients were distributed almost uniformly in the cerebral hemispheres. The head of the caudate nucleus, the putamen, claustrum, amygdala and hippocampus were lesioned in some cases. The thalamus was always spared.

Method of Stimulation and Psychophysical Procedure

The subjects were at no time allowed to see the stimuli. They were free to palpate the stimuli as they wished - but not to use their nails. The way the subjects palpated or manipulated the stimuli was recorded in detail. Control subjects and patients were instructed by the same standardized instructions and the number of times a subject needed it read to comprehend it,was noted. All patients had a full clinical neuropsychological examination including a test for aphasia. Their behavior was recorded continuously during the discriminations.

The subjects discriminated the stimuli after a two-alternative foced-choice procedure. Two stimuli (S1 and S2) were selected after a randomized schedule, and presented successively to the subjects. They were allowed to palpate each stimulus a maximum of five times and then they had to make a decision. Four patients had so severe pareses of their hands that active palpation was impossible. Their hands were then moved around the stimuli in a way that imitated active palpation. Each combination of stimuli was presented at least five times. In the patients the hand ipsilateral to the lesion was tested first.

For each type of stimuli a number of age matched normal control subjects were examined (see tables). All testing was done by an assistent with no knowledge of the site of lesion.

The response matrices were analysed for response bias. A maximum likelihood estimate of the probability of correct response (PC) was calculated. We (Roland and Mortensen, 1984) developed an energy detection model to describe the relation between signal energy

TWO ALTERNATIVE FORCED CHOICE DISCRIMINATION:

SIGNAL TO NOISE RATIO:

$$\text{INPUT} \quad \frac{S_I}{N_I} = \frac{\sqrt{\Sigma \Delta^2}}{\sqrt{S_1 + S_2}}$$

$$\text{OUTPUT} \quad \frac{S_O}{N_O} = \frac{\sqrt{\Sigma \Delta^2}}{K_O \sqrt{S_1 + S_2}}$$

PROBABILITY OF CORRECT RESPONSE

$$PC = \frac{\exp\left(\frac{S_O}{N_O}\right)^2}{1 + \exp\left(\frac{S_O}{N_O}\right)^2}$$

Fig. 1. Energy discrimination model. Δ^2 is the difference in stimulus energy for corresponding points of the two stimuli, S1 and S2. Ko is the noise figure.

and behavioral binary output. An ultrashort description of the model is presented in Fig. 1. The signal for discrimination is the difference in energy (power) between the two stimuli S1 and S2. Empirically, the means ratio of signal to noise (So/No) of the output was a linear function of the means ratio of signal to noise (Si/Ni) of the input for both brain injured and normal controls. Si/Ni divided by So/No is called the noise figure, Ko. Since the human somatosensory system amplifies the input signal, Ko is usually below one in normals. The noise figure, Ko, increases as a result of noisy discrimination,for example due to damage of the somatosensory demodulators in the cerebral cortex. The criteria for deviation from normality were two output signal to noise ratios at .75 probability of correct response exceeding the 99% confidence levels of normal individuals, or three So75/No ratios exceeding the individual 95% confidence levels.

DISCRIMINATION OF MICROGEOMETRIC SURFACE DEVIATIONS

Two examples of the surfaces used as stimuli for discrimination of object microgeometry are shown in Fig. 2. The profile of the surfaces were known exactly and a full discription of the theory and manufacturing of these stimuli has been published earlier (Roland, 1975). The stimulus energy and signal energy was

$$S_1 = \tfrac{1}{2}\, Ru_1^2\, w_1;\; S_2 = \tfrac{1}{2}\, Ru_2^2\, w_2;\; \Delta_2^2 = S_1 - S_2$$

The mean output signal to noise ratio (Fig. 1) of normal control subjects is listed in Table 1 for a probability of correct response PS = 0.75. The mean noise figure, Ko, which is independent of the

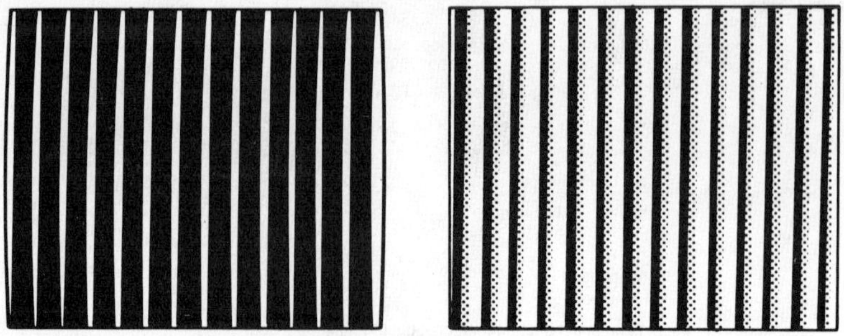

Fig. 2. Drawings of microscope images of two surfaces used for microgeometric discrimination. Left: Peak to mean line amplitude, Ru = 4.50μ, wavelength w = 215μ. Right: Ru = 13.1μ, w = 353μ.

.75 probability of correct response is also listed. For amplitudes, Ru, between 1.3μ and 16.6μ the noise figure was almost constant.

With one minor exception, all microgeometric and macrogeometric discrimination loss was confined to lesions of the contralateral hemisphere. In general the patients had to have a certain proportion lesioned of a cortical area participating in somatosensory discrimination to elicit a discrimination loss. Fig. 3 shows the area of maximal overlap of any two lesions associated with discrimination loss in roughness or microgeometric discrimination.

Thirty patients had lesions of the postcentral gyrus (SI). In 19 of these, the lesion invaded the SI-hand area. The SI-hand area was defined as the middle third of the postcentral gyrus on the lateral brain surface. Only three of these 19 patients had microgeometric discrimination loss (Table 2). In one of these patients the afferent fibres to SI were damaged and no discrimination thresholds could be measured. A total lesion of the SI-hand area raised the So/No ratio to 34 times that of normals. Lesions of the overt part of the SI-hand area had a more moderate effect. In the rest of the patients whose lesions invaded up to 80% of the SI-hand area the noise figure was normal. In the patients with discrimination loss the noise was uniformly distributed over the full stimulus bandwidth although in Table 2 only the bandwidth from 1.3μ to 16.6μ was shown.

The supplementary sensory area was defined as the cortical area on the superolateral border and the upper mesial surface of the superior parietal lobule immediately behind the SI-leg area (Roland et al., 1982). The area was situated between the sections F8 and F10 in Fig. 3. Three patients with lesions of this area plus the adjacent cortex down to the intraparietal sulcus had a slight, non bandlimited noisy roughness discrimination (Table 2). The patient with the most complete lesion (98%) had the largest noise figure

TABLE 1. DISCRIMINATION OF MICROGEOMETRIC SURFACE DEVIATIONS, NORMATIVE DATA (61 SUBJECTS).

Amplitude Ru μ	Wavelength W μ	$\frac{So75}{No}$	Noise figure Ko	Individual Upper 95% confidence limits of Ko
.99	50	.346	.330	.700
1.31	109	.130	.123	.260
3.83	200	.108	.103	.215
4.50	215	.116	.111	.230
4.94	256	.150	.143	.290
6.87	297	.140	.133	.290
7.85	312	.151	.144	.290
10.3	353	.136	.129	.290
10.7	488	.159	.151	.310
16.6	453	.152	.145	.310
22.0	488	.298	.284	.600

(.370). One of these patients had ipsilateral discrimination loss. Fourteen patients with partial, but smaller lesions of the cortical zone occupied by the lesions of the patients with discrimination losses, had normal noise figures.

The retroinsular cortex was arbitrarily defined as the upper half of the cortex lining the rear end of the Sylvian fissure from 1 cm beneath the surface from F8 to midway between F9 and F10 (Fig. 3). Nineteen patients had lesions invading this zone. Five of the seven patients with more than 50% of this zone destroyed had increased noise figures. Those with from 25% to 50% of the zone destroyed had mean noise figures in the bandwidth from 1.0μ to 16.6μ ranging from 0.193 to 0.287. None of the six patients with lesions exclusively of the inferior parietal cortex covering the retroinsular cortex had any discrimination loss. The noise after retroinsular cortex lesions was limited to the large amplitude (7-22μ) long wavelength (200-400μ) bandwidth (Table 2).

Lesions of the left middle frontal gyrus caused a non bandlimited slightly noisy discrimination. These patients rarely reexamined the surfaces for additional cues. Four patients with equiterritorial lesions in the right hemisphere were normal. Finally two lesions of the superior frontal gyrus caused sporadic exceedings of the 99% confidence limits of normals.

DISCRIMINATION OF MACROGEOMETRIC OBJECT DIFFERENCES

The size and shape stimuli have been extensively described ear-

TABLE 2. MICROGEOMETRIC DISCRIMINATION LOSS ACCORDING TO LOCUS OF LESION.

Lesion Area	No of Subjects	Amplitude range μ	Noise Figure, Ko range	Noise Figure, Ko mean
S_I	3	1.30-16.6	0.819-4.476 (∞)	2.467
Supplement. sensory	3	1.30-16.6	0.187-0.370	0.259
Retroinsular	5	1.30-7.00	0.120-0.133	0.126
		7.00-16.6	0.151-0.481	0.342
		16.7-22.0	0.284-0.484	0.380
Mid.front.gy.	3	1.30-16.6	0.161-0.205	0.186
Sup.front.gy.	2	0.99	0.964-1.064	1.014
		1.30-16.6	0.134-0.187	0.160

lier (Roland, 1975). All objects were in hard aluminium and had the same weight. All other physical properties except size and shape were identical. The size stimuli were 31 spheres with a diameter range from 20.0 mm to 50.0 mm with the diameter increased in steps of 1.0 mm. The shape stimuli were 15 rectangular parallelepipeda with square bases and 15 rotational symmetrical ellipsoids all of the same volume 11500 mm^3. The side lengths of the parallelepipeda ranged from 22.57 mm to 44.50 mm in small steps. The long axis of the ellipsoids ranged from 14.01 mm to 29.99 mm in small steps. The stimulus energy and signal energy for the rectangular parallelepipeda was

$$S_1 = 4A_1+8B_1 \quad S_2 = 4A_2+8B_2, \ \Delta^2 = 4(A_1-A_2)^2+8(B_1-B_2)^2$$

in which A was the side length of the long side and B the side length of the square bases. The stimulus energy and signal energy for the ellipsoids was

$$S_1 = \Sigma\, K_1xyz, \ S_2 = \Sigma\, K_2xyz, \ \Delta^2 = \Sigma\, \Delta Kxyz^2$$

in which Kxyz was the curvature in every point of the surface. Each subject or patient had a minimum of 1000 discriminations in each test.

Normal subjects could with a probability of correct response of 0.75 distinguish spheres with a diameter difference of 0.74 mm. This threshold was constant throughout the stimulus bandwidth (upper 95% confidence limit: 1.70 mm). Table 3 shows the output signal to noise ratio and the mean noise figure of normals. The oblongness referred to in the table was the objects deviation from spherical shape and cubic shape, respectively. This was the physical entity to be discriminated by the subjects. This figure can be calculated

TABLE 4. SHAPE DISCRIMINATION LOSS. ACCORDING TO LOCUS OF LESION.

Lesion area	No of subjects	Oblongness	Noise figure, Ko range	mean
		Ellipsoids		
SI-hand	7	0.00-2.00	0.031-0.142 (∞)	0.0953
		0.200-4.700	0.043-0.109 (∞)	0.0804
Supplement. sensory	1	0.00-2.00		0.0152
		0.200-4.700		0.0181
Postcentral sulcus	2	0.00-200	0.017-0.023	0.020
		0.200-4.700	0.016-0.019	0.018
		Rectangular parallelepipeda		
SI-hand	7	0.000-2.000	0.507-2.393 (∞)	1.328
		2.000-47.000	0.694-1.796 (∞)	1.329
Supplement. sensory	1	0.00-2.000		0.281
		2.000-47.000		0.194
Postcentral sulcus	2	0.00-2.000	0.208-0.269	0.239
		2.000-47.000	0.163-0.219	0.191
Retroinsular	1	0.000-2.000		0.239
		2.000-47.000		0.134
Middle front. gyrus	2	0.00-2.00	0.179-0.306	0.241
		2.00-47.000	0.173-0.180	0.177
Sup.front. gyrus	2	0.00-2.00	0.083-0.199	0.141
		2.00-47.000	0.153-0.257	0.205

shape discrimination loss after the criteria, and the mean noise level of the third was at both bandwidths increased from that of the normal controls.

These three areas the cortex lining the postcentral sulcus, the supplementary sensory area and adjacent part of cortex down to the intraparietal sulcus, and the retroinsular cortex was regarded as somatosensory association areas. The noise in shape discrimination arising from lesions of these areas was always uniformly distributed over the whole bandwidth. Eleven patients had lesions involving the second somatosensory area, but all had normal macrogeometric (and microgeometric) discrimination.

The rectangular parallelepipeda were the most difficult to discriminate for both normals and patients. As was the case for microgeometric discrimination, selected lesions of the prefrontal cortex caused macrogeometric discrimination losses of the parallelepipeda. The two patients with most of the midpart of the midfrontal gyrus lesioned seldom resampled information about the objects and their noise figures were quite varying although their average noise level in parallelepipeda discrimination was close to the mean of normals. Strangely enough, four patients with lesions encompassing the midfrontal gyrus were normal, as were 14 patients with partial lesions of the midpart of the gyrus. Finally, lesions of the superior frontal gyrus in its prefrontal division caused sporadic exceedings of the 95% confidence limits of normals.

None of the patients with either microgeometric or macrogeometric discrimination loss had any troubles comprehending or memorizing the instructions. There were many variants of deviant manipulation and palpation of the objects. However, 23 out of 30 patients with abnormal palpation had normal noise figures. Even lesions of the primary motor hand area, which did not intrude into the SI-hand area, were associated with normal discrimination capacity.

DISCUSSION

Theoretically discrimination loss after brain lesions could be due to suboptimal sampling of information, noisy transmission, destruction of somatosensory demodulators and destruction of the discriminator. To this list may be added more unspecific factors such as suboptimal decision strategies, defects of attention, intention, memory or comprehension. There were, among the patients with discrimination loss, no defects in their comprehension and memorizing of the instructions and no defects in their maintenance of intention which could give rise to erroneous measurements of a noisy discrimination.

Motor disturbances of the sampling of information was likewise of subordinate importance since size, shape and roughness discrimination could be normal despite paralysis arising from lesions of the motor areas. This was in accordance with the findings that the regional cerebral blood flow did not increase in the motor areas during discrimination of macrogeometric object differences when the hand or fingers did not move (Roland and Larsen, 1976). Furthermore these findings indicate that the motor areas were of no crucial importance for the demodulation of complex somatosensory information.

The response criteria were measured in all subjects and patients. Since both patients and normal control subjects had no difficulties maintaining the symmetrical response criterion necessary for optimal two alternative forced choice discrimination, suboptimal response criteria in the sense of asymmetrical response bias could not explain the discrimination loss.

The discrimination loss was therefore naturally ascribed to

the destruction of somatosensory demodulators and disturbances of the transmission to the SI-hand area and from the SI-hand area to the somatosensory association cortex. Furthermore, transmission disturbances between these areas and areas participating in discrimination and control of sensory demodulation might have caused additional discrimination loss. Since discrimination of auditory and visual signals has been found to give rise to regional cerebral blood flow increases in the superior and midfrontal cortex, and since these two areas also participate in many other non sensory types of brain work (Roland, 1984) the demodulation and decoding of the complex somatosensory signals must have taken place in the SI-cortex and the somatosensory association areas.

There was no correlation between the noise figures for microgeometric surface discrimination and those for macrogeometric discrimination. Neither had the patients any correlation between the noise figures for size and those for shape. One of the reasons for this lack of correspondence might be that the information was differently represented in the different stimuli. All information in the microgeometric surfaces was represented in a line orthogonal to the orientation of the ridges. Since the profile curve along this line consisted of a replication of the same ridge profile, the microgeometric surface was highly redundant. Similarly, the size of a sphere is the inverse of the curvature in every point of the surface. Since the curvature in a sphere is the same in every point of the surface, the spheres were also highly redundant stimuli. In the ellipsoids the curvature of the surface was identical in only four symmetric points of the surface, and in the parallelepipeda the dimensions of at least three sides had to be known for the subjects in order for them to detect the object. The shape stimuli thus were highly compact or non redundant.

The whole SI-hand area or the afferent fibres to SI had to be lesioned to cause a major impairment in microgeometric discrimination. Since lesion of the overt part of the SI-hand area caused an eight times rise in noise figures this part of the SI might have been of particular importance for the central reconstruction of the surface characteristics. However, even about 20% preservation of the overt part seemed sufficient for a normal roughness discrimination. Similarly, the whole SI-hand area had to be destroyed in order to raise the diameter threshold above 20 mm in size discrimination. Full lesion of either the anterior half of the gyrus or the whole overt part only raised the threshold to six mm. That is, for these redundant stimuli, the mechanoreceptive demodulation provided by a small part of either the anterior or overt part of the SI-hand area gave sufficient information to reconstruct or discriminate these stimuli.

Destruction of the anterior half of the SI-hand area including both the overt and deep part, made it impossible to distinguish edged objects from rounded - the shape recognition was totally abolished. Destruction of only the overt part destroyed the discrimination between different degrees of curvature and shape, but the edge - round discrimination was intact. This indicated that the mechanoreceptive demodulation provided by the deep anterior part of the SI-hand area was necessary information for the neuronal population in the whole overt part. This part in turn provided the further information necessary for the reconstruction or discrimination of different degrees of shape. That is the overt part seemed to provide quantitative information about curvature. Since large lesions of the cortex lining the postcentral sulcus caused a moderate microgeometric and macrogeometric discrimination loss. This part of the cortex should also be included in the somatosensory association cortex. The noise level was only slightly and uniformly raised over the stimulus bandwidth so it was difficult to assess the contribution of this area, unless it was further filtering.

The supplementary sensory area has been activated by movements of the fingers in intrapersonal space (Roland et al., 1982). Damage to the region caused a slight noisy discrimination of both microgeometric and macrogeometric object variables. The supplementary sensory area, perhaps in concert with the cortex just lateral to it, could therefore be participating in demodulation of sensory information retrieved during movements in particular.

Since only one patient with a retroinsular lesion had a slightly noisy discrimination of parallelepipeda, whereas many patients with equiterritorial lesions or greater lesions of the retroinsular cortex had normal discrimination of shape, this region was probably not crucial for shape reconstruction. The retroinsular cortex was important for the demodulation of microgeometric surface deviations, because all patients with more than 50% of the area lesioned had noise figures significantly different from normals. The noise was limited to the high amplitude low frequency surfaces. If the spatial frequency of the microgeometric surface was somewhat lower than the spatial frequency of the epidermal ridges the surface would be difficult to discriminate unless information about phase was preserved. It could therefore be that the retroinsular cortex may contain phase sensitive neurons which were especially important for the central reconstruction of large amplitude low frequency surfaces. Neurons that are phase sensitive probably are able to fire with very high frequencies. Such neurons have been found in retroinsular cortex of the monkey (Robinson and Burton, 1980). Murray et al. (1980) found impaired discrimination of different grades of sandpaper after retroinsular lesions in the rhesus monkey.

No matter what modality is discriminated in a two alternative forced choice procedure, the posterior superior prefrontal cortex and another prefrontal area around the midfrontal gyrus has always been activated together with the sensory cortex and the sensory association cortex of the relevant modality (Roland, 1984). The participation of the midfrontal gyrus in the present task was probably best reflected by the observation that lesions here decreased the probability of resampling of sensory information. The role of this area as a search area for sensory information was reviewed recently (Roland, 1984). The paradox, that lesions of the midfrontal gyrus caused suboptimal somatosensory discrimination, whereas big lesions that entirely precluded communication with the entire midfrontal cortex and most of the rest of the prefrontal cortex were without effect, could probably be explained that the communication between the somatosensory association cortex and the prefrontal cortex in the latter case followed an alternative and therefore not noisy route to the midfrontal gyrus in the other hemisphere.

It was evident, that minor lesions of the SI-hand area or the somatosensory association areas were without any effect on the discrimination of microgeometric and macrogeometric object differences. When the lesion of any of these areas became larger first the demodulation and discrimination of the most compact objects - the shape stimuli suffered. When the lesion of either the SI-hand area or the retroinsular cortex got even larger the demodulation of the redundant stimuli suffered. This could indicate that the information about the object or surface in both SI and in any of the somatosensory association areas was represented in a relatively big field. If such a field was moderately lesioned representation of the most compact or non redundant objects would suffer first, because an integration of the information contained in the remaining undamaged field would be insufficient to reconstruct the object with great accuracy. On the other hand,representations of redundant objects would survive such a lesion because an integration of the information contained in the remaining undamaged field would be sufficient to reconstruct the object with great accuracy. In support of this idea Juliano et al. (1981) found that vibration of the finger tip in the monkey provoked widespread metabolic activation in the SI-hand area, areas 2 and 5 and the retroinsular cortex and S II. When a cortical area participates in a particular brain work, for example sensory discrimination, the regional cerebral blood flow increases always over an area of 3-7 cm^2. Within these areas of activation the rCBF always had a coefficient of variation usually less than 7% in the individuals (Roland and Larsen, 1976, Roland 1984). This indicates that small areas of 3-7 cm^2 functions as units.

REFERENCES

Hoffmann, H. (1884). Stereognostische Versuche, Dtsch. Arch. Klin. Med., 35, 529-561.

Juliano, S.W., Hand, P.J. and Whitsel, B.L. (1981). Patterns of increased metabolic activity in somatosensory cortex of monkeys Macaca fascicularis, subjected to controlled cutaneous stimulation: a 2-deocyglucose study. J. Neurophysiol., 46, 1260-1284.

Murray, E.A., Nakamura, R.K. and Mishkin, M. (1980). A possible cortical pathway for somatosensory processing in monkeys. Soc. Neurosci. Abstr., 6, 654.

Robinson, C.J. and Burton, H. (1980). Somatic submodality distribution within the second somatosensory (SII), 7b, retroinsular, postauditory and granular insular, cortical areas of M. fascicularis. J. Comp. Neurol., 192, 93-108.

Roland, P.E. (1975). Some principles and new methods of tactile stimulation. Behav. Res. Meth. Instrum., 7, 333-338.

Roland, P.E. (1976). Astereognosis. Arch. Neurol., 33, 543-550.

Roland, P.E. and Larsen, B. (1976). Focal increase of cerebral blood flow during stereognostic testing in man. Arch. Neurol., 33, 551-558.

Roland, P.E., Meyer, E., Shibasaki, T., Yamamoto, Y.L. and Thompson, C.J. (1982). Regional cerebral blood flow changes in cortex and basal ganglia during voluntary movements in normal human volunteers. J. Neurophysiol., 48, 467-480.

Roland, P.E. (1984). Cortical activity in man during discrimination of extrinsic patterns and retrieval of intrinsic patterns. Exp. Brain Res. Suppl., 9 (in press).

Roland, P.E. and Mortensen, E. (1984). Somatosensory discrimination of macrogeometric object differences in normal subjects (submitted).

Talairach, J., Szikla, G., Tournoix, P., Prossalentis. A., Bordas-Ferrer, M., Covello, L., Iacob, M. and Mempel, E. (1976). Atlas d'Anatomie Stéréotaxique du Télencéphale. Masson, Paris.

Sessions III and IV

PERIPHERAL AND INTEGRATIVE MECHANISMS OF SOMATOSENSATION

WHY ARE SELECTIVELY RESPONSIVE AND MULTIRECEPTIVE NEURONS BOTH PRESENT IN SOMATOSENSORY PATHWAYS?

EDWARD R. PERL

University of North Carolina at Chapel Hill, NC 27514, USA

It is a particular privilege to join in this symposium honoring Yngve Zotterman. His gracious hospitality and enthusiasm for science was an inspiration to me as to so many others. My invitation asked for an overview for the presentations to follow in this session, a rather difficult task because of the scope of the topics involved. I have chosen instead to offer a theory about a feature of the organization of the somatosensory system that touches on the several topics, with the realization that while speculation in science is a precarious game, it would have pleased Yngve Zotterman; he loved to practice it.

The theory draws in large part from the work of others and is fairly simple. Its testing and application will not, but in my view the eventual evaluation will, depend on ingenuity of experimental design. The question partially concerns mechanisms of pain, an interest Professor Zotterman and I shared, but its implications are broader.

The information a mammal gathers from its environment leads to constant adjustments of a large range in behavior; these include modifications of body position or motion, search for nourishment, rest, procreation, protection, etc. These changes may be automatic, that is, reflexive; they may involve affect and thus represent behavioral drives related to preservation; or, they may be cognitive. Environmentally shaped behavior demands substantial changes in levels of consciousness from sleep to heightened awareness or alertness. In particular, alerting may necessitate such focusing of attention that information from channels or sources outside this focus would be ignored. Reflection on the many challenges faced by an animal and their behavioral outcomes indicates that the types of information necessary or sufficient to accomplish the outcomes differ considerably. For example, alerting may require no more than signals indicating environmental change, whereas perceiving and discriminating demand a wealth of detail with minimal ambiguity. Is it efficient or even possible to use a single system of information transfer to accomplish tasks demanding different levels of detail?

The largest amount of somatosensory information probably would be needed for cognitive purposes, in which not only a stimulus is detected, but also its nature is distinguished, its intensity indicated, and its location delineated. For example, the skin is a somatic tissue which is a major source of afferent information. There is much evidence demonstrating that the afferent input from mammalian skin provides the organism with the capacity to differentiate several features of mechanical disturbance, to detect temperature change and direction, and to reliably distinguish damaging from innocuous stimulation. Although the full story is far from known, it is generally agreed that neural apparati provide distinctive signals appropriate for the various discriminative functions associated with the cognitive aspects of cutaneous mechanoreception and thermoreception. On the other hand, the issue remains clouded for nociception and for how some of the reactions, other than discrimination, triggered by or dependent upon sensory input, are mediated.

While the concept of specific nerve function was suggested prior to Johannes Müller (1840, 1842), and while he actually proposed something less than the usual present-day understanding of those terms, it was his codification of the concept that fell upon a fertile environment, stimulating thought and experimentation. The most controversial view of specific function from the standpoint of ideas about somatosensory organization appeared in the writings of von Frey (1897). Von Frey extended Blix's discovery of punctate cutaneous sensibility to a generalization. He maintained that each of the several cutaneous sensory experiences began with activation of a particular afferent organ. Questions arose about the validity of this concept when some of the predicted relationships between sense organ and sensory experience could not be confirmed (Melzack and Wall, 1962; Sinclair, 1955).

To disengage the present discussion from the old arguments, let me avoid the term "specific" in favor of what might be a more realistic description of the relationship between a sensory system's activity and the stimuli that engage it. One or another aspect of the physical and chemical environment of a mammal varies as time passes. When a sense organ or the neural system to which it is connected is much more responsive to one kind of environmental change or event than to others, it can be considered selectively responsive. It is selective rather than specific because whenever the energy level of a type of change increases, the spectrum of kinds of events that can excite a particular receptive structure often broadens. In other words, biological transducers may be adapted to respond to a particular format of change in the status of the tissue, but they are never absolute in their sensitivity. Although this discussion will concentrate on the afferent signals that stem from the skin, essentially the same arguments can be mustered for other somatic tissues.

Selectivity of Cutaneous Sense Organs

Much has been written on the responsiveness of cutaneous sense organs; however, not all commentators appreciate the limits and differences in the selectivity of response exhibited by them. I suggest that selectivity of responsiveness can only be properly judged when one considers both the natural history of the organism and the range of experiences evocable from a tissue. Only when the selectiveness of behavior of the population of individual cutaneous receptors is understood can the behavior of central neurons in turn be properly evaluated for selectivity in responsive patterns. For example, a variety of mechanoreceptors respond to quite gentle disturbances of the skin's surface or body

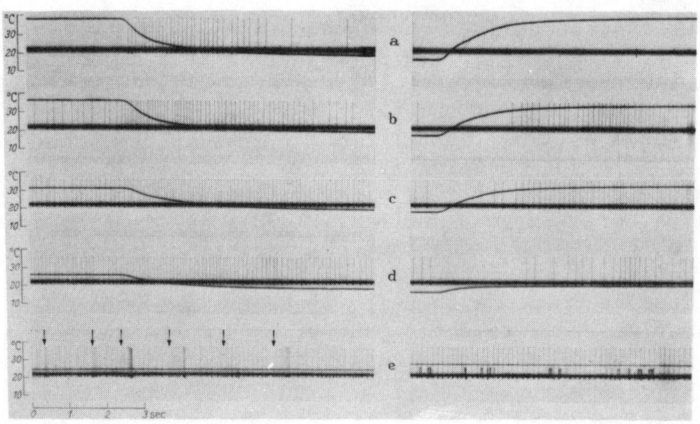

Figure 1: Activity recorded from a single type II tactile receptor of cat after cooling and warming of the skin (a-d). Temperature of the receptive field was controlled by a contact thermode of $6 cm^2$. In *e* on the left at each arrow the skin of the receptive field was lightly touched with the flat side of a dissecting needle. (Reproduced with permission from Witt and Hensel, 1959).

hairs. Often they are simply described as "rapidly adapting" or "slowly adapting". However, this use of adaption refers to a maintained contact with or deflection of structure which, on careful consideration of the ordinary variety of mechanical events impinging upon a mammal's integument, is too simple to describe or to differentiate either stimuli or sense organs (Burgess and Perl, 1973). The selectivity with which mechanoreceptors of the skin respond turns out to be based not only on the magnitude of a mechanical event but also on the components of that event's motion, a point that several of this symposium's participants have demonstrated (e.g., Brown and Iggo, 1967). Looked at in this way mechanoreceptors are tuned to one or another aspect of motion from maintained displacement through higher derivatives of displacement, and although the tuning represents a substantial focusing at threshold amplitudes, the spectrum of responsiveness broadens greatly as stimuli become more intense. Thus, at higher suprathreshold intensities of stimulation, stimulus features must be indicated by the dynamics of a population of neurons with overlapping ranges of responsiveness. As we have already heard in this symposium, features of complex events or those differentially affecting an extent of tissue, such as the movement of objects across the skin surface, must be coded by the activity of a population of sense organs. This is hardly surprising since the same principles apply also for other organs of sense such as the eye and the ear.

In attempting to decipher the combinations of sense organs activating central neurons, it is necessary to keep in mind that some cutaneous mechanosensitive sensors are also excited by sudden temperature changes. One of the earlier accounts of this appeared in Witt and Hensel's (1959) description of what is now known as the Type II slowly adapting receptor (Fig. 1). The Type II receptor's discharges signal degrees of skin pressure-tension with a reproducible pattern (Burgess and Perl, 1973); its responsiveness to temperature changes is thought to be spurious and unrelated to temperature sense because characteristics of its afferent fiber, of its quantitative response to temperature changes, and of its central projections fail to correspond with data on the psychophysics and the functional organization of systems important for temperature sensations.

Cutaneous mechanoreceptors belonging to the rapidly adapting group also are activated by temperature changes; they include C-fiber units (C-mechanoreceptor) (Fig. 2) and a sensory unit with a thinly myelinated afferent fiber (D-hair or δ mechanoreceptor). It certainly would seem reasonable to consider such elements nonselective regarding cutaneous stimuli, yet the characteristics of each type suggest a substantial selectivity in terms of the features of mechanical stimuli for which they have the lowest threshold. It is generally conceded that these sense organs also are truly mechanoreceptors rather than thermoreceptors, principally because their responses to mechanical stimuli are more vigorous and consistent than their responses to alterations of temperature.

Other cutaneous afferent units exhibit a greater selectivity. Thermoreceptors respond to very small changes in surface temperature in a consistent and predictable fashion, and do not respond to ordinary mechanical manipulations. (The thermoreceptors and the mechanoreceptors with thin afferent fibers proved something of a problem some decades ago in terms of the afferent basis of pain when Zotterman (1939) noted discharges from thin fibers evoked by stimuli that ordinarily do not cause pain. Yet, his own work and that of others had previously shown that activity in the thin fibers was associated with pain.)

The low threshold mechanoreceptors served by the thicker myelinated fibers appear to include the most selective of the cutaneous sense organs. Pacinian corpuscles and some of the hair follicle receptors (G1) do not respond to cutaneous temperature

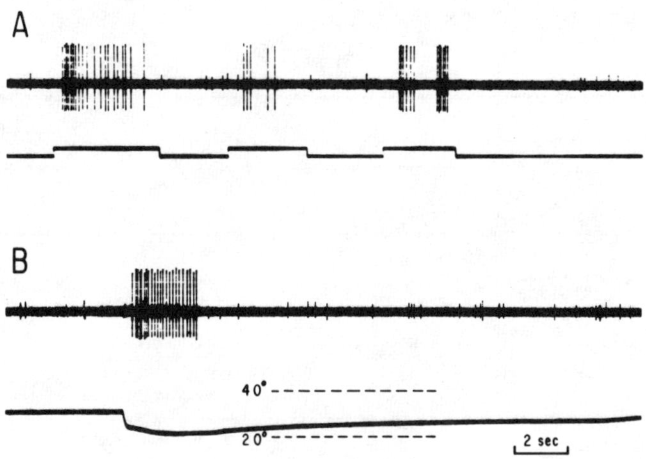

Figure 2: Discharges recorded from a cat C-fiber cutaneous mechanoreceptor. Afferent fiber conduction velocity 1 m/sec. A: Mechanical contact of receptive field by von Frey-type stimulator bending at 0.04 g at each upward deflection of lower trace. B: Receptive field cooling by evaporation of volatile liquid; temperature changes indicated by a thermistor on the receptive field surface. (With permission from Bessou and Perl, 1969).

changes, are inactivated by noxious stimulation of the skin and look upon the world of mechanical changes through a relatively narrow window. They are effectively excited by transient disturbances or disturbances with a substantial component of acceleration.

The cutaneous nociceptors provide another dimension. Their selectivity results in part from elevated thresholds for all varieties of modifications of the cutaneous environment; this is particularly evident when their responses are compared with those of the other kinds of sensory elements in the skin. But they have selectivity to stimulus type as well; one type with myelinated fibers responds promptly only to strong mechanical distortion, and the polymodal C-fiber type is unique in being excited by low concentrations of irritant chemicals. Do they respond only to noxious stimuli? For a part of the population making up each type the answer is no, since a fraction of each kind of nociceptor has thresholds that may be below the intensity necessary to damage the skin; e.g., see Figure 3A for mechanical stimuli. However, nociceptors differ from the low threshold sense organs that might also be excited by strong stimuli, in that they provide progressively augmented activity to progressively increased stimulation at damaging intensities; vigorous responses are regularly reserved for unambiguously noxious stimulation.

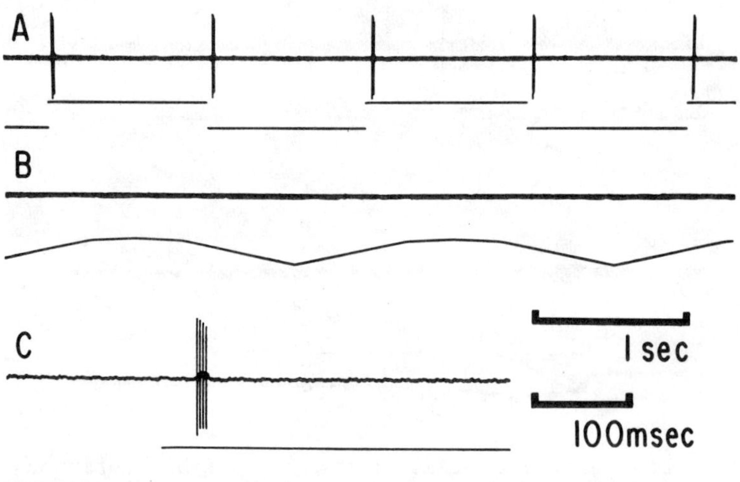

Figure 3: Responses of a gracilis nucleus neuron to hair movement. Upper trace of each pair: recording from pipette microelectrode. Lower trace: analogue indication of displacement of a fine probe attached to an electromechanical stimulator moving a small group of hairs. Upper time mark - A and B. Lower time mark - C. (With permission from Perl, Whitlock and Gentry, 1961.)

In summary, the breadth of a sense organ's vision of the cutaneous environment varies from one kind of afferent unit to another. However, in terms of threshold and the minimal intensity of stimulation that provokes maximal activity, different types of sensory units exhibit considerable selectivity for a) one of the various attributes of a mechanical disturbance, b) small temperature changes, and c) stimuli whose nature or intensity have the potential to damage the tissue.

Why is there so great a variety of cutaneous sense organs, each with its distinctive characteristics? Based upon their selectivity, it would appear that, when viewed as a member of a population, each kind of sensory unit is adapted to have a maximal sensitivity to a particular set of environmental modifications of the skin; in other terms, each sensory unit is best suited to signal, for example, minimal movement at particular velocities, or the degree of maintained displacement, or a direction of temperature change, etc. Therefore, it seems reasonable to view sense organ diversity as reflecting a developmental compromise in which sensitivity is maximized for a limited set of stimulus features.

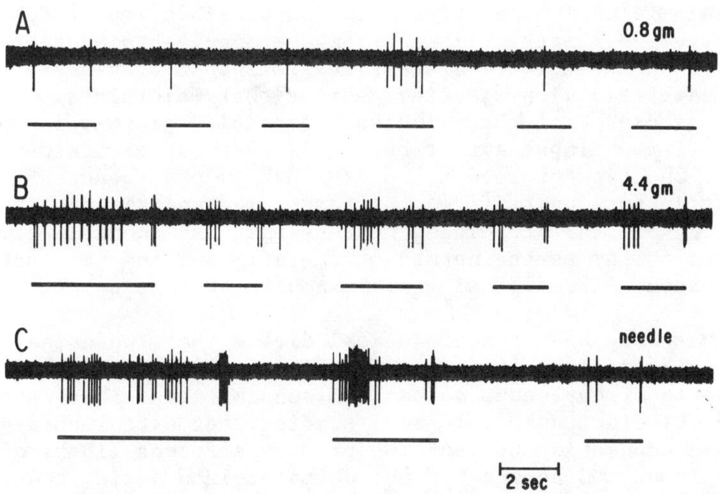

Figure 4: Discharges recorded from a C-fiber polymodal nociceptor to graded mechanical stimuli. Bar under each trace indicates approximate time of skin contact at one point. A: von Frey stimulator bending with 0.8 g. B: von Frey stimulator bending at 4.4 g. C: Pressure with a sharp needle sufficient to cause the point to penetrate the skin. (With permission from Bessou and Perl, 1969.)

Selectivity in Somatosensory Projection Neurons

Thus, when information from the skin enters the central nervous system it is partially segregated by the relative selectivity of the transduction apparatus of the periphery. What subsequently happens to this first order segregation?

Some neurons that stand as intermediaries between primary afferent input and motor output within the spinal cord or lower medulla may be independent of those that send axons rostrally. My discussion is not aimed at such elements. On the other hand, it appears likely that at least certain connections between primary afferent activity and reflex output are made by neurons that are also part of an organization related to ascending afferent transmission, as is the case for some of the class excited by the flexor reflex afferents (FRA) (Lundberg and Oscarsson, 1962; Rastad et al, 1977).

It is well established that in mammals at least three different ascending spinal pathways project somatosensory information rostrally as far as the midbrain or diencephalon: the dorsal column system, the dorsolateral tract-lateral cervical nucleus (spinocervicothalamic) system, and the ventrolateral tract (spinothalamic). The relative size and possible importance of these systems apparently varies from one species to another. The literature on these three systems suggests that they share two common features: 1) projecting neurons that maintain a substantial degree of the stimulus selectivity present in the primary afferent input either through a dominant excitatory input from only one type of receptor or through a convergence of congruent sensory units (i.e., different mechanical velocity detectors); and, 2) projecting neurons that have nonselective properties as far as the nature of the stimulus and its location are concerned. The following are examples to this point.

It has long been a neurological dictum that the dorsal column system has major importance for certain cognitive and discriminative aspects of cutaneous mechanical sensibility. This is partly borne out by electrophysiological studies that established some time ago a connection between the primary afferent fibers of the dorsal columns and neurons of the dorsal column nuclei that retain substantial mechanoreceptive selectivity (Perl et al, 1961). An example of this kind of selective neuron in dorsal column nucleus (DCN) appears in Figure 4; it is activated by relatively rapid displacement of skin hair. The peripheral receptive field activating such a unit is a discrete, sharply limited, ipsilateral area. Contact with the cutaneous surface or hairs surrounding the excitatory field suppresses evoked activity. Noxious pressure within the receptive field evokes discharge only on first contact. In addition to neurons driven by moving skin contact, other groups of DCN cells are excited in a similarly selective manner by vibration-causing stimuli in a fashion suggesting that they receive a dominant input from Pacinian corpuscles. However, the selective cutaneous projection to DCN cells is joined by a nonselective projection. One type of non-selective dorsal column projection consists of the second order fibers described by Uddenberg (1968) and Angaut-Petit (1975a,b). The second order dorsal column elements articulate with ascending neurons at the dorsal column nuclei and have distinct multi-receptive characteristics (Angaut-Petit 1975a,b; Rustioni, 1973). They respond to activation of hair follicle receptors, cutaneous displacement (pressure) receptors and heat sensitive nociceptors. They give maximal responses to noxious stimulation of the skin. Their receptive fields are large and may be discontinuous.

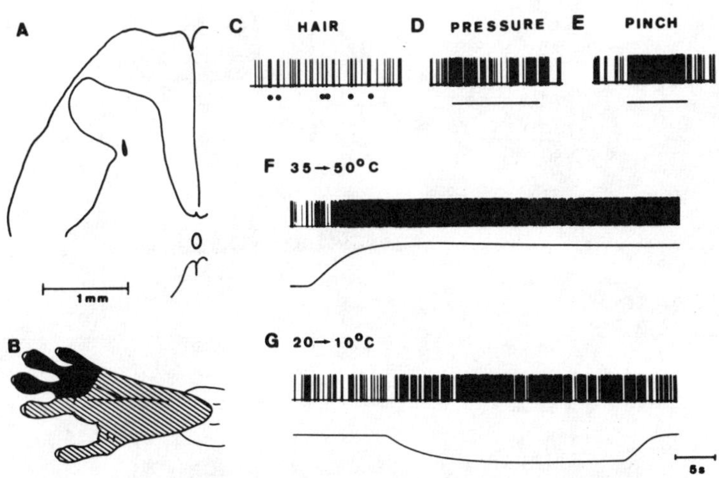

Figure 5: Discharges recorded from a multireceptive spinal neuron with an axon projecting to the contralateral thalamus. Cell location diagrammed in A. B: receptive field from which innocuous mechanical stimuli was effective is indicated by solid area; this was surrounded by an area from which responses were evoked only by strong mechanical stimuli (hatched). C: hair movement; D: pressure; E: pinch. Stimuli times in C-E indicated by dots. F & G: thermode-controlled temperature changes of receptive field. (With permission from Chung *et al*, 1979.)

The spinocervicothalamic system is particularly well developed in cat. Its projection in the dorsolateral column of the spinal cord (spinocervical tract) consists of second order fibers with several categories of neurons, including one class that is maximally excited only by moving skin contact or hair displacement from restricted ipsilateral receptive fields. Another type of spinocervical tract neuron is multireceptive, responding not only to moving distortion of the cutaneous surface but also to maintained skin pressure, and giving still more vigorous response to noxious mechanical stimulation and noxious heat (Brown, 1973; Brown et al, 1975; Brown, 1981). Thus, in the carnivore much of the spinocervical tract organization seems concerned with rostrally conveying information about light mechanical disturbances of

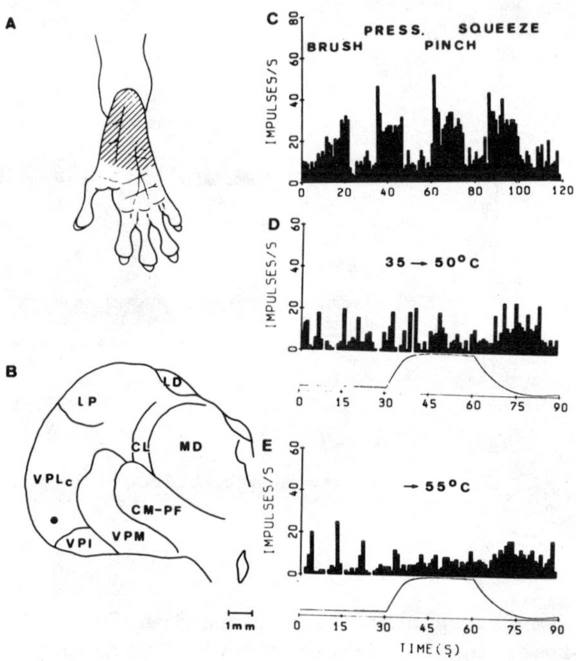

Figure 6: Discharges evoked from monkey VPL neuron responsive to gentle tactile stimuli on the glabrous skin of the foot.
A: receptive field indicated by hatched area. B: approximate cell location indicated by the dot B. C: graphic indication of discharges of the cell for graded mechanical stimulation of the receptive field. D-E: graphic indication of discharges of the cell to heating of the skin from 35° to 50 or 55°C. (With permission from Kenshalo et al, 1980.)

the body surface, but this system contains a sizeable population of multireceptive, nonselective neurons as well.

The ventrolateral spinal tracts are best known for their spinothalamic component. The fibers that make up the ventrolateral tracts, as in the spinocervicothalamic system, are from second or higher order neurons. For the better part of a century this system has been known to carry information important for pain and temperature sensation as well as for an aspect of tactile sense. The ventrolateral spinal projection consists in part of neurons whose activity is selectively related to noxious stimulation of the skin and of neurons whose responses are thermoreceptive in nature (Kumazawa et al, 1975). The thermoreceptive and nocireceptive neurons of the ventrolateral tracts have restricted receptive fields on the contralateral body surface (Trevino et al, 1973). The ventrolateral tracts also contain neurons that project rostrally as far as the diencephalon and have a distinctly multireceptive excitation; they are nonselective in terms of the kinds of stimuli to which they respond. These multireceptive neurons of the ventrolateral system respond weakly to light touch, a little more vigorously to pressure, and still more vigorously to pinch or to noxious heat (Fig. 5).

The ascending systems passing through the spinal cord in the dorsal columns, in the dorsolateral tracts, and in the ventrolateral tracts converge in part upon the ventrolateral thalamus. There, one finds a prominent pattern of a somatotopic organization of neurons with selective tactile responsiveness (Poggio and Mountcastle, 1960). As illustrated in Figure 6, these tactile neurons do not exhibit altered responses to increasing intensities of stimuli. Another population of neurons of the ventrolateral thalamus, located outside the tactile zone, receives a dominant input from joint and muscle receptors (Andersson et al, 1965; Poggio and Mountcastle, 1963). Furthermore, recent studies (Honda et al, 1983; Kniffki and Mizamura, 1983) have reaffirmed the presence of a rim of neurons surrounding the large tactile aggregation which include highly selective nocireceptive neurons excited from restricted contralateral receptive fields (Fig. 7). Again, in addition to neurons selective for tactile, position and nocireceptive input, nonselective or multireceptive neurons are present in the ventrolateral thalamic region at loci similar to those occupied by the nocireceptive types. The multireceptive neurons receive excitation from several different classes of cutaneous receptors located in bilateral receptive fields and feature the pattern of maximal responses to noxious stimuli seen at caudal levels of somatosensory systems (Fig. 8).

Finally, in the SI cortical zone of somatosensory projection, selective neurons for tactile, vibratory and musculoskeletal

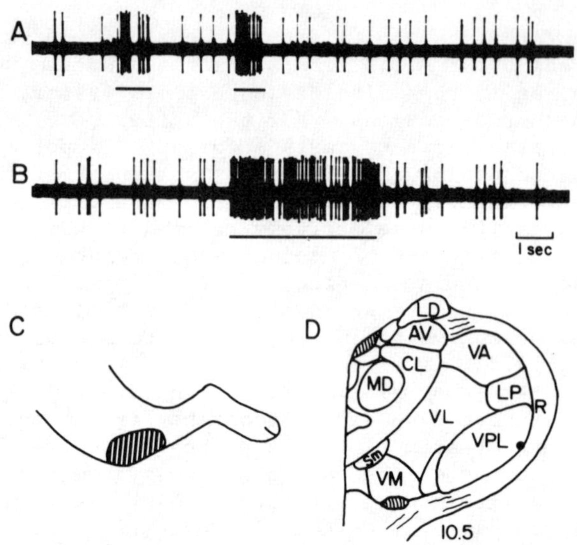

Figure 7: Responses from a selective nociceptive neuron of the cat lateral thalamus. A-δ component had to be present in afferent volleys to excite the neuron. No responses evoked by gentle mechanical manipulations or by noxious thermal stimuli. A: bar indicates time of forceful prick to receptive field (shading in diagram of contralateral hindlimb). B: pinching of receptive field with sharp dissecting forceps. Recording site indicated by dot in the outline drawing of the thalamic cross section. (With permission from Honda et al, 1983.)

stimulation are well-established features. However, as recently reported by Kenshalo and Isensee (1983) for monkey, and by Guilbaud (in this volume) for rat, selective neurons of both the selective nocireceptive type (Fig. 9) and the multireceptive type (Fig. 10) also are a part of the SI organization. The nocireceptive type of Figure 9 is unresponsive to gentle cutaneous stimuli, while the multireceptive type of Figure 10 responds to skin brushing, but more vigorously to pinch and to noxious heat to bilateral locations.

Other Central Connections

So far only some of the central projections of the somatosensory pathways have been mentioned. In addition to the termination in the lateral thalamus, dorsal and ventrospinal pathways are both known to contribute importantly to midline regions of the medulla, the midbrain and the diencephalon, and to the thalamic posterior nuclear group (Po) (Boivie and Perl, 1975). It is possible to summarize the available information in a general way by stating that the midline brain stem and diencephalic somatosensory projections are mostly nonselective. They usually are described as being activated by the kinds of stimuli that effectively excite several classes of receptors and are influenced from complex, often bilateral receptive fields. One of the properties is augmentation of the level of activity evoked by a weak stimulus when noxious intensities of stimulation are employed.

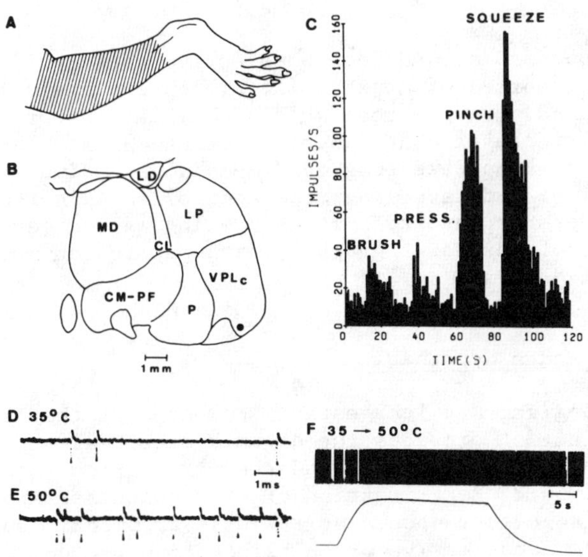

Figure 8: Responses of a multireceptive neuron of monkey lateral thalamus to mechanical and thermal stimuli. A: receptive field on contralateral hindlimb indicated by hatched area. B: location of recording site (dot) in the VPL nucleus. C: responses evoked by brush, pressure, pinch, and squeeze - progressively more intense mechanical stimuli. D & E: discharges (rapid time base) at thermode-controlled temperatures of receptive field as indicated. F: activity of cell upon heating skin to 50°C. from adapting temperature of 35°C. (With permission from Kenshalo et al, 1980.)

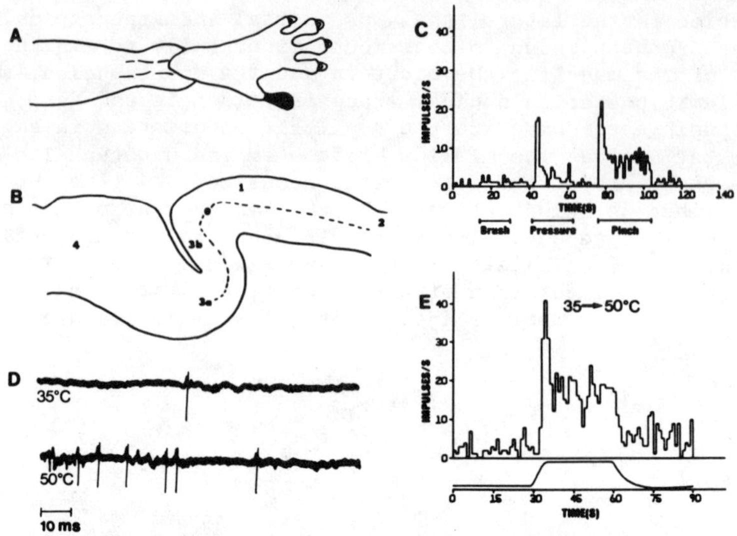

Figure 9: Responses evoked from a selective nociceptive neuron of the SI cerebral cortex of monkey. A: receptive field locations indicated by shading - glabrous skin of hallux. B: recording location (dot) in layer 4 at junction between area 3b and 1. C: histogram of responses to graded mechanical stimuli. D: discharges at indicated thermode-controlled skin temperatures. E: histogram of response to noxious heating as indicated by the analogue trace of the lower graph. (With permission from Kenshalo and Isensee, 1983.)

Some Comments and a Hypothesis

It is a logical and long-established concept that the selective projection provides the signals essential for cognitive, discriminative mechanoreception and thermoreception (Rose and Mountcastle, 1959). Reaffirmation of this postulate about functional arrangements should stir little argument. But what is the role of the nonselective or multireceptive neuron? It is commonplace in the current literature to describe such multireceptive neurons as "wide dynamic range." This terminology Iggo once decried at another symposium (1980) as reflective of the patter used in a shop selling stereophonic sound equipment. Is the dynamic range of the neuron with multireceptive characteristics greater than that of the neuron without such characteristics? Not necessarily. If the frequency of discharge from no primary afferent input to the maximal obtainable is used as a gauge, many neurons of the selective type have the same range as those with multireceptive characteristics.

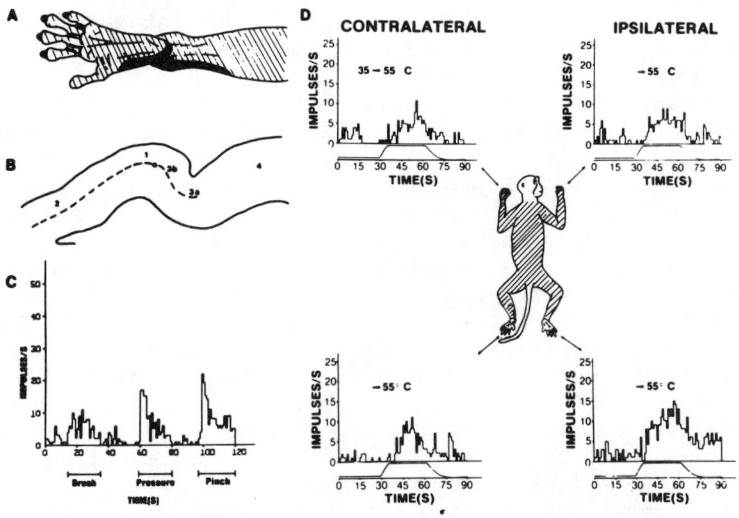

Figure 10: Responses of a nonselective SI neuron in monkey. A: receptive field locations for innocuous mechanical stimulation (dark area) and for noxious stimulation (hatch) on contralateral hindlimb. B: recording location (dot) at junction of area 3b and 1. C: histogram of responses to graded mechanical stimuli to the contralateral hindlimb. D: histograms of discharge to noxious heating (temperature increase from 35° to 55°C as indicated) of the distal part of each of the four limbs. (With permission from Kenshalo and Isensee, 1983.)

It is currently popular to describe multireceptive neurons as nocireceptive. In the sense that their maximal activity is provoked by noxious stimulation, they assuredly are nocireceptive. On the other hand, they lack the capacity to signal differences between noxious and innocuous stimulation without ambiguity, a property which certainly is characteristic of the projection of selectively nociceptive neurons to the thalamus and the cerebral cortex. To consider multireceptive neurons as nocireceptive and as serving the cognitive and discriminative purposes of modality or place recognition would be to ignore the existence of an arrangement with a surer capacity. Furthermore, multireceptive neurons are found in the dorsal column pathway, a projection which has never been shown to have a primary part in discriminative pain sense.

Does the activity of multireceptive neurons provoke pain? Mayer _et al_ (1975) argued that fibers with these features in the ventrolateral tracts do, on the basis of cross species comparisons of the excitability in the ventrolateral spinal tracts in monkey and human beings. Multireceptive neurons were shown by Kenshalo _et al_ (1979) to provide, under controlled experimental conditions, discharge patterns quantitatively reflecting the intensity of noxious heat. On the other hand, the receptive field organization of multireceptive neurons, particularly at thalamic and cortical levels, does not match the discreteness of the reference of pain experienced by normal individuals after injuring a portion of the body. Moreover, multireceptive neurons are reported to lose their nocireceptive properties as a result of descending modulation (Willis, 1982). It would be remarkably redundant to have selective neurons conveying detailed and unquestioned information on the nature of the stimulus, on the stimulated region, and on quantitative features about different events representative of cutaneous environment, without utilizing such information for discriminative and cognitive purposes. Given the existence of the selective system, it is artifice to presume that the nonselective projections serve the same purposes. What then might be the function of nonselective neurons since the necessary signals about type and location of stimulus are conveyed by another projection?

The following are hypotheses on this question. I propose that the nonselective projections to the midline reticular regions of the brainstem and perhaps also those to the Po region of the thalamus are involved in the initiation of alerting functions and the effects of somatosensory input upon states of consciousness; this would be an extension of concepts proposed by Magoun and Rhines (1947). Still to be accounted for, however, are the seemingly parallel projections of selective and nonselective information along the several ascending pathways which converge upon the somatosensory part of the lateral thalamus. Therefore, a second and more novel part of this proposal is that the nonselective projection to the ventrolateral thalamus and to the somatosensory cerebral cortex would represent part of a sensitivity-modifying mechanism related to each form of somatosensory input. In such a case, nonselective neurons would underlie mechanisms serving to adjust the level of neuronal sensitivity (up or down) at the upstream nuclei so as to shape the relationship between the selective signal and background activity. There must be some process at central neural stations to provide for an adjustment of the conditions upon which new or pervasive signals fall to account for the remarkable changes of responsiveness under differing states. The responsiveness of the group of cells receiving selective signals does systematically vary as a function of immediate history, as does the level and influence of intercurrent activity. This theory would also have the nonselective projections to both medial brainstem and cerebral centers activating descending projections, which could aid in sharpening the distinction, between incoming signals that must be attended to (because of novelty persistence or intensity) and messages that are less important. The nonselective input is also presumed to be a part of processes which suppress competing activity, either from other sensory systems or from the internal workings of the brain, at stations where crucial information must enter into the apparatus of cognition.

In summary, then, I support the century-old idea suggesting that the selective projections over the somatosensory system are the components from which discrimination, quantification and localization are taken, whether that be for various aspects of tactile sense, for vibration, for temperature or for pain. Such an arrangement would parallel the one existing for audition and vision. I add to that concept the idea that the nonselective projections are the origins of (1) changes in states of consciousness and awareness, and (2) adjustments that modify the sensitivity of the receptive neurons of higher centers to best manage selective information. This does not mean that the nonselective projection by itself cannot, under special circumstances, generate a form of sensory experience. The very fact that the selective and nonselective projections run rostrally in parallel over the several somatosensory pathways makes it possible that their terminal regions may in part overlap. The present proposal argues that there is an overlap in connectivity and that its function is to modulate. This overlap suggests that under pathological conditions nonselective projections could be the source of abnormal somatosensory experience. For instance, if selective input is prevented from reaching target neurons of higher centers due to lesions of the nervous system, an excitatory nonselective connection, as a consequence of enhanced sensitivity from denervation, may become suprathreshold and thus activate neurons controlling the transition to perception. Similarly, massive stimulation, particularly if prolonged, could cause subliminal synaptic connections from nonselective neurons to become effective enough to reach suprathreshold levels.

ACKNOWLEDGMENTS

I am grateful for the substantial assistance given by Ms. Caterri Miller, Ms. Marianna Chambless and Mr. David Maloof in the preparation of this manuscript. Support was provided by a grant (NS 10321) from the NINCDS of the U.S. Public Health Service.

REFERENCES

Andersson, S.A., Landgren, S. and Wolsk, D. (1966). The thalamic relay and cortical projection of group I muscle afferents from the forelimb of the cat. J. Physiol., Lond., 183, 576-591.

Angaut-Petit, D. (1975a). The dorsal column system: I. Existence of long ascending postsynaptic fibres in the cat's fasciculus gracilis. Expl. Brain Res., 22, 457-470.

Angaut-Petit, D. (1975b). The dorsal column system: II. Functional properties and bulbar relay of the postsynaptic fibres of the cat's fasciculus gracilis. Expl. Brain Res., 22, 471-493.

Bessou, P. and Perl, E.R. (1969). Response of cutaneous sensory units with unmyelinated fibers to noxious stimuli. J. Neurophysiol., 32, 1025-1043.

Boivie, J.J.G. and Perl, E.R. (1975). Neural substrates of somatic sensation. In Neurophysiology I. (ed. C.C. Hunt). University Park, Baltimore.

Brown, A.G. (1973). Ascending and long spinal pathways: dorsal columns, spinocervical tract and spinothalamic tract. In Handbook of Sensory Physiology. Somatosensory System. (ed. A. Iggo). Springer-Verlag, Berlin.

Brown, A.G. (1981). Organization in the Spinal Cord. Springer-Verlag, Berlin.

Brown, A.G., Hamann, W.C. and Martin, H.F., III. (1975). Effects of activity in non-myelinated afferent fibres on the spinocervical tract. Brain Res., 98, 243-259.

Brown, A.G. and Iggo, A. (1967). A quantitative study of cutaneous receptors and afferent fibres in the cat and rabbit. J. Physiol., Lond., 193, 707-733.

Burgess, P.R. and Perl, E.R. (1973). Cutaneous mechanoreceptors and nociceptors. In Handbook of Sensory Physiology. Somatosensory System. (ed. A. Iggo). Springer-Verlag, Berlin.

Chung, J.M., Kenshalo, D.R., Jr., Gerhart, K.D. and Willis, W.D. (1979). Excitation of primate spinothalamic neurons by cutaneous C-fiber volleys. J. Neurophysiol., 42, 1354-1369.

Frey, M. von. (1897). Beiträge zur Sinnesphysiologie der Haut (Part IV). Koenigl. Saechs. Ges. Wiss., Math.-Phys. Klasse, 49, 462-468.

Guilbaud, G. (1983). Organization of noxious and non-noxious inputs in SmI cortex: comparison in normal and in arthritic rats. Presentation at symposium on Somatosensory Mechanisms held at Stockholm, Sweden, June 8-10, 1983. Published in this volume.

Honda, C.N., Mense, S. and Perl, E.R. (1983). Neurons in ventrobasal region of cat thalamus selectively responsive to noxious mechanical stimulation. J. Neurophysiol., 49, 662-673.

Iggo, A. (1980). Presentation at Satellite Symposium of the International Congress of Physiological Sciences held at Keszthely, Hungary, July 9-12, 1980.

Kenshalo, D.R., Jr., Leonard, R.B., Chung, J.M. and Willis, W.D. (1979). Responses of primate spinothalamic neurons to graded and to repeated noxious stimuli. J. Neurophysiol., 42, 1370-1389.

Kenshalo, D.R., Jr., Giesler, G.J., Jr., Leonard, R.B. and Willis, W.D. (1980). Responses of neurons in primate ventral posterior lateral nucleus to noxious stimuli. J. Neurophysiol., 43, 1594-1614.

Kenshalo, D.R., Jr., and Isensee, O. (1983). Effects of noxious stimuli on primate SI cortical neurons. In Advances in Pain Research and Therapy. (eds. J.J. Bonica, U. Lindblom and A. Iggo). Raven Press, New York.

Kniffki, K.-D. and Mizumura, K. (1983). Responses of neurons in VPL and VPL-VL region of the cat to algesic stimulation of muscle and tendon. J. Neurophysiol., 49, 649-661.

Kumazawa, T., Perl, E.R., Burgess, P.R. and Whitehorn, D. (1975). Ascending projections from marginal zone (lamina I) neurons of the spinal dorsal horn. J. Comp. Neurol., 162, 1-11.

Lundberg, A. and Oscarsson, O. (1962). Two ascending spinal pathways in the ventral part of the cord. Acta physiol. scand., 54, 270-286.

Magoun, H.W. and Rhines, R. (1947). Spasticity - The Strength Reflex and Extra-Pyramidal Systems. Charles C. Thomas, Springfield, Ill.

Mayer, D.J., Price, D.D. and Becker, D.P. (1975). Neurophysiological characterization of the anterolateral spinal cord neurons contributing to pain perception in man. Pain, 1, 51-58.

Melzack, R. and Wall, P.D. (1962). On the nature of cutaneous sensory mechanisms. Brain, 85, 331-356.

Müller, J. (1840). Handbuch der Physiologie des Menschen. J. Hölscher, Coblenz.

Müller, J. (1842). Elements of Physiology. Taylor and Walton, London.

Perl, E.R., Whitlock, D.G. and Gentry, J.R. (1961). Cutaneous projection to second-order neurons of the dorsal column system. J. Neurophysiol., 25, 337-358.

Poggio, G.F. and Mountcastle, V.B. (1960). A study of the functional contributions of the lemniscal and spinothalamic systems to somatic sensibility. Bull. Johns Hopkins Hosp., 106, 266-316.

Poggio, G.F. and Mountcastle, V.B. (1963). The functional properties of ventrobasal thalamic neurons studied in unanesthetized monkeys. J. Neurophysiol., 26, 775-806.

Rastad, J., Jankowska, E. and Westman, J. (1977). Arborization of initial axon collaterals of spinocervical tract cells stained intracellularly with horseradish peroxidase. Brain Res., 135, 1-10.

Rose, J.E. and Mountcastle, V.B. (1959). Touch and kinesthesis. In Handbook of Physiology, Section 1 - Neurophysiology, v. 1. (eds. J. Field and H.W. Magoun). American Physiological Society, Washington, D.C.

Rustioni, A. (1973). Non-primary afferents to the nucleus gracilis from the lumbar cord of the cat. Brain Res., 51, 81-95.

Sinclair, D.C. (1955). Cutaneous sensation and the doctrine of specific energy. Brain, 78, 584-614.

Trevino, D.L., Coulter, J.D. and Willis, W.D. (1973). Location of cells of origin of spinothalamic tract in lumbar enlargement of the monkey. J. Neurophysiol., 36, 750-761.

Uddenberg, N. (1968). Functional organization of long, second-order afferents in the dorsal funiculus. Expl. Brain Res., 4, 377-382.

Willis, W.D. (1982). Control of nociceptive transmission in the spinal cord. In Progress in Sensory Physiology 3. (ed. D. Ottoson). Springer-Verlag, Berlin.

Witt, I. and Hensel, H. (1959). Afferente Impulse aus der Extremitätenhaut der Katze bei thermischer und mechanischer Reizung. Pflügers Arch. ges. Physiol., 268, 582-596.

Zotterman, Y. (1939). Touch, pain and tickling: an electrophysiological investigation on cutaneous sensory nerves. J. Physiol., Lond., 95, 1-28.

TACTILE SENSATION RELATED TO ACTIVITY IN PRIMARY AFFERENTS WITH SPECIAL REFERENCE TO DETECTION PROBLEMS

Å.B. VALLBO

Nobel Institute of Neurophysiology, Karolinska Institutet, S-104 01 Stockholm, Sweden

An ultimate goal in the study of sensory mechanisms is to arrive at an understanding of how the brain produces useful sensations on the basis of afferent information from sense organs. An attractive approach which bridges the gap between physiology and psychology is to analyse the correlations between neuronal events and psychophysical test data. This field was first explored, with the resolution of single unit discharges, by Mountcastle and co-workers about 20 year ago when the now classical papers on correlations between psychophysical data extracted from human subjects were related to activity in first order afferents studied in monkeys (Werner & Mountcastle, 1965, 1968; Talbot et al. 1968; Mountcastle et al. 1969). One conclusion drawn from their studies on vibrotactile sensation was "that the detection capacity of the alert and attending subject appears to be set by the peripheral neuronal threshold; it is not elevated by any higher central threshold" (Mountcastle, 1975).

This conclusion is in agreement with the psychophysical study by Hecht et al. (1942) on the absolute threshold in vision. They found that human subjects were able to detect a flash of light when a small number of rods (5-14) absorbed only one light quantum each. On the other hand, the notion that the threshold is set by the properties of the sense organs clashes with prominent threshold theories held among psychologists, notably the signal detection theory which was elaborated by Swets (1964) and Green and Swets (1966). This theory states that the threshold is set by central mechanisms and not by the sensitivity of the sense organs. Moreover, it predicts that the noise prevailing in central structures of the sensory system accounts for the general finding in detection studies that the psychometric function, i.e. the threshold curve, is S-shaped and not a step function. However, already Hecht et al. (1942) argued against this interpretation on the basis of their own findings: "The results clarify the nature of the fluctuations shown by an organism in response to a stimulus. The general assumption has been that the stimulus is constant and the organism variable. The present considerations show, however, that at the threshold it is the stimulus which is variable and that the properties of its variation determine the fluctuations found between response and stimulus".

The present communication will be much concerned with this dichotomy between the concept of a threshold set by the sensitivity of the sense organs and a threshold set by central mechanisms. When these matters are discussed some of the approaches to the study of psychoneural correlates will be presented which have emerged since Mountcastle and his coworkers presented their pioneer work.

Recording and stimulation

With the development of the method of recording impulses from peripheral nerves of alert human subjects the two facets of psychoneural correlative studies were allowed to be combined at the single unit level in man. With one exception (Hensel and Boman, 1960) investigations of this nature have been based on the non-traumatic method of recording impulses with percutaneously inserted tungsten needle electrodes (Vallbo & Hagbarth, 1967, 1968).

In the studies of Mountcastle and co-workers on vibrotactile sensibility the subject was facing the task to decide whether the stimulus had a vibratory quality or not (Talbot et al. 1968; Mountcastle et al. 1969). The focus of the present report, on the other hand, will be on the absolute threshold, i.e. the subject was asked to report whether he noticed any stimulus at all. A series of studies on tactile sensibility in the human hand, largely performed by Vallbo and Johansson, constitutes the basis of the account.

In order to study the absolute detection threshold to tactile stimuli a stimulation technique was developed that would excite a minimal number of afferents and avoid the complications that might arise in a detection analysis from lateral inhibition when a stabilizing surround close to the target area is used (Westling et al. 1976). The essence of the stimulation technique was that the effective skin indentation was controlled with high precision, not just the amplitude of the probe movement. A very small probe was made to indent the skin at a low speed to minimize excitation of remote units sensitive to travelling waves which are particularly effective when they contain high frequencies. Analysis of single afferents' receptive fields suggested, in fact, that the technique was adequate to excite single end organs within the receptive field of the unit (Johansson, 1978).

Psychoneural group data

When this method of stimulation was used to analyse the detection capacity in the human hand it was found that the psychophysical thresholds clustered around 10 µm in large areas of the glabrous skin, whereas it was strikingly higher, about 30 µm, in the center of the palm and some other regions of the hand.

The functional characteristics of the four kinds of sensitive mechanoreceptors present in the glabrous skin of the hand have been described in previous reports (Knibestöl & Vallbo, 1970; Knibestöl, 1973, 1975; Vallbo & Johansson, 1977, Johansson, 1978; Johansson & Vallbo, 1980; Johansson et al. 1980). The two types of slowly adapting units, SA I and SA II, were originally described in the hairy skin of other species by Iggo and co-workers (Iggo & Muir, 1969; Chambers et al. 1972) who demonstrated that the afferents are connected to Merkel cells and Ruffini endings respectively. The fast adapting units, FA I and FA II, which presumably are connected to Meissner corpuscles and Pacini or Golgi-Mazzoni endings (Iggo, 1974) have previously been denoted RA or QA and PC. However, these terms are inadequate, often confusing, and they are getting more and more loosely used in papers dealing with peripheral as well as central mechanisms. For instance, often it is not clear whether RA refers to activity related to a particular kind of afferent units or just to any rapidly adapting response regardless of origin. It would therefore be an advantage to adopt a semantic that strictly refers to the functional properties of the primary afferents and clearly differentiates between the separate unit types.

A comparison between group data on psychophysical thresholds and thresholds of the afferent units revealed that the two types of slowly adapting units could not account for the subjects' detection capacity because the thresholds of the SA units were too high (Johansson & Vallbo, 1979b). However, the thresholds of the two types of fast adapting units were clustered around 10 µm for all parts of the glabrous skin area. These findings demonstrate that the psychophysical detection threshold was based on one single nerve impulse elicited in one or a few of the fast adapting mechanoreceptive afferents. Hence the detection capacity is set by the sensitivity of the sense organs and not by central mechanisms. However, this was not true for the whole glabrous skin area. In the center of the palm the relation between psychophysical and afferent thresholds indicates that central mechanisms might set the detection capacity when the stimuli were delivered here. Thus, the nature of the absolute threshold is apparently different within different sections of the somatosensory system and is not universally set either by the sense organs, or by central mechanisms.

Psychoneural data pairs

The previous conclusions are based on group data of two kinds, extracted from human species when identical stimuli were delivered. However, the method of recording single unit impulses in man offers the possibility to extract more intimately related data, e.g. neural and psychophysical responses to the same individual tests. Such data are highly pertinent because the subjects are alert and fully able to attend to a psychophysical task during the nerve recording and, as shown in a separate study, the single unit impulses recorded by the needle electrode do pass the recording site and therefore may account for the detection (Vallbo, 1976).

When data pairs from the same tests were collected and the percentage of detection was plotted against stimulus amplitude to produce psychometric functions as well as threshold curves for FA I units, it was found that these two curves not only were centered around the same stimulus amplitude (Vallbo & Johansson, 1976). Also their slopes were identical indicating that the inter-trial variation in stimulus amplitude required to produce an impulse in the afferent unit was identical with the variation required to produce a psychophysical response from the attending subject. The implication being that there is no central noise added to the afferent signal at the higher levels of the somatosensory system. Obviously these findings are totally consonant with the study of Hecht et al. (1942) which was quoted above.

Interestingly this was not true for the whole glabrous skin area, because in the center of the palm the two kinds of detection curves differed drastically. The psychometric curve was centered at larger amplitudes and its slope was less steep compared to the threshold curve of the most sensitive afferents. A substantial noise must therefore be postulated at central levels of the somatosensory system because the variation in stimulus amplitude required to produce a pychophysical response was much larger than the variation required to produce a uniform afferent response.

Population response

When an almost exact agreement between psychophysical and neural threshold is present, as found e.g. for the finger pads, the inevitable conclusion is that the very unit that was recorded also accounted for the psychophysical detection. However, it may be argued that there remains the possibility that not only the recorded afferent was excited but also one or a few additional units with endings close to the target area. This seems unlikely considering the spatial accuracy of the stimulation technique and the weak stimulus employed. However, to cover this possibility an analysis of the population response was undertaken. Since it is technically not possible to record accurately the discharge from the whole population of afferents at a time in order to assess whether one or several units were excited, an estimate of the number of stimulated units was done. This estimate was based on histological data concerning the number of nerve fibres distributed to the glabrous skin area, and on data describing a sample of units collected in experiment which were designed to avoid sampling bias (Johansson and Vallbo, 1976, 1979a; Vallbo & Johansson, 1978). The estimate indicated that on the average, only one FA I unit or less was activated at the finger tip in the range of stimulus amplitude where 50 per cent of the psychophysical thresholds fell in our sample (Johansson & Vallbo, 1979b). Thus an estimate of the population response clearly supported the interpretation that one nerve impulse in a single FA I afferent often accounted for the detection.

Electrical microstimulation

A totally independent approach has also been employed to confirm and extend this conclusion. With the method of microstimulation it is possible to electrically excite the recorded and identified afferent unit in alert human subjects (Torebjörk & Ochoa, 1980; Konietzny et al. 1981; Vallbo, 1981; Schady, 1983). When the train parameters of the electrical stimulation were varied and the subject was asked to report whether he felt the stimulus or not, it turned out that a sensation was reported when only one single pulse was delivered with several units of the FA I type. With the SA I units on the other hand, not only a long train of pulses was required but also the rate had to exceed at least 10 pulses/s before a sensation was reported. Thus the SA I units not only have a higher threshold to mechanical stimulation; their perceptive power seems to be much lower when defined in terms of number of impulses required to produce a sensaiton.

Concluding remarks.

The present report has exhibited three approaches to the study of psychoneural correlates useful in experiments with alert and attending human subjects.

The recording of single afferent impulses has been combined with psychophysical tests to directly relate neuronal events to perceptive experience.

Because afferent impulses from only one unit can be properly discriminated at a time an estimate of the population response might be necessary to cover particular aspects of a problem when neuronal activity responsible for a psychophysical response is evaluated. A model of the sensory innervation of the appropriate skin area may constitute a basis for estimates of population responses.

Finally, the method of electrical microstimulation of identified units which has recently been developed, offers the possibility to inject, in one single afferent channel, a train of impulses of controlled time pattern. When the subject is asked to report his sensation during such stimulation, particular aspects of an afferent's perceptive role may be explored which can not be analysed with other methods.

The use of these three approaches to explore the nature of the psychophysical detection threshold in the glabrous skin of the human hand has been reviewed. The findings indicate that the threshold is not universally set either by central mechanisms as implied by most threshold theories based on psychophysical investigations, or by the sensitivity of the sense organs. Rather there seem to be differences between separate sections of the somatosensory system related to skin area and unit type. In certain sections there is no significant noise in the central part of the somatosensory system because an indivisible neural quantum, i.e. a single impulse from one FA I unit gives rise to a psychophysical response. However, this is not true for all skin areas in the hand. In regions which have less tactile significance the threshold seems to be set by central mechanisms rather than by the sensitivity of the endings. Moreover, a considerably noise is apparently present in central structures which handle the information from these skin areas.

References

Chambers, M.R., Andres , K.H., von Duering, M. and Iggo, A. (1972). The structure and function of the slowly adapting type II mechanoreceptors in hairy skin. Q. Jl exp. Physiol., 57, 417-445.

Green, D.M. and Swets, J.A. (1966). Signal detection theory and psychophysics. Wiley, New York.

Hecht, S., Schlaer, S. and Pirenne, M.H. (1942). Energy, quanta and vision. J. Gen. Physiol., 25, 819-840.

Hensel, H. and Boman, K.A. (1960). Afferent impulses in cutaneous nerves in human subjects. J. Neurophysiol., 23, 564-578.

Iggo, A. (1974). Cutaneous receptors. In The peripheral nervous system. (ed. J.I. Hubbard) Plenum Press, New York, pp. 347-404.

Iggo, A. and Muir, A.R. (1969). The structure and function of a slowly adapting touch corpuscle in hairy skin. J. Physiol., Lond., 200, 763-796.

Johansson, R.S. (1978). Tactile sensibility in the human hand: receptive field characteristics of mechanoreceptive units in the glabrous skin area. J. Physiol., Lond., 281, 101-123.

Johansson, R.S. and Vallbo. Å.B. (1976). Skin mechanoreceptors in the human hand: An inference of some population properties. In Sensory functions of the Skin in Primates, with special reference to Man. (ed. Y. Zotterman). Pergamon Press, Oxford, pp. 71-184.

Johansson, R.S. and Vallbo, Å.B. (1979a). Tactile sensibility in the human hand: relative and absolute densities of four types of mechanoreceptive units in glabrous skin. J. Physiol., Lond., 286, 283-300.

Johansson, R.S., and Vallbo, Å.B. (1979b). Detection of tactile stimuli. Thresholds of afferent units related to psychophysical thresholds in the human hand. J. Physiol., Lond., 297, 405-422.

Johansson, R.S., and Vallbo, Å.B. (1980). Spatial properties of the population of mechanoreceptive units in the glabrous skin of the human hand. Brain Res., 184, 353-366.

Johansson, R.S., Vallbo, Å.B. and Westling, G. (1980). Threshold of mechanosensitive afferents in the human hand as measured with von Frey hairs. Brain Res., 184, 343-351.

Knibestöl, M. (1973). Stimulus-response functions of rapidly adapting mechanoreceptors in the human glabrous skin area. J. Physiol., Lond., 232, 427-452.

Knibestöl, M. (1975). Stimulus-response functions of slowly adapting mechanoreceptors in the human glabrous skin area. J. Physiol., Lond., 245, 63-80.

Knibestöl, M. and Vallbo, Å.B. (1970). Single unit analysis of mechanoreceptor activity from the human glabrous skin. Acta physiol. scand., 80, 178-195.

Konietzny, F., Perl, E.R., Trevino, D., Light, A. and Hensel, H. (1981). Sensory experiences in man evoked by intraneural electrical stimulation of intact cutaneous afferent fibres. Exp. Brain Res., 42, 219-222.

Mountcastle, V.B. (1975). The view from within: Pathways to the study of perception. The Johns Hopkins Medical Journal, 136, 109-131.

Mountcastle, V.B., Talbot, W.H., Sakata, H. and Hyvärinen, J. (1969). Cortical neuronal mechanisms in flutter-vibration studied in unanesthetized monkeys neuronal periodicity and frequency discrimination. J. Neurophysiol., 32, 452-484.

Schady, W. (1983). Projections of sensory fibres in human median nerve. Acta Univ. Upsaliensis, Abstr. Uppsala Diss.

Swets, J.A. (1964).(ed.) Signal detection and recognition by human observers: Contemporary readings. Wiley, New York.

Talbot, W.H., Darian-Smith, I., Kornhuber, H.H. and Mountcastle, V.B. (1968). The sense of flutter-vibration: Comparison of the human capacity with response patterns of mechanoreceptive afferents from the monkey hand. J. Neurophysiol., 31, 301-334.

Torebjörk, H.E. and Ochoa, J.L. (1980). Specific sensations evoked by activity in single identified sensory units in man. Acta Physiol. Scand. 110, 445-447.

Vallbo, Å.B. (1976). Prediction of propagation block on the basis of impulse shape in single unit recordings from human nerves. Acta physiol. Scand., 97, 66-74.

Vallbo, Å.B. (1981). Sensations evoked from the glabrous skin of the human hand by electrical stimulatin of unitary mechanosensitive afferents. Brain Res., 215, 359-363.

Vallbo, Å.B. and Hagbarth, K.-E. (1967). Impulses recorded with microelectrodes in human muscle nerves during stimulation of mechanoreceptors and voluntary contractions. Electroen. Neurophysiol., 23, 392.

Vallbo, Å.B. and Hagbarth, K.-E. (1968). Activity from skin mechanoreceptors recorded percutaneously in awake human subjects. Exptl. Neurol., 21, 270-289.

Vallbo, Å.B. and Johansson, R.S. (1976). Skin mechanoreceptors in the human hand: Neural and psychophysical thresholds. In Sensory Functions of the Skin in Primates, with special reference to Man. (ed. Y. Zotterman). Pergamon Press, Oxford, pp. 185-199

Vallbo, Å.B. and Johansson, R.S. (1978). The tactile sensory innervation of the glabrous skin of the human hand. In Active Touch: The Mechanisms of Recognition of Objects by Manipulation. A Multi-disciplinary approach. (ed. G. Gordon) Pergamon Press, Oxford, pp. 29-54.

Werner, G. and Mountcastle. (1965). Neural activity in mechanoreceptive cutaneous afferents: Stimulus-response relations, Weber functions, and information transmission. J. Neurophysiol., 28, 359-397.

Werner, G. and Mountcastle, V.B. (1968). Quantitative relations between mechanical stimuli to the skin and neural responses evoked by them. In The skin senses (ed. D.R. Kenshalo) C.C. Thomas, Springfield, pp. 112-137.

Westling, G., Johansson, R.S. and Vallbo, Å.B. (1976). A method for mechanical stimulation of skin receptors. In Sensory functions of the Skin in Primates, with special reference to Man. (ed. Y. Zotterman) Pergamon Press, Oxford, pp. 151-153.

ROLE OF SINGLE MECHANORECEPTOR UNITS IN TACTILE SENSATION

H.E. TOREBJÖRK, *J.L. OCHOA, and W.J.L. SCHADY

Department of Clinical Neurophysiology, University Hospital, S-751 85 Uppsala, Sweden
**Department of Neurology, University of Wisconsin Medical Scool, Madison, WI 53792, USA*

A fundamental task of the hand is to extract information about surface structure, shape and weight of objects during manual exploration and manipulation. Much of this information is signalled from low-threshold mechanoreceptive units in the skin. Deep receptors of various types are also involved in the complex feedback mechanisms for motor control. We know from microneurographic studies in man that there are four basic types of low-threshold mechanoreceptive units in the glabrous skin of the hand. Extrapolation from animal experiments suggests that RA units are connected to Meissner corpuscles, PC units to Pacinian corpuscles, SA I units to Merkel disks, and SA II units to Ruffini end-organs (Lindblom and Lund, 1966; Iggo and Muir, 1969; Lynn, 1969; Chambers et al., 1972; Iggo and Ogawa, 1977). Detailed information is at present available about the stimulus-response functions, receptive field characteristics and innervation densities of these four unit types in the fingers and palm (Knibestöl and Vallbo, 1970; 1980; Knibestöl, 1973; 1975; Vallbo and Johansson, 1978; Johansson, 1978; Johansson and Vallbo, 1979; Johansson et al., 1982). Much less is known about the role of these mechanoreceptive units in sensory experience. Here we describe a new technique for studying sensory correlates of single mechanoreceptive unit activation and discuss the contribution of these units to tactile sensation and motor control.

METHODS

Microneurography and intraneural microstimulation (INMS) were carried out in cutaneous fascicles of the median and ulnar nerves at wrist or elbow level in healthy subjects. A lacquer-insulated tungsten electrode (tip diameter 1-5 μm, shaft diameter 200 μm, impedance 100-200 kΩ at 1000 Hz) was inserted percutaneously into the nerve, and a low impedance electrode of similar shape served as reference in the subcutaneous tissue 1-2 cm outside the nerve. After identifying a single mechanoreceptor unit by recording, and mapping its receptive field in the skin with a von Frey hair 5 times threshold, electrical pulses at rising amplitudes were delivered through the electrode and the subject was asked to describe quality, magnitude and projection area of the evoked sensation. Alternatively, electrical stimuli were first delivered through the intraneural electrode and psychophysical tests were carried out, before the electrode was used for recording neural activity at the site of intraneural stimulation. Square wave pulses of 0.25 ms duration were delivered at 1-300 Hz in regular trains of 1-3 s duration, repeated at intervals of 30-60 s. Amplitudes usually ranged from 0 to 0.3 V, rarely up to 1 V. Further details about methodology have been reported elsewhere (Torebjörk and Ochoa, 1980; Ochoa and Torebjörk, 1983).

RESULTS

Recruitment of Elementary Sensations

When the amplitude of intraneural stimulation at 10 or 100 Hz was gradually raised above threshold for sensation (usually 0.2-0.3 V), even naïve subjects in their first experiment could detect successive recruitment of several sensations of distinct quality, described as tapping, vibration, tickling, pressure, sharp pain, dull pain or itch. Such sensations had different temporal profiles, either intermittent as for tapping and vibration, or sustained as for pressure and pain. Furthermore, the sensations were projected to focal and usually quite separate fields in the glabrous skin of the hand (Fig. 1). For a particular electrode position, the order of recruitment of sensations was reproducible and the threshold in terms of stimulus amplitude was typical for each sensation. For a given stimulus frequency, the subjective attributes of each sensation did not change as the stimulus amplitude was increased; rather, there was discontinuous recruitment of new sensations, each having its own typical threshold, quality, temporal profile, magnitude and projected field. We have used the term "elementary sensations" for such all-or-nothing sensory phenomena (Ochoa and Torebjörk, 1983).

Eventually, after identifying up to seven discrete elemen-

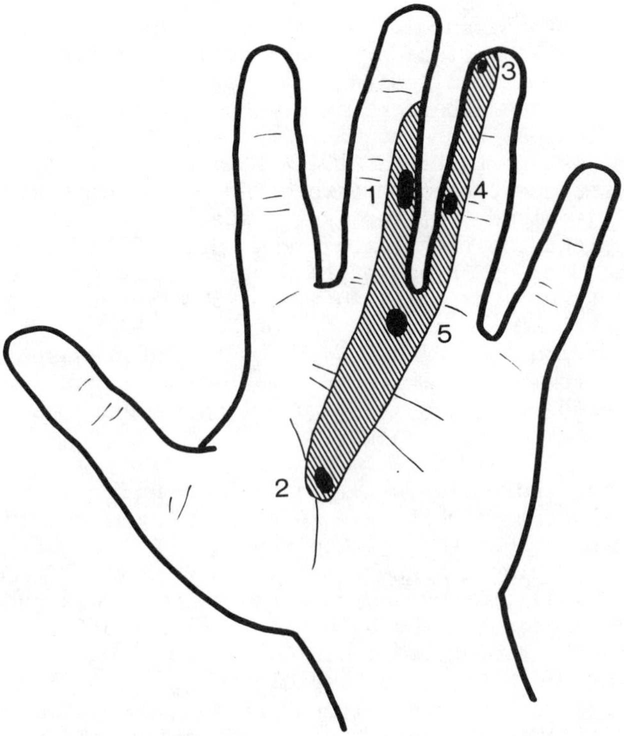

Fig. 1. Example of recruitment of elementary sensations (black areas) evoked by INMS (10 Hz) at increasing amplitudes (0.2-0.3 V) in a cutaneous fascicle of the median nerve at the elbow. Sensations of different quality (1 = dull pain, 2 = pressure, 3 = tapping, 4 = tapping, 5 = pressure) were projected to separate fields. Eventually, at higher amplitudes, an "electric" sensation was projected to an area corresponding to the fascicular innervation territory (hatched zone).

tary sensations, subjects reported bridging of projected fields and mixed sensations as the stimulus amplitude was increased, and at 1 V a strong "electric" sensation was projected to the entire innervation territory of the stimulated fascicle (Schady et al., 1983a; 1983c).

Specific Sensations from Specific Types of Mechanoreceptive Units

The combination of intraneural microstimulation and microneurography allowed direct correlation between quality of elementary sensation and basic unit type, as well as correlation between site of sensory projection and location of the receptive field of the stimulated unit. Such correlations were made when a first elementary sensation was recruited at a low stimulus amplitude (0.20-0.28 V) and subsequent recording revealed single unit waveforms indicative of impalement of the myelin sheath of a particular fibre by the electrode tip (Torebjörk et al., 1970; Vallbo, 1976). Under these conditions it was often possible to predict correctly, from the quality and temporal profile of the sensation, what type of unit to anticipate on recording. Thus, intermittent tapping predicted recording from RA units and sustained pressure predicted recording from SA I units. Furthermore, the sensation was usually projected very close to or overlapping the actual receptive field of the recorded unit.

The quantal properties of elementary sensations, and their specific correlation to unit type in terms of quality and projection, suggested that such evoked sensations were the result of stimulation of single sensory units (Torebjörk and Ochoa, 1980; Vallbo, 1981). Further evidence to support this notion was obtained by recording the afferent impulse traffic in a sensory fascicle with a proximal electrode while stimulating through a distal electrode in the same fascicle (Fig. 2). A close time-lock was observed between recruitment of single unit potentials and reports of elementary sensations, and quality and projection of each sensation correlated with type and receptive field of each recorded unit (Ochoa and Torebjörk, 1983).

More than 100 low-threshold mechanoreceptive units innervating mainly the glabrous skin of the hand have been classified and activated by INMS. RA units gave rise to intermittent tapping at low frequency (<10 Hz) and oscillatory flutter-vibration at higher frequencies (up to 100 Hz). Each tap was described as a brief superficial contact on the skin confined to a small area of 10 mm^2 or less. For RA units innervating the finger pulps, a single impulse was often enough for sensory detection and quality discrimination, as well as localization and demarcation of the field of sensory projection. A minimum frequency of 5 Hz or more was usually required for detection of a sensation from RA units innervating the palm. Increases in the stimulus frequency were perceived as increases in the frequency component of the tactile sensation in a linear fashion up to about 100 Hz, but did not lead to obvious increase in the subjective intensity of sensa-

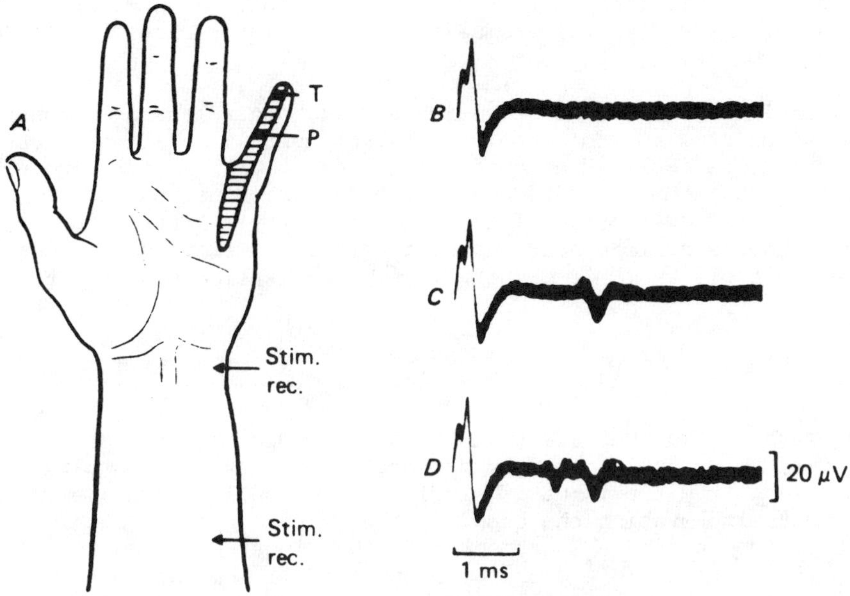

Fig. 2. Recording with a proximal electrode while stimulating through a distal electrode in the same cutaneous fascicle of the ulnar nerve, innervating the hatched area in A. No response was observed below threshold for sensory detection (B). Recruitment of the first single unit potential (C) coincided with the subjective report of an elementary sensation of tapping, projected to T in A. Further recruitment of a second potential (D) coincided with the report of an additional sensation of pressure, projected to P in A. 10 sweeps superimposed in each record. (Reproduced with permission from J. Physiol. 342, 633-654, 1983).

tion (Torebjörk and Ochoa, unpublished).

Intraneural stimulation of SA I units gave rise to sensations of sustained skin pressure above detection thresholds around 5-10 Hz. Single impulses were never felt. Increases in frequency increased the magnitude of pressure in an almost linear fashion (Torebjörk and Ochoa, unpublished), but subjects could not detect frequency per se, in contrast to RA units. Mean projection areas were somewhat larger than for RA units, being smallest in the fingertips and largest in the palm.

Sensations from stimulation of PC units were described as vibration or tickling, detected above 10-80 Hz, and projected to areas which were typically smaller than the large receptive fields of these units.

SA II units gave no sensation at all, regardless of frequency, when stimulated in isolation. Proof that an SA II unit was indeed activated was obtained by changing the excitability of the recorded fibre by INMS at 200-300 Hz for 5-10 min. After such stimulation, natural activation of the SA II receptor elicited high frequency bursts of impulses, indicating that the fibre had become hyperexcitable by INMS. Despite this, no sensation was reported.

DISCUSSION

The present data indicate that the human brain has exquisite capacity to detect, localize, delineate, qualify and quantify sensations from the input of single low-threshold mechanoreceptor units innervating the hand. In the extreme case, a single impulse from a single RA receptor in the finger pulp conveys all this information. This does not mean that all the information that comes up from the exploring hand in everyday situations reaches consciousness. Voluntary finger movements can activate large numbers of mechanoreceptive units (Hulliger et al., 1979), and yet little sensation is projected to the skin. Much of this information is probably used as sensory feedback for motor control, and is prevented from reaching conscious levels by various inhibitory mechanisms. Thus, voluntary movement of a finger or briefly touching the skin of the hand can temporarily abolish the sensation from a single unit stimulated intraneurally (Schady and Torebjörk, 1983). A likely candidate for proprioceptive function is the SA II unit, which gives no conscious sensation when stimulated in isolation. It is tempting to speculate that these units, which are often found close to joints and which respond directionally to lateral movements of the skin, might be involved in motor control (Knibestöl, 1975; Knibestöl and Vallbo, 1980). However, our findings do not completely rule out a possible role of SA II units for sensation, since it is conceivable that they might give a conscious percept through spatial coactivation.

The observation that PC units signal vibration is in line with previous conclusions of Talbot et al. (1968). However, a contribution of these units to motor control as well is suggested by studies in man (Torebjörk et al., 1978; Eklund et al.,

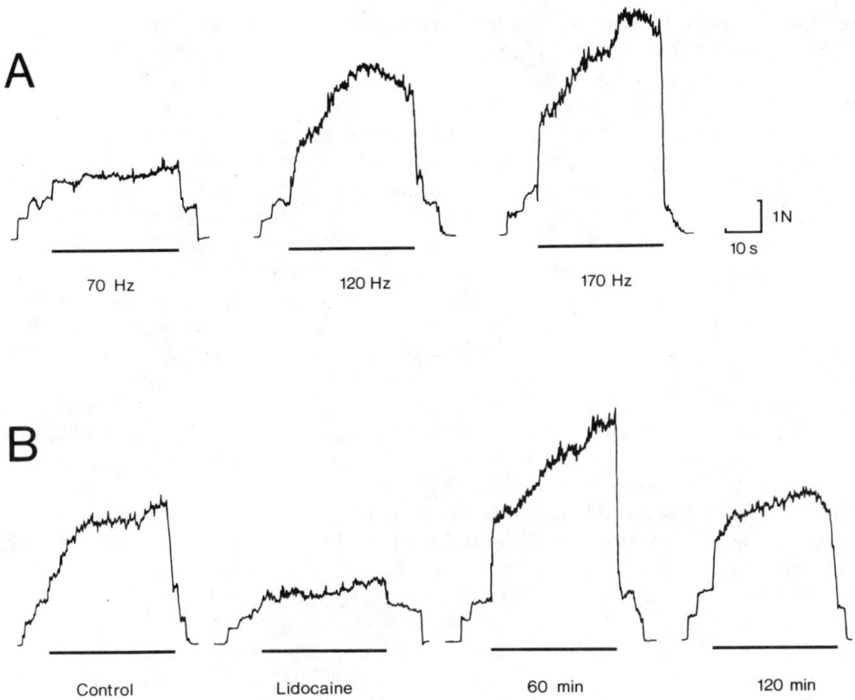

Fig. 3. Effect of index pulp vibration (horizontal bars) on grip force between thumb and index fingers. The grip force increased involuntarily with increasing frequency of vibration (A). Lidocaine block of the skin of the vibrated finger reduced the finger flexor reflex, which was subsequently increased during cutaneous paraesthesiae 60 min after the block (B). These experiments indicate that the finger flexor reflex is induced by cutaneous receptors, particularly those responsive to high frequency vibration like PC units.

1978), where high frequency vibration on the skin of a finger elicited an exteroceptive finger flexion reflex, possibly by contribution from vibration-sensitive PC afferents (Fig. 3).

RA and SA I units are probably of great importance for spatial discrimination, particularly in the fingertips, where their innervation density is higher (Johansson and Vallbo, 1976; 1979) and their receptive fields are smaller than in the proximal parts of the hand (Schady and Torebjörk, 1983). Furthermore, the present results point out several central factors that enhance the discriminative capacity of the fingertips. Thus, the capacity to detect impulses is higher for RA units in the fingertips than in the palm (Vallbo and Johansson, 1976; Ochoa and Torebjörk, 1983), the sizes of projected sensations are smaller for SA I units in the fingertips than in the palm (Schady and Torebjörk, 1983), and the ability to localize a projected sensation relative to the activated unit's receptive field is also more accurate in the fingertips than the palm (Schady et al., 1983b; 1984, this volume). In conclusion, both the peripheral and central neural mechanisms are propitious for spatial discrimination in the fingertips.

While both types of units are involved in spatial analysis, their respective roles in different sensory tasks may differ. The main difference between the RA and SA I sensory systems is the ability of the RA system to faithfully reproduce frequency changes, whereas the SA I system signals magnitude of sensation without frequency detection. When this is coupled with the distinctly different qualities of sensation they convey - skin contact for the RA system and pressure for the SA I system - it appears that the dynamic RA system is best equipped to recreate a spatio-temporal image of the textured surface (sandpaper, cloth) being explored by the moving hand. The static SA I system is better equipped to recreate a spatio-intensity image of the contour and force of objects compressing the skin. Obviously this dichotomy is artificial in everyday situations, when the brain compiles the complex input from many receptors of several types to recreate all combinations of sensations, but awareness of the basic differences between sensory submodalities at the single unit level is of importance for our understanding of the fundamental laws that govern somatosensory processing.

ACKNOWLEDGEMENTS

This work was supported by grants from the Swedish Medical Research Council (Nos. 5206-06, 6153-02 and 5822-02) and the National Institutes of Health (No. 1R01NS18315-01).

REFERENCES

Chambers, M.R., Andres, K.H., von Düring, M. and Iggo, A. (1972). The structure and function of the slowly adapting type II mechanoreceptor in hairy skin. Q. J. Exp. Physiol., 57, 417-455.

Eklund, G., Hagbarth, K.E. and Torebjörk, H.E. (1978). Exteroceptive vibration-induced finger flexion reflex in man. J. Neurol. Neurosurg. Psychiat., 41, 438-443.

Hulliger, M., Nordh, E., Thelin, A.-E. and Vallbo, Å.B. (1979). The response of afferent fibres from the glabrous skin of the hand during voluntary finger movements in man. J. Physiol., 291, 233-249.

Iggo, A. and Muir, A.R. (1969). The structure and function of a slowly adapting touch corpuscle in hairy skin. J. Physiol., 200, 763-796.

Iggo, A. and Ogawa, H. (1977). Correlative physiological and morphological studies of rapidly adapting mechanoreceptors in cat's glabrous skin. J. Physiol., 266, 275-296.

Johansson, R.S. (1978). Tactile sensibility in the human hand: receptive field characteristics of mechanoreceptor units in the glabrous skin. J. Physiol., 281, 101-123.

Johansson, R.S., Landström, U. and Lundström, R. (1982). Sensitivity to edges of mechanoreceptive afferent units innervating the glabrous skin of the human hand. Brain Res., 244, 27-32.

Johansson, R.S. and Vallbo, Å.B. (1976). Skin mechanoreceptors in the human hand: an inference of some population properties. In Sensory Functions of the Skin in Primates (ed. Y. Zotterman). Pergamon Press,Oxford, pp 171-184.

Johansson, R.S. and Vallbo, Å.B. (1979). Tactile sensibility in the human hand: relative and absolute densities of four types of mechanoreceptive units in the glabrous skin area. J. Physiol., 286, 283-300.

Knibestöl, M. (1973). Stimulus-response functions of rapidly adapting mechanoreceptors in the human glabrous skin area. J. Physiol., 232, 427-452.

Knibestöl, M. (1975). Stimulus-response functions of slowly adapting mechanoreceptors in the human glabrous skin area. J. Physiol., 245, 63-80.

Knibestöl, M. and Vallbo, Å.B. (1970). Single unit analysis of mechanoreceptor activity from the human glabrous skin. Acta Physiol. Scand., 80, 178-195.

Knibestöl, M. and Vallbo, Å.B. (1980). Intensity of sensation related to activity of slowly adapting mechanoreceptive units in the human hand. J. Physiol., 300, 251-267.

Lindblom, U. and Lund, L. (1966). The discharge from vibration-sensitive receptors in the monkey foot. Exp. Neurol., 15, 401-417.

Lynn, B. (1969). The nature and location of certain phasic mechanoreceptors in cat's foot pad. J. Physiol., 201, 768-773.

Ochoa, J.L. and Torebjörk, H.E. (1983). Sensations evoked by intraneural microstimulation of single mechanoreceptor units innervating the human hand. J. Physiol., 342, 633-654.

Schady, W.J.L., Ochoa, J.L., Torebjörk, H.E. and Chen, L.S. (1983a). Peripheral projections of fascicles in the human median nerve. Brain. 106, 745-760.

Schady, W.J.L. and Torebjörk, H.E. (1983). Projected and receptive fields: a comparison of projected areas of sensations evoked by intraneural stimulation of mechanoreceptive units, and their innervation territories. Acta Physiol. Scand., 119, 267-275.

Schady, W.J.L., Torebjörk, H.E. and Ochoa, J.L. (1983b). Cerebral localisation function from the input of single mechanoreceptive units in man. Acta Physiol. Scand., 119, 277-285.

Schady, W.J.L., Torebjörk, H.E. and Ochoa, J.L. (1983c). Peripheral projections of nerve fibres in the human median nerve. Brain Res., 277, 249-261.

Talbot, W.H., Darian-Smith, T., Kornhuber, H.H. and Mountcastle. V.B. (1968). The sense of flutter-vibration: comparisons of the human capacity with response patterns of mechanoreceptive afferents from the monkey hand. J. Neurophysiol. 31, 301-334.

Torebjörk, H.E., Hagbarth, K.E. and Eklund, G. (1978). Tonic finger flexion reflex induced by vibratory activation of digital mechanoreceptors. In Active Touch (ed. G. Gordon). Pergamon Press, Oxford, pp. 197-203.

Torebjörk, H.E., Hallin, R.G., Hongell, A. and Hagbarth, K.E. (1970). Single unit potentials with complex waveform seen in microelectrode recordings from the human median nerve. Brain Res., 24, 443-450.

Torebjörk, H.E. and Ochoa, J.L. (1980). Specific sensations evoked by activity in single identified sensory units in man. Acta Physiol.Scand., 110,445-447.

Vallbo, Å.B. (1976). Prediction of propagation block on the basis of impulse shape in single unit recordings from human nerves. Acta Physiol. Scand., 97, 66-74.

Vallbo, Å.B. (1981). Sensations evoked from the glabrous skin of the human hand by electrical stimulation of unitary mechanosensitive afferents. Brain Res., 215, 359-363.

Vallbo Å.B. and Johansson R.S. (1976). Skin mechanoreceptors in the human hand: neural and psychophysical thresholds. In Sensory Functions of the Skin in Primates (ed. Y. Zotterman). Pergamon Press, Oxford, pp 185-199.

Vallbo, Å.B. and Johansson, R. (1978). The tactile sensory innervation of the glabrous skin of the human hand. In Active Touch (ed. G. Gordon). Pergamon Press, Oxford, pp 29-54.

SIZE PERCEPTION AND LOCALISATION OF SENSATIONS FROM SINGLE MECHANORECEPTIVE UNITS IN MAN

W.J.L. SCHADY, H.E. TOREBJÖRK, and *J.L. OCHOA

Department of Clinical Neurophysiology, University Hospital, S-751 83 Uppsala, Sweden
**Department of Neurology, University of Wisconsin Medical School, Madison WI 53792, USA*

In assessing the functional role played by individual mechanoreceptive unit types in sensation, it is not sufficient to study the organisation and physiology of their cutaneous endings. Such characteristics as receptor response properties and unit innervation density define the quality and quantity of the raw inflow from cutaneous mechanoreceptors. But all perceptual experience is the result of cortical processing of afferent signals, and central rather than peripheral factors may determine which signals are given priority. Some inputs may be enhanced at successive relay stations, some may be inhibited and some may be entirely prevented from reaching consciousness, while still playing a role e.g. in proprioception. Psychophysical methods in combination with microneurography are therefore necessary in order to establish how the human brain deals with single unit somatic inputs: whether, for instance, signals from all mechanoreceptor types and from all skin regions are given equal treatment.

Natural stimulation of the skin, however slight, results in a "complex" sensation due to coactivation of a number of sensory units with overlapping receptive fields. The contribution of each unit type cannot be determined, nor can the number of activated units be assessed. Attempts have been made to control these variables by recording somatosensory evoked potentials while delivering threshold stimuli to the skin (Yamauchi et al., 1981; Soininen & Järvilehto, 1982), but the amplitude of SEPs is, at best, only a rough guide to the number of activated primary sensory neurons. On the other hand, the technique of intraneural microstimulation, introduced in 1980 (Torebjörk & Ochoa, 1980) and described in detail in the preceding chapter, allows selective stimulation of single identified sensory units, or groups of them, at will. The attributes of evoked sensations can then be correlated with the electrophysiological characteristics of the excited units, and conclusions may be drawn about their role in tactile sensibility.

In the following we shall describe the "spatial" attributes of sensations evoked by intraneural stimulation of single mechanosensitive units supplying the hand and forearm: the size, shape and location of projected sensations with respect to the receptive fields of activated units. The finding of gradients in these properties for different mechanoreceptive unit types and for different skin regions allow certain conclusions to be drawn about the processing priorities of the central nervous system. Furthermore, they are the basis of observed proximodistal gradients in clinical tasks such as point localisation and two point discrimination tests.

METHODS

Microneurography and intraneural microstimulation were undertaken in the median nerve or in the medial cutaneous nerve of the forearm of 22 healthy human subjects aged 21 to 37 years. Lacquer insulated tungsten microelectrodes with a tip diameter of 5-10 µm and an exposed tip length of 50-200 µm were inserted percutaneously into the appropriate nerve in the upper arm or at the wrist. The electrode was then advanced manually in small steps until a single cutaneous mechanoreceptive unit could be clearly discriminated. At this point, the electrode was connected to a stimulator delivering square wave electrical pulses of 0.25 ms duration at frequencies between 1 and 200 Hz in the form of short trains of 2-5 s duration. The stimulus amplitude was gradually increased from zero to the liminal point for conscious detection, and the threshold and quality of the first sensation were determined. The area to which this first sensation was referred (projected field, or PF) was drawn by the subject on actual size photographs of his hand or, in the case of the forearm, directly on the skin. Intraneural microstimulation (INMS) was then discontinued and recording was recommenced. The discriminable mechanoreceptive unit was classified as either RA, PC, SA I or SA II, according to criteria laid down by Knibestöl and Vallbo (1970). In hairy skin, RA and PC units could not easily be distinguished and were thus grouped under fast adapting (FA) units. The unit's receptive field (RF) was mapped on the subject's skin using a von Frey hair 4-5 times threshold.

At the end of each session another photograph of the subject's hand was made, containing the mapped RFs of one or more sensory units. The outlines of the hand, skin creases and RFs were traced and superimposed upon the photographs containing the projected fields drawn by the subjects, so that the borders of the hand and the skin creases matched as accurately as possible. The distance between the centre of each receptive field and its corresponding projected field was measured. When necessary, oblique views of the fingers or forearm were taken in order to show receptive fields face-on. The areas of receptive and projected fields were determined by tracing their margins onto a digitiser connected to a computer. Results were grouped according to RF location on either the fingertips (limited by a line drawn across the whorl of the skin ridges), the main body of the fingers or the palm. Finally, the distance between the centre of each RF and the nearest fingertip was measured.

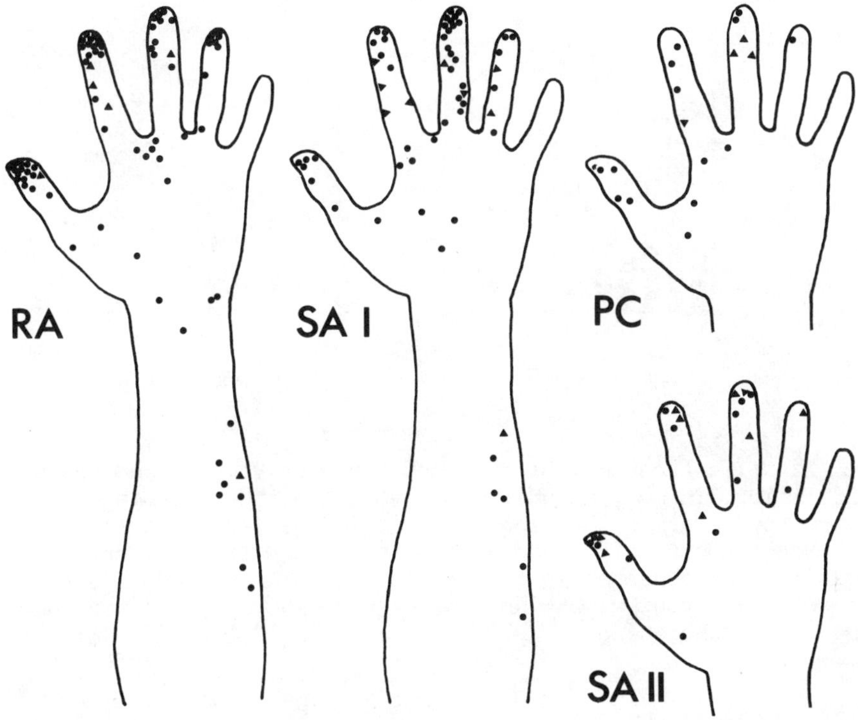

Fig. 1. Location of the receptive fields of sampled mechanoreceptive units. Dots represent units on the volar aspect, triangles units on the dorsal aspect of the hand and forearm. Symbols indicate approximate position of the centres of the receptive fields but do not reflect their size. Note concentration of RA and SA I units in the fingertips.

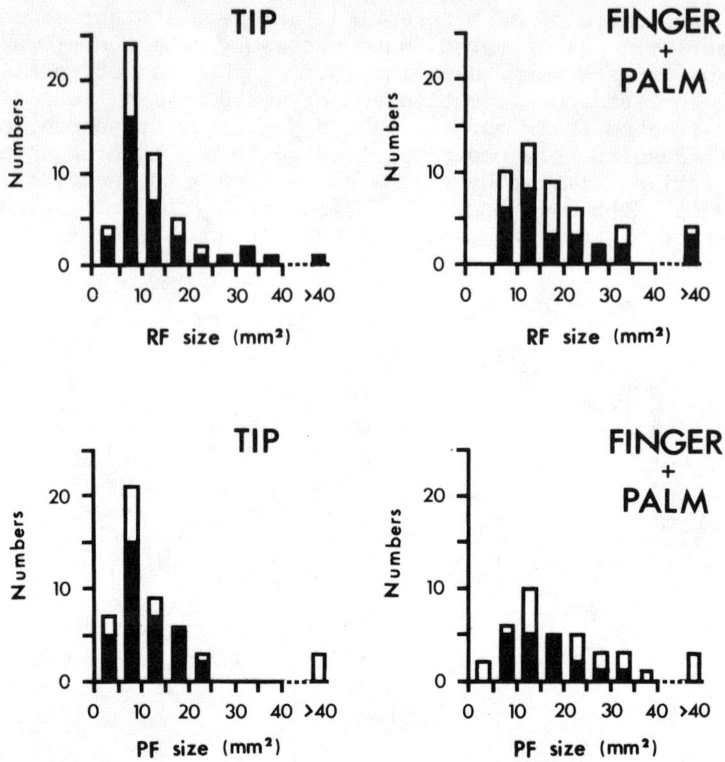

Fig. 2. Size-frequency histograms of receptive (RF) and projected (PF) fields, in mm^2, of units innervating either the fingertips or the remainder of the fingers and palm. Filled-in and empty bars represent numbers of RA and SA I units, respectively. There is overall correspondence in the distributions of RFs and PFs for the same skin region, but not in the contribution of either unit type to extreme values.

RESULTS

Observations were made on 134 mechanoreceptive units projecting to the glabrous skin of the hand, and 38 units projecting to the hairy skin of the dorsum of the fingers or the forearm (Fig. 1). Based on a sample of 34 units in whom unfiltered action potential waveforms were studied, it was estimated that over 80% of recordings were intracellular - if not intraaxonal, at least from within the myelin sheath. This suggests that in most instances the fibre that was being recorded from was also the one most likely to be activated first by anodal pulses along the low impedance route from electrode to axon (Zealear & Crandall, 1982). Sensations induced by INMS at liminal amplitudes were therefore accepted as arising from excitation of the recorded unit if the following two criteria were met: (i) the character of the evoked tactile sensation was intermittent for RA and PC units, and sustained for SA I units (see preceding chapter and Ochoa & Torebjörk, 1983), and (ii) the receptive field of the discriminable unit was "reasonably" close to the projected field of the evoked sensation, i.e. at least on the same finger. When these criteria were not met, it was assumed, on the basis of previous work (Torebjörk & Ochoa, 1980; Ochoa & Torebjörk, 1983; Schady & Torebjörk, 1983), that the sensation reported by the subject during INMS resulted from activation of a unit other than the one recorded from, and the unit was rejected. In practice, only two units were rejected for non-fulfilment of criterion (ii) alone. SA II units are not thought to convey a conscious sensation when activated individually (Torebjörk & Ochoa, 1980; Ochoa & Torebjörk, 1983).

Receptive fields were usually rounded or oval, and when oval their long axis tended to be parallel to the axis of the finger. Mean RF sizes for RA, SA I, PC and SA II units in glabrous skin were 17.5 mm^2, 15 mm^2, 137 mm^2 and 81 mm^2, respectively. A significant ($p < 0.05$) proximodistal RF size gradient existed for RA units (13 mm^2 in the fingertips vs. 27 mm^2 in the palm) and SA I units (11 mm^2 vs. 18 mm^2). The projected fields of sensations induced by INMS also varied depending upon the type of mechanoreceptive unit activated, and on the region of skin it projected to. The smallest PFs were those of sensations mediated by RA units (11 mm^2 in the fingertips vs. 16 mm^2 in the palm, $p < 0.05$), followed by PC units (14 vs. 22 mm^2) and SA I units (19 vs. 44 mm^2, $p < 0.05$). Comparable size gradients from fingertip to palm were thus found for receptive and projected fields of RA and SA I units (Fig. 2). For FA units in hairy skin, RFs and PFs also increased by fourfold from the dorsum of the fingers to the forearm. When all populations of units from glabrous and hairy skin were pooled, their projected field sizes correlated fairly well (r = 0.58, Pearson) with distance to the limb tip (Fig. 3).

The next step was to measure the distance separating RF from PF for each unit (interfield distance, or IFD) in order to obtain an index of the brain's localisation capacity from the input of a single unit. It was found that IFDs, like RFs and PFs, tended to increase the more proximally located they were: the smallest mean values were those in the fingertips (2.6 mm for SA I units, 4.4 mm for RA units) and the largest those in the forearm (15 mm for FA units, 23 mm for SA I units). As can be seen in Fig. 4, significant

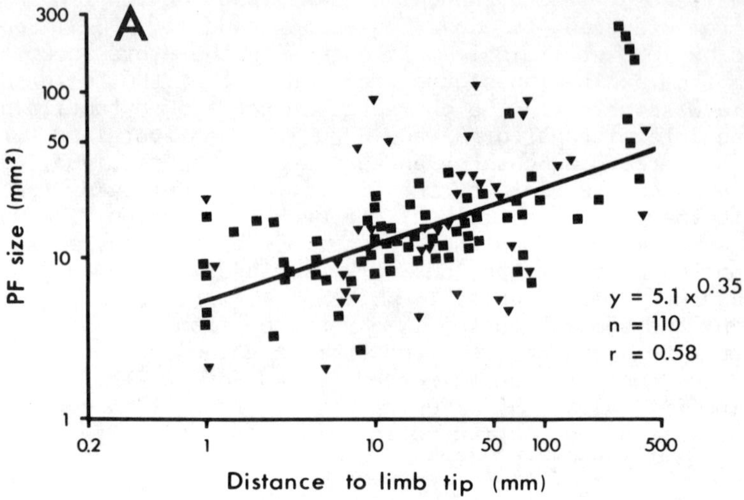

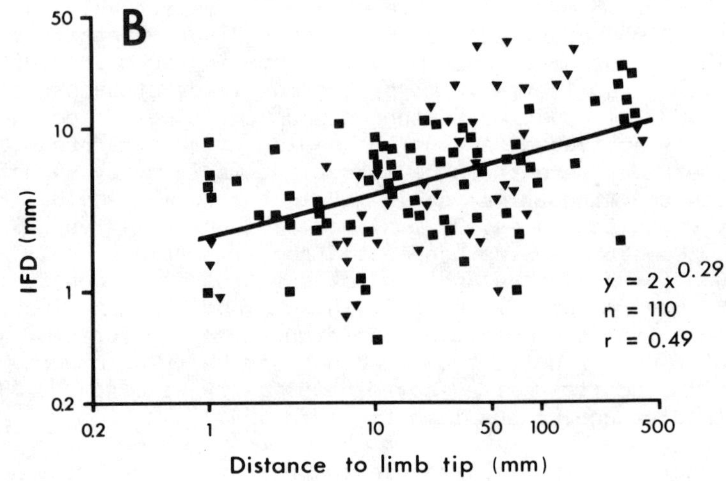

Fig. 3. Scatter plots of projected field size (A) and interfield distance (B) against distance from receptive field to limb tip. Only RA + FA units (squares) and SA I units (triangles) are included. Both sets of data fit a power function with an exponent smaller than one.

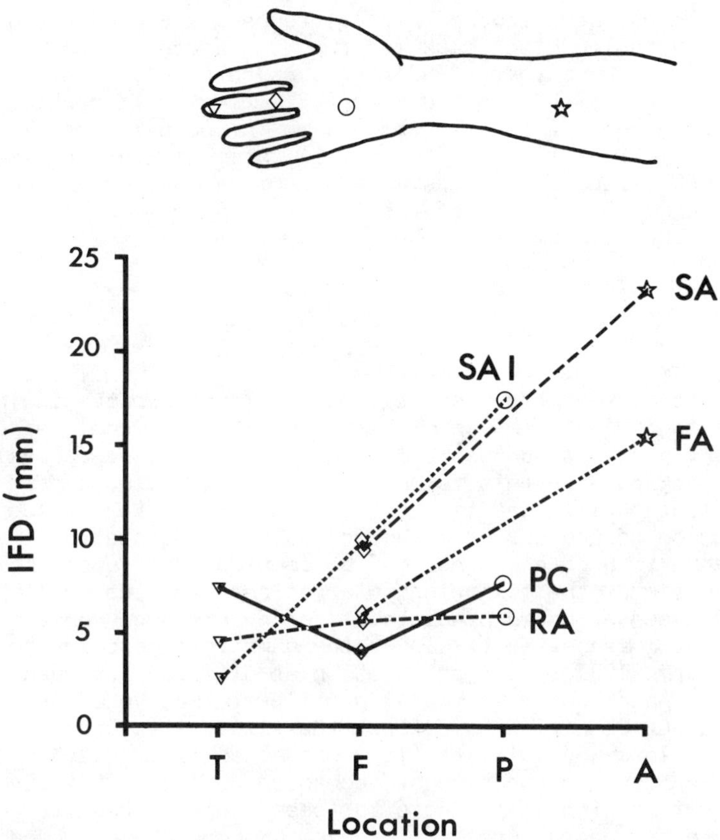

Fig. 4. Schematic representation of gradients in interfield distance between fingertips (T), bodies of the fingers (F), palm (P) and forearm (A) for each mechanoreceptive unit type. SA and FA refer to units in hairy skin.

proximodistal IFD gradients existed for SA I units in glabrous skin, and for FA and SA I units in hairy skin.

Since it seemed of interest to compare IFD values with the results of conventional point localisation tests based on natural stimulation of the skin, seven subjects underwent testing by a method introduced by Noordenbos (1972). Briefly, red dots were marked at standard points in the skin of the hand and forearm, and were touched with a 20 mN von Frey hair. The subject, wearing red goggles, was instructed to make a black mark at the spot where he thought he had been touched. The distance between each red dot and its corresponding black mark was thus a measure of the subject's "locognosia" (Hamburger, 1980) for a given skin region. Mean values obtained by this method were 3.2 mm in the fingertips, 5.7 mm in the remainder of the finger, 6.4 mm in the palm and 17 mm in the forearm. Such figures are only slightly smaller than mean pooled IFD values for the same skin regions: 4.2 mm in the fingertips, 7.4 mm in the rest of the finger, 9.9 mm in the palm and 18 mm in the forearm.

DISCUSSION

If it is accepted that it is possible to excite single sensory units through INMS in awake human subjects (see preceding chapter), a host of unresolved issues in sensory neurophysiology may be tackled with this technique. Activation of one primary afferent neuron will lead to excitatory and inhibitory discharges in a number of neurons at each successive relay station in a sensory channel. Such a channel may be considered a functional unit on the basis of the following observations: (i) the quality of the resulting sensation is predictable for each sensory unit type and consistent in repeated trials at the same intraneural site (Ochoa & Torebjörk, 1983); (ii) stimulation at different frequencies will alter the magnitude or temporal profile of the evoked sensation but does not significantly affect the size of its projection (Schady & Torebjörk, 1983); (iii) projected field size is critically dependent on the type of mechanoreceptive unit activated and on the skin region it projects to (see Results). PFs may therefore be considered characteristic for each sensory channel, and may ultimately be linked to some property of the neurons within the corresponding cortical excitation zone.

Since natural stimulation of the skin coactivates several overlapping mechanoreceptors, one would have expected the resulting sensation to be projected to an area equivalent to the sum of the PFs of the activated units. However, this is not so. It is common experience that a punctiform stimulus is felt in a tiny spot and not over an area 5-10 mm across. What, then, is the relevance of the projected fields obtained by our method in terms of perception of natural stimuli? We believe the results of intraneural and natural stimulation are firmly linked, and apparent discrepancies such as the one mentioned above can be explained by the effects of surround inhibition. Mountcastle & Powell (1959) have shown that the excitation zone in the somatosensory cortex induced by a cutaneous stimulus is surrounded by a ring of cortical neurons which are inhibited by the same stimulus. In the case

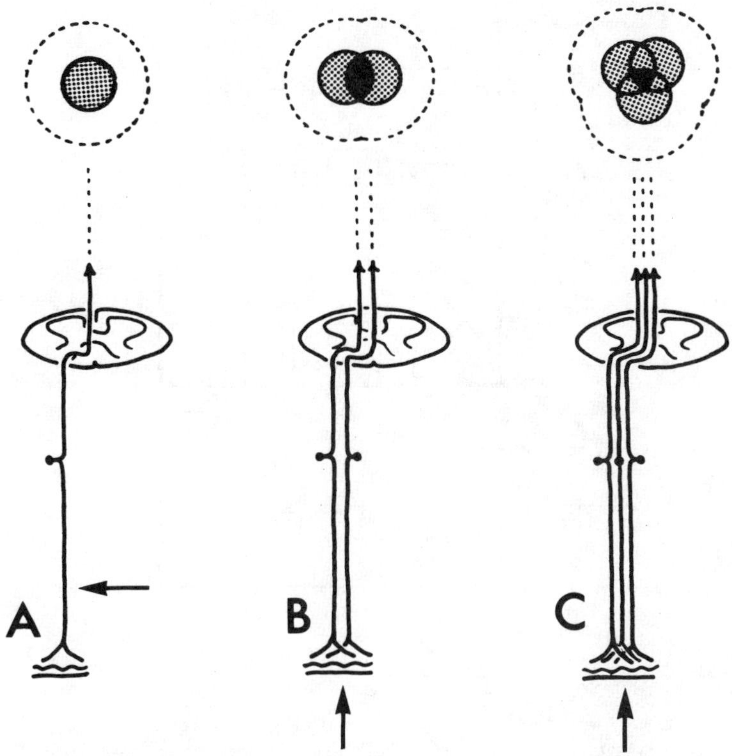

Fig. 5. Diagram illustrating cortical excitatory (stippled) and inhibitory zones resulting from activation of one (A), two (B) or three (C) mechanoreceptive units with overlapping receptive fields. Arrows indicate site of stimulation: intraneural in A, and cutaneous in B and C.

of single unit activation by means of INMS, surround inhibition would have no apparent effect (Fig. 5A). But in the hypothetical case of activation of two neighbouring mechanoreceptive units (Fig. 5B), their cortical inhibitory surrounds would be expected to suppress discharges in each other's excitation zone, except where both excitation zones overlap. At this point there should be summation of impulses from both sources with no inhibition. Such funneling (Gardner & Spencer, 1972) would be multiplied when 3 (Fig. 5C) or more mechanoreceptive units are coactivated. The resulting concentration of convergent input should both enhance weak signals and enable the brain to judge more precisely the size of the stimulus during natural as opposed to intraneural stimulation.

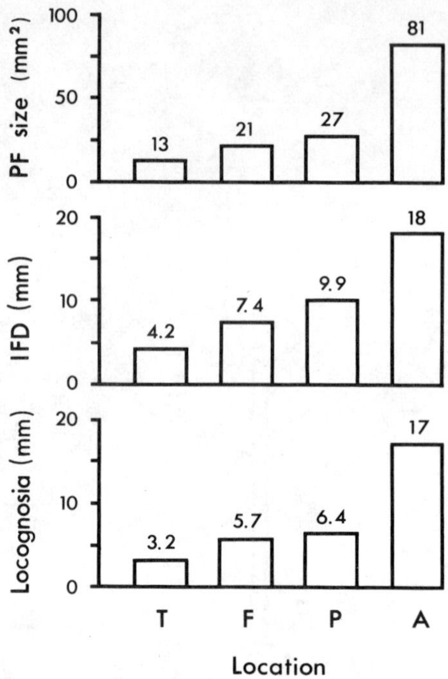

Fig. 6. Gradients in PF size, IFD and locognosia between fingertips (T), bodies of the fingers (F), palm (P) and forearm (A). Numbers above each bar refer to mean values for individual skin regions. In the upper two panels, results from RA, SA I and PC units have been pooled.

The present work has demonstrated proximodistal gradients for all functions studied (Fig. 6). Of particular interest is the finding that PF size and IFD, both functions resulting from single unit inputs, are graded: channels projecting distally to the fingertips are endowed with greater resolution and localisation capacity than those projecting proximally to the palm and forearm. It follows, therefore, that functional sensitivity gradients, for instance in locognosia and two point discrimination, are not only the result of a different innervation density in the skin regions concerned, but are, at least partly, due to differential processing of inputs from individual channels.

Interfield distances are in our view a true measure of the most elementary localisation capacity (Schady et al., 1983). After all, locating the source of a cutaneous stimulus is equivalent to identifying and localising the receptive fields of the sensory units activated by the stimulus. Taken at face value, the mean IFDs given in Fig. 6 may appear large, but they should be judged relative to the locognostic values given in the lower panel. In fact, there are reasons to believe that true IFD values may be smaller than those given. Firstly, the only firm criterion for exclusion of a unit, as long as RF and PF were located on the same finger, was incompatibility between sensory unit type and quality of sensation, as laid down by Ochoa & Torebjörk (1983). By chance it is likely that a few sensations were included which resulted from activation of a unit other than the one recorded from, but whose quality happened to match that which would have been expected from activation of the recorded unit. Secondly, in mapping PFs on two-dimensional photographs of their hands, subjects may have misplaced laterally located PFs medially relative to their RFs. Both these factors would tend to give a falsely elevated mean IFD value. It is all the more striking that IFDs and locognostic values are so similar. Obviously, our localising abilities do not improve substantially by coactivating many overlapping mechanoreceptive units. The force, size and shape of the delivered stimulus may be better judged by the added information, but not its location.

In conclusion, the accuracy of localisation of a cutaneous stimulus depends primarily on the region of skin to which it is delivered. The intensity of the stimulus is relatively unimportant, as also shown by the results of testing locognosia with different stimulus forces (Hamburger, 1980). Single mechanoreceptive units are capable of providing sufficient information to elicit a percept, to specify its quality and to locate the stimulus. In this context, the degree of mechanoreceptive unit overlap demonstrated by Johansson & Vallbo (1980) may seem wasteful, but of course the brain is normally engaged in much more complex tasks than the highly artificial performance of a point localisation test.

ACKNOWLEDGEMENTS

This work was supported by grants from the Swedish Medical Research Council (Nos. 5206-06, 6153-02 and 5822-02) and the National Institutes of Health (No. 1R01NS18315-01).

REFERENCES

1. Gardner, E.P. & Spencer, W.A. (1972). Sensory funneling. II. Cortical neuronal representation of patterned cutaneous stimuli. J. Neurophysiol., 35, 954-977.

2. Hamburger, H.L. (1980). Locognosia. MD Thesis, University of Amsterdam.

3. Johansson, R.S. & Vallbo, Å.B. (1980). Spatial properties of the population of mechanoreceptive units in the glabrous skin of the human hand. Brain Res., 184, 353-366.

4. Knibestöl, M. & Vallbo, Å.B. (1970). Single unit analysis of mechanoreceptive activity from the human glabrous skin. Acta Physiol. Scand., 80, 178-195.

5. Mountcastle, V.B. & Powell, T.S. (1959). Neural mechanisms subserving cutaneous sensibility, with special reference to the role of afferent inhibition in sensory perception and discrimination. Bull. Johns Hopkins Hosp., 105, 201-232.

6. Noordenbos, W. (1972). The sensory stimulus and the verbalisation of the response: the pain problem. In Neurophysiology studied in man. (ed. J.J. Somjen). Excerpta Medica, Amsterdam.

7. Ochoa, J.L. & Torebjörk, H.E. (1983). Sensations evoked by intraneural microstimulation of single mechanoreceptor units innervating the human hand. J. Physiol., 342, 633-654.

8. Schady, W.J.L. & Torebjörk, H.E. (1983). Projected and receptive fields: a comparison of projected areas of sensations evoked by intraneural stimulation of mechanoreceptive units, and their innervation territories. Acta Physiol. Scand. , 119, 267-275.

9. Schady, W.J.L., Torebjörk, H.E. & Ochoa, J.L. (1983). Cerebral localisation function from the input of single mechanoreceptive units in man. Acta Physiol. Scand., 119, 277-285.

10. Soininen, K. & Järvilehto, T. (1982). Are there somatosensory evoked responses to subliminal tactile stimuli? In Soviet-Finnish Symposium on Psychophysiology, Helsinki.

11. Torebjörk, H.E. & Ochoa, J.L. (1980). Specific sensations evoked by activity in single identified sensory units in man. Acta Physiol. Scand., 110, 445-447.

12. Yamauchi, N., Fujitani, Y. & Oikawa, T. (1981). Somatosensory evoked potentials elicited by mechanical and electrical stimulation of the skin. Tohuku J. Exp. Med., 133, 81-92.

13. Zealear, D.L. & Crandall, W.F. (1982). Stimulating and recording from axons within their myelin sheaths. J. Neurosci. Methods, 5, 27-54.

CAN SYMPATHETIC OUTFLOW INFLUENCE AFFERENT ACTIVITY IN MAN?

ZSUZSANNA WIESENFELD-HALLIN and ROLF G. HALLIN

Department of Clinical Neurophysiology, Huddinge University Hospital, S-141 86 Huddinge, Sweden

INTRODUCTION

The total afferent barrage of signals traveling centripetally through the CNS input systems far exceeds the number of impulses that can be analysed by the brain. Various inhibitory mechanisms exist which filter out irrelevant signals so that the organism can concentrate on important infomation vital for its survival. The mechanisms of presynaptic inhibition at the afferent terminals and postsynaptic inhibition at the earliest synaptic levels in the CNS have been extensively studied (see Eccles, 1964) and it is generally accepted that these forms of inhibition are predominant in the vertebrate nervous system. Direct influence on afferent activity at the receptor level is common in the somesthetic and kinesthetic system of invertebrates, but not in vertebrates (Schmidt, 1973). However, some physiological studies in animals have yielded data suggesting that even in vertebrates a peripheral modulation of the afferent input may exist in the somesthetic system (see Table 1). Futhermore, clinical observations initially made over a hundred years ago (Mitchell, 1872) on some patients with peripheral nerve lesions are most readily explained by modulation of certain afferent signals at the peripheral level. A number of findings suggests that the postganglionic sympathetic fibers which travel to the periphery could be involved in such a modulatory function.

It has previously been difficult to study the possible role of the sympathetic outflow on peripheral afferent inflow in man due to the methodological difficulties of recording neural activity in conscious humans. With the introduction of percutaneous recording techniques it became possible to study evoked asynchronous activity in single myelinated axons in human nerves

(Vallbo and Hagbarth, 1968). This procedure also allows recording of unitary activity in the thinnest unmyelinated C afferents, as well as sympathetic efferent C fibers, in alert man (Hallin and Torebjörk, 1970, 1974; Torebjörk and Hallin, 1970, 1974). We have started to examine the question of sympathetic and afferent interaction in healthy, conscious human subjects since we recently developed a new type of electrode for percutaneous neurography in man (Hallin and Wiesenfeld, 1981) which has proven to be particularly suitable for simultaneously recording unitary activity in single cutaneous afferents and high amplitude mass discharges in sympathetic cutaneous efferent fibers.

A brief summary of some previous clinical and experimental observations suggesting a role for the sympathetic system in the modulation of cutaneous mechanoreceptor sensitivity is presented below. Our own recent findings in conscious, healthy human subjects seem to further substantiate the existence of a functional link between sympathetic and afferent fibers in the periphery. Some of these results, as well as experimental details, have been published previously (Hallin and Wiesenfeld-Hallin, 1983).

SOME CLINICAL AND EXPERIMENTAL OBSERVATIONS SUPPORTING THE EXISTENCE OF A FUNCTIONAL SYMPATHETIC-AFFERENT COUPLING IN THE PERIPHERY

Clinical data

After damage to one of the major peripheral nerves, such as the median or sciatic nerves, some patients develop a condition characterized by a severe burning pain which often arises very soon after the injury and may persist for years (Nathan,1947; Livingston, 1976; Bonica, 1979). The most severe form of this painful condition is called causalgia and is relatively rare. A number of less severe forms of this syndrome are more common and are generally grouped under the name of reflex sympathetic dysthrophy (Bonica, 1979). Apart from the pain, which is of a burning character, these patients state that even the slightest touch stimulus to the affected skin area can be experienced as excruciatingly painful, a symptom termed allodynia. The patients can also be hyperesthetic and hyperalgesic. Furthermore, dysfunctions in the sympathetic nervous system seem to be closely coupled to the maintenance of the sensory abnormalities. Thus, all emotional stimuli, which increase the sympathetic outflow to the skin (Delius et al, 1972; Hallin and Torebjörk, 1970, 1974), greatly exacerbate the burning pain. In addition, there is increased sweating and trophic changes within the affected body region. Most importantly, the majority of patients are dramatically improved or totally relieved of the burning pain by a peripheral sympathetic block or sympathectomy (Spurling, 1930).

In a small group of patients with causalgia-like pain condition temporarily relieved from their pain the hyperalgesia could be reproduced by injection of small amounts of noradrenaline or adrenaline intracutaneously in the affected area (Wallin et al, 1976). There was sometimes a delay of up to 30 min. before the pain reappeared. Intracutaneous injections of saline had no effect. These data provide additional evidence suggesting that the sympathetic outflow may be involved in the generation of some abnormal sensations in this condition.

Experimental data from animal studies

The effect of sympathetic activation on afferent input from peripheral receptors has been studied in acute experiments in anaesthetized normal animals or in vitro (see Akoev, 1981 for review of earlier literature). Some results are summarized in Table 1. All these studies were done on mechanoreceptors in cats. Most authors have found that the responses of the rapidly adaptin Pacinian corpuscles (PC) are facilitated by sympathetic activation (Loewenstein and Altamirano-Orrego, 1956; Schiff, 1974; Akoev, 1981) although in some receptors inhibition was also observed (Freeman and Rowe, 1981). Both slowly and rapidly adapting hair receptors are desensitized by electrical sympathetic activation (Nilsson, 1972; Pierce and Roberts, 1981; Roberts and Levitt, 1982). Slowly adapting type II (SA II) mechanoreceptors are characterized by an ongoing discharge in the absence of intentional stimulation and give a direction selective response to stretching the skin. SA II units selectively sensitive to stretching the skin perpendicular to the long axis of the leg showed an increase in the rate of ongoing activity during sympathetic stimulation, whereas those sensitive to stretch parallel to the long axis of the leg were unaffected (Pierce and Roberts, 1981). The effects of electrical stimulation of sympathetic fibers on responses in periodontal mechanoreceptors were examined in two recent studies (Cash and Linden, 1982; Passatore and Filippi, 1983). Although the experimental procedures used seem to be rather similar, the results were quite inconsistent. Afferents, presumably nociceptor fibers, in the dental pulp responded to electrical sympathetic stimulation with an initial excitation followed by inhibition (Edwall and Scott, 1971).

In several of the studies cited above attempts were made to differentiate between direct and indirect effects of sympathetic stimulation on the receptor. Some investigators ascribed the functional changes to indirect effects due to vasoconstriction (Edwall and Scott, 1971; Freeman and Rowe, 1981), whereas in other cases there was no relationship noted between changes in receptor responsiveness and variations in blood pressure (Nilsson, 1972; Pierce and Roberts, 1981; Passatore and Filippi, 1983). The latter results were interpreted to suggest a direct effect on receptor

Table 1. Summary of recent studies on sympathetic effects of evoked or ongoing activity in mechanoreceptors in the cat. In the case of SA II units ⊥ stands for units activated by skin stretch stimuli applied perpendicular to the long axis of the leg, ∥ indicate units excited by stimuli exerted parallel to the long axis of the leg. +, - and 0 symbolize facilitation, inhibition and unchanged responsiveness to an applied stimulus respectively.

PACINIAN CORPUSCLE	
+ evoked, ongoing	Loewenstein and Altamirano-Orrego, 1956
+ evoked	Schiff, 1974
+ evoked	Akoev, 1981 (review)
+, - evoked	Freeman and Rowe, 1981
HAIR RECEPTORS	
- vibrissae	Nilsson, 1972
- evoked, + ongoing G hair	Pierce and Roberts, 1981
- D hair	Roberts and Levitt, 1982
SA II	
⊥, 0 evoked, + ongoing	Pierce and Roberts, 1981
∥, no effect	" "
PERIODONTAL MECHANORECEPTORS	
- ongoing and evoked	Cash and Linden, 1982
+ ongoing, 0 evoked	Passatore and Filippi, 1983
TOOTH PULP	
+ followed by -	Edwall and Scott, 1971

structures by sympathetic activation. Indeed, histological studies have shown that small noradrenaline containing fibers may be found in close connection to some skin receptors (Fuxe and Nilsson, 1965; Santini et al, 1971; Roberts and Levitt, 1982), although the evidence is somewhat controversial (Spencer and Schaumburg, 1973). However, all the physiological studies cited above were done on anaesthetized cats or in vitro. In addition, the sympathetic fibers were electrically stimulated in most studies, a clearly non-physiological method of stimulation. Therefore the results may not be directly applicable to conscious human subjects. Nevertheless most of the above results suggest an influence of sympathetic outflow on the firing properies of certain receptors in normal skin.

Experimental peripheral nerve injuries in rats and mice can induce abnormal behaviors (Wall et al, 1979a; Wiesenfeld and Lindblom, 1980) which at least in part seem to depend on the level of sympathetic outflow (Wall et al, 1979b; Wiesenfeld and Hallin, 1980, 1981; Devor and Jänig, 1981). The injuries induce the formation of a neuroma which contains intermingled sprouts from sympathetic and afferent fibers. Many of the afferent sprouts exhibit an abnormal ongoing activity and increase their firing to applied α-adrenergic agents (Wall and Gutnick, 1974; Korenman and Devor, 1981; Scadding, 1981). As the clinical data presented above, these results also suggest that peripheral autonomic nerve mechanisms may be of pathophysiological significance, possibly being involved in the symptomatology described by patients with causalgia or reflex sympathetic dystrophy.

MICRONEUROGRAPHIC FINDINGS SUGGESTING A FUNCTIONAL SYMPATHETIC-AFFERENT LINKAGE AT THE PERIPHERY IN MAN

Pacinian corpuscles are highly sensitive mechanoreceptors that respond phasically and are most sensitive to high frequency vibratory stimuli (see Burgess and Perl, 1973 for review). We encountered on several occasions an arrhythmic, ongoing firing in PC afferents in the absence of intentional stimulation. An example of such a unit recorded from the median nerve at the wrist is shown in Fig. 1 B. The large, positive, irregularly occurring spikes in the PC afferent did not evoke any conscious percept in the subject. The receptive field of this unit was large, comprising almost half of the vola manus with the area of maximal sensitivity situated on the thenar eminence. The unit was exquisitely sensitive to light mechanical skin stimulation (Fig. 1 A) and responded even to slight taps on the platform supporting the subject´s arm. Mainly negative polarity multiunit bursts in cutaneous sympathetic C fibers, as judged from their discharge pattern (Delius et al, 1972; Hallin and Torebjörk, 1974), regularly coincided with increased activity in the PC afferent (Fig 1 B). The waveform of the PC afferent showed no irregularities indicative of fiber damage (Fig. 1 C), suggesting that the ongoing activity was not likely to be injury discharges (Torebjörk et al, 1970; Vallbo, 1976). However, it should be ruled out that the effects observed were due to some abnormal connection between the sympathetic and the afferent fibers at the electrode site. If abnormal connections were present, one would expect the sympathetic and afferent activity to be closely coupled with a delay of only a few ms (Granit et al, 1944; Granit and Skoglund, 1945). If the effects were due to the sympathetic impulses traveling to the periphery and affecting the afferent near the receptor stuctures, there should be a suitable time delay between sympathetic and afferent activity. Indeed, such a delay was usually seen in the

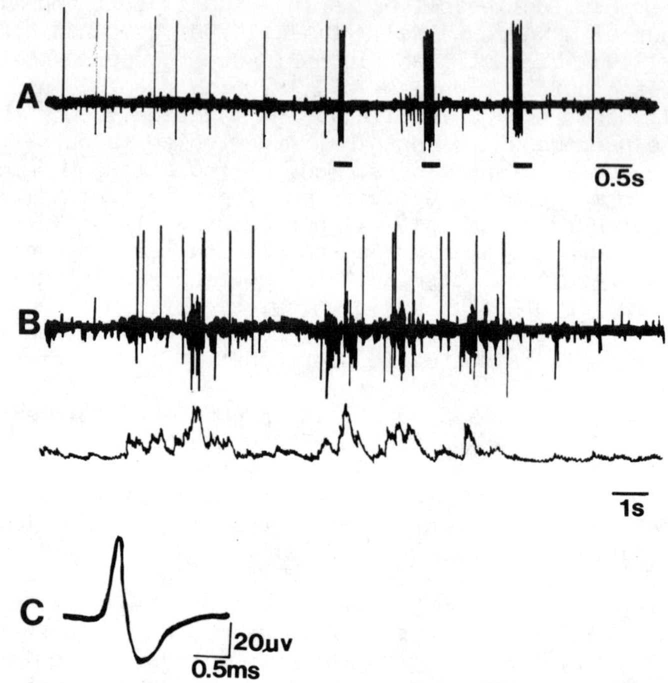

Fig. 1. Ongoing activity in an afferent fiber innervating a PC-receptor (predominantly positive large amplitude spikes) and multi-unit sympathetic activity (predominantly negative spikes). Underneath trace shown in B is the integrated sympathetic neurogram. A. Single spikes in the PC-afferent appeared irregularly in the absence of skin stimulation (left). When gently stroking the skin within a fairly large palmar area as underlined, the unit fired repetitively at a high frequency (right). B. In sequences where sympathetic activity was also distinctly recorded there was a clear relationship between ongoing PC-activity and sympathetic outflow. C. Shape of the action potential of the PC unit.

integrated neurogram trace. A spontaneous (Fig. 2 A) or evoked (Fig. 2 B) sympathetic burst generally preceded firing of several action potentials in the PC afferent. However, single spikes in the PC afferent sometimes occurred without clear sympathetic activity being recorded (Fig. 2 B, last positive potential). The delay was 300-400 ms and the distance between the recording site in the nerve and location of the area of maximal sensitivity of the PC receptor in the skin was about 7 cm. The above calculated values would correspond to an estimated conduction velocity of the peripherally propagated sympathetic impulses of around 0.2 m/s, including any eventual delay near the receptor

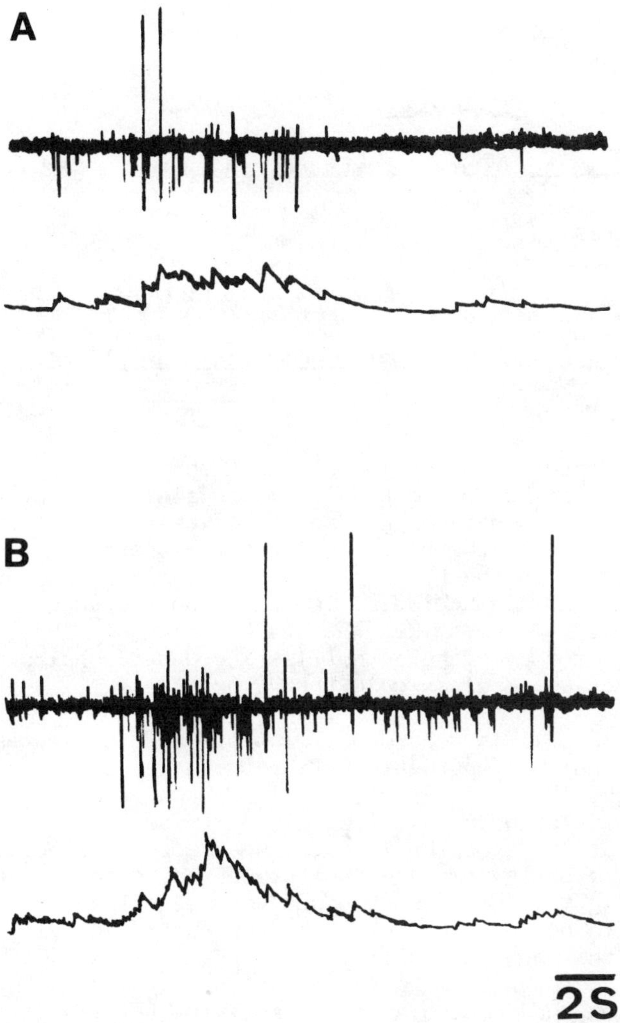

Fig. 2. Time relation between a spontaneously occurring (A) and an induced sympathetic burst (B) and accompanying afferent PC activity. The sympathetic bursts preceded the PC spikes by 300-400 ms.

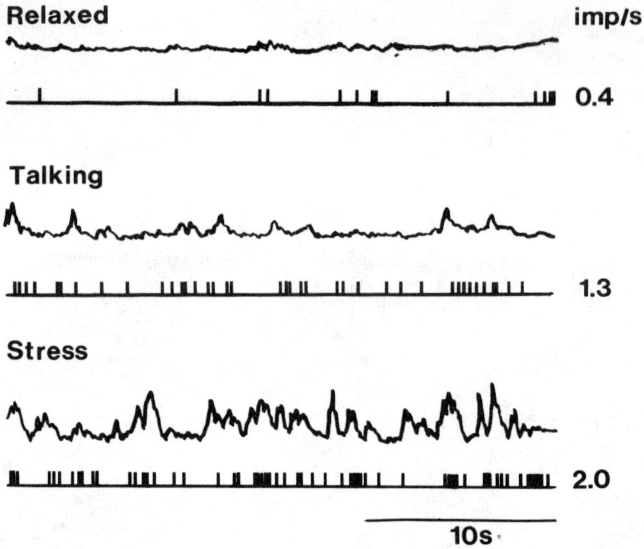

Fig. 3. Integrated sympathetic multiunit activity (top) and ongoing activity in PC afferent (bottom) in alert subject during different states of attention. The PC spikes were uniformly rectified. The numbers to the right indicate the average number of ongoing impulses/s in the PC unit. Top: The subject was totally relaxed in a silent room. Middle: There was an increase in sympathetic activity and PC spikes when talking to the subject. Bottom: During stress the sympathetic outflow increased considerably as did the number of impulses/s in the PC unit.

site, which is well in accordance with propagation rates of 0.5-1 m/s in human sympathetic C fibres studied previously (Hallin and Torebjörk, 1974).

In a few stable unitary recordings it was possible to study how changes in the sympathetic outflow affected the ongoing PC firing. Fig. 3 shows some representative recording sequences. Initially (Fig. 3, top) when the subject was relaxed, quietly reclining with her eyes closed, the sympathetic activity level was low and the ongoing activity in the PC was about 0.4 imp/s. When the subject was talking there was an increase in both sympathetic outflow and ongoing afferent activity (Fig. 3, middle). When she was stressed by performing mental arithmetic, the sympathetic outflow increased substantially (Fig. 3, bottom). There was a clear relationship between the increased number of deflections in the integrated sympathetic neurogram and increase of "spontaneous" firing in the PC, with the highest number of im-

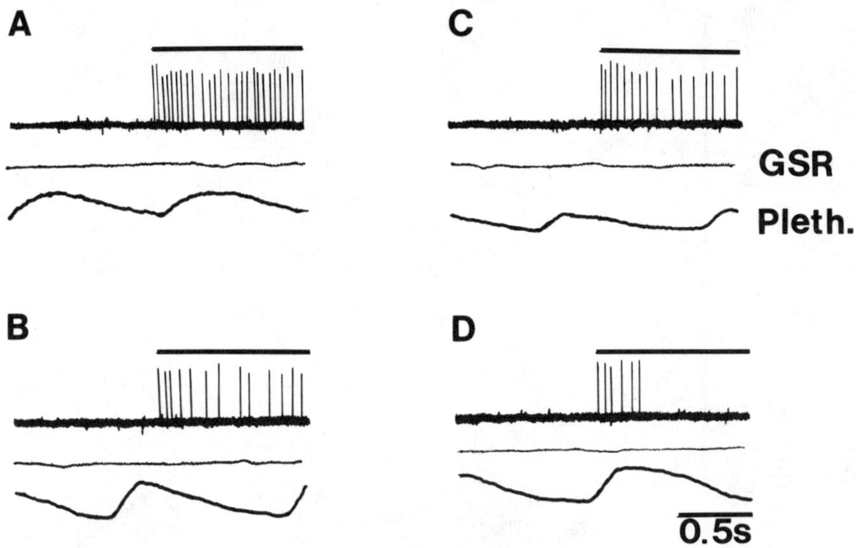

Fig. 4. Modulation of mechanical responsiveness in a SAI unit by cutaneous vasomotor (pleth) and sudomotor (GSR) outflow. The vasomotor and sudomotor changes were induced by cooling respectively warming the subject. The response of the unit to stimulation with von Frey´s hairs exerting a force of 16 mN (A and B) and 6.9 mN (C and D) are shown. The responses were more pronounced during vasoconstriction (A and C) than during vasodilation (B and D).

pulses coinciding with the largest amplitude bursts. We have seen similar effects of sympathetic activation on PC receptors in 4 of 20 units examined in detail.

Sympathetic modulation of slowly adapting type I (SAI) receptor responses to mechanical stimuli have also been observed. The response of these receptors is characterized by a maintained, irregular discharge to a continuous mechanical stimulus with no ongoing activity in the absence of stimulation (see Burgess and Perl, 1973 for review). The activity from the SAI unit shown in Figs. 4 and 5 was recorded from the median nerve. The receptive field of the unit was on the thumb. No sympathetic activity was recorded simultaneously with the afferent unit response. The level of sympathetic activation was monitored by recording the galvanic skin response (GSR) from the skin area near the receptive field of the SAI unit and the finger plethysmogram along with the neural recordings. The sudomotor and vasoconstrictor sympathetic fibers to the skin were selectively activated by warming or cooling the subject, respectively (Bini et al, 1980).

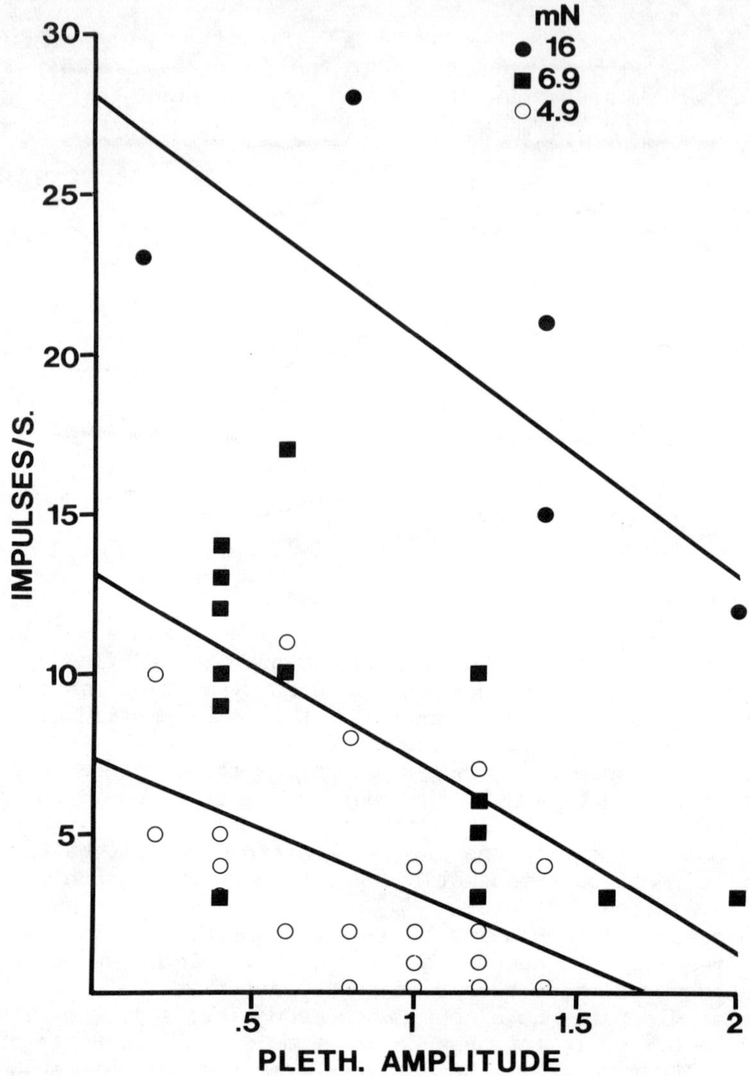

Fig. 5. Relationship between unitary SAI responses to defined mechanical stimuli (Y axis) and level of vasoconstriction as measured by the finger plethysmogram (X axis). Same experiment as in Fig. 4. Both threshold (4.9 mN) and supratheshold (6.9 mN and 16 mN) stimuli induced larger number of impulses during the first second of stimulation under periods of relative vasoconstriction. The regression coefficients for the three stimulus strengths were statistically significant: 4.9 mN r= -.48, p<.0.5; 6.9 mN r= -.67, p<.01; 16 mN r= -.82, p<.05.

A clear relationship was found between the pulse amplitude and the responsiveness of the unit to stimulation with von Frey´s hairs. Fig. 4 shows some responses during the first second after applying two different suprathreshold stimuli during relative vasoconstriction (A and C) and vasodilation (B and D). The responses during vasoconstriction were clearly more vigorous than during vasodilation. No ongoing activity in the absence of stimulation was observed for this unit. The skin temperature was measured and was not found to vary by more than 1.5° C, which is not enough to account for the large observed changes in sensitivity (Tapper, 1965). The quantitative relationship between plethysmogram amplitude and response strength for the same SAI unit is shown in Fig. 5. A significant relationship was found for all three stimulus intensities used; plethysmogram amplitude was negatively correlated to the number of action potentials in the response. Considerable variability in responsiveness was observed, especially for the threshold stimulus (4.9 mN). Some of the variability was undoubtably due to the rather crude method of stimulation. No relationship was found between response size and changes in GSR. We have seen similar modulatory effects during vasoconstriction on response magnitude in 2 of 13 SAI units studied in such detail.

DISCUSSION

The microneurographic results presented above indicate that in healthy, awake human subjects mechanoreceptor function in some receptors in the skin can be modulated by the level of sympathetic activity. These data cannot resolve the question of whether the effects are direct at the site of the generator potential or indirect, due to changes in, for example, the blood supply to the skin during sympathetic activation. Whether this effect is direct or indirect is a problem of considerable interest. Functionally, however, in either case the effects are mediated via the sympathetic nervous outflow.

We considered several possible sources of error which might have contributed to or distorted the data. It has been shown that cutaneous receptors in glabrous skin of the hand respond vigorously to movements of the hand (Hulliger et al, 1979). These effects are, however, only seen with relatively large movements, which never occurred in our studies. In fact, even the slightest unobservable tremors were accompanied by clearly discernible electromyographic (EMG) potentials. We observed no afferent activation during such periods of EMG activity. Furthermore, the records of the mechanoreceptor responses that were influenced by sympathetic activation were not contaminated by EMG potentials.

Another possible source of artefact could be the formation of "ephapses" or "artificial synapses" (Granit et al, 1944; Granit and Skoglund, 1945) between sympathetic and afferent fibers at the recording site, due to acute injury caused by the electrode. There are three lines of evidence against this possibility. First, the waveform of the afferent spikes had no signs of fiber injury (see Fig. 1 C). Secondly, no typical injury discharges were recorded and the units were responsive for long periods, up to hours. Thirdly, in those experiments where sympathetic outflow and afferent activity were recorded simultaneously, the estimated peripheral conduction velocities of the sympathetic bursts accorded well with the idea that the sympathetic impulses were conducted peripherally to the skin before affecting the afferent unit.

The fact that serious clinical pain states following nerve trauma can be permanently relieved following sympathetic block or sympathectomy suggests that the sympathetic system may exert a modulatory influence on the responsiveness of some peripheral afferents. Results from histological, pharmacological and physiological studies in animals lend additional support to such an idea of a peripheral sympathetic sensory coupling. The evidence is especially strong in the case of PC receptors (Akoev, 1981). In view of the presented considerations and results it is therefore tempting to suggest that activity in at least a subpopulation of normal human PC and SAI receptors may become modified by changes in sympathetic outflow. It can not be excluded that the peripheral interaction between afferent and sympathetic nerve fibers may become more pronounced or changed in certain sympathetic-dependent pain states. This and other questions of pathophysiological significance may be examined by performing intraneural recordings from patients with well defined peripheral nerve lesions.

ACKNOWLEDGEMENT

Supported by Swedish Medical Research council grant K 79-14 V -5318-02, Harald Jeanssons Stiftelse, Harald and Greta Jeanssons Stiftelse, Ciba Geigy AB, Folksam and Skandia Insurance companies and research funds of the Karolinska Institute.

REFERENCES

Akoev, G.N. (1981). Catecholamines, acetylcholine and excitability of mechano-receptors. Prog. Neurobiol. 15, 269-294.

Bini, G., Hagbarth, K.-E., Hynninen, P. and Wallin, B.G. (1980). Thermoregulatory and rhythm-generating mechanisms governing the sudomotor and vasoconstrictor outflow in human cutaneous nerves, J. Physiol. (London), 306, 537-552.

Bonica, J.J. (1979). Causalgia and other reflex sympathetic dystrophies. In Advances in Pain Research and Therapy, Vol. 3. (eds. J.J. Bonica and V. Ventafridda). Raven Press, New York.

Burgess, P.R. and Perl, E.R. (1973). Cutaneous mechanoreceptors and nociceptors. In Handbook of Sensory Physiology, Vol. 2.

Somatosensory System. (ed. A. Iggo). Springer, Berlin-Heidelberg-New York.

Cash, R.M. and Linden, R.W.A. (1982). Effects of sympathetic nerve stimulation on intra-oral mechanoreceptor acitvity in the cat. J. Physiol. (London), 329, 451-463.

Delius, W., Hagbarth, K.-E., Hongell, A., and Wallin, B.G. (1972). Manoeuvres affecting sympathetic outflow in human skin nerves. Acta Physiol. Scand., 84, 177-186.

Devor, M. and Jänig, W. (1981). Activation of myelinated afferents ending in a neuroma by stimulation of the sympathetic supply in the rat. Neurosci. Lett., 24, 43-47.

Eccles, J.C. (1964). The Physiology of Synapses. Springer, Berlin-Göttingen-Heidelberg-New York.

Edwall, L. and Scott, D. (1971). Influence of changes in microcirculation on the excitability of the sensory unit in the tooth of the cat. Acta Physiol. Scand., 82, 555-566.

Freeman, B. and Rowe, M. (1981). The effect of sympathetic nerve stimulation on responses of cutaneous Pacinian corpuscles in the cat. Neurosci. Lett., 22, 145-150.

Fuxe, K. and Nilsson, B.Y. (1965). Mechanoreceptors and adrenergic nerve terminals. Experientia, 21, 641-642.

Granit, R., Leksell, L. and Skoglund, C.R. (1944). Fibre interaction in injured or compressed regions of nerve. Brain, 67, 125-140.

Granit, R. and Skoglund, C.R. (1946). Facilitation, inhibition and depression at the "artificial synapse" formed by the cut end of a mammalian nerve. J. Physiol. (London), 103, 435-448.

Hallin, R.G. and Torebjörk, H.E. (1970). Afferent and efferent C units recorded from human skin nerves in situ. Acta Soc. Med. Upsal., 75, 277-281.

Hallin, R.G. and Torebjörk, H.E. (1974). Single unit sympathetic activity in human skin nerves during rest and various manoeuvres. Acta Physiol. Scand., 92, 303-317.

Hallin, R.G. and Wiesenfeld, Z. (1981). A standardized electrode for percutaneous recording of A and C fibre units in conscious man. Acta Physiol. Scand., 113, 561-563.

Hallin, R.G. and Wiesenfeld, Z. (1983). Does sympathetic activity modify afferent inflow at the receptor level in man? J. Autonom. Nerv. Syst., 7, 391-397.

Hulliger, M., Nordh, E., Thelin, A.-E. and Vallbo, Å.B. (1979). The responses of afferent fibres from the glabrous skin of the hand during voluntary finger movements in man. J. Physiol. (London), 291, 233-249.

Korenman, E.M.D. and Devor, M. (1981). Ectopic adrenergic sensitivity in damaged peripheral nerve axons in the rat. Exp. Neurol., 72, 63-81.

Livingston, W.K. (1976). Pain Mechanisms. A Physiologic Interpretation of Causalgia and Its Related States. Plenum Press, New York.

Loewenstein, W.R. and Altamirano-Orrego, R. (1956). Enhancement of activity in a Pacinian corpuscle by sympathomimetic agents. Nature, 178, 1292-1293.

Mitchell, S.W. (1872). Injuries of Nerves and Their Consequences. Lippincott, Philadelphia.

Nathan, P.W. (1947). On the pathogenesis of causalgia in peripheral nerve injuries. Brain, 70, 145-170.

Nilsson, B.Y. (1972). Effects of sympathetic stimulation on mechanoreceptors on cat vibrissae. Acta Physiol. Scand., 85, 390-397.

Passatore, M. and Filippi, G.M. (1983). Sympathetic modulation of periodontal mechanoreceptors. Arch. Ital. Biol., 121, 55-65.

Pierce, J.P. & Roberts, W.J. (1981). Sympathetically induced changes in the response of guard hair and type II receptors in the cat. J. Physiol. (London), 314, 411-428.

Roberts, W.J. and Levitt, G.R. (1982). Histochemical evidence for sympathetic innervation of hair receptor afferents in cat skin. J. Comp. Neurol., 210, 204-209.

Santini, M., Ibata, Y. and Pappas, G.D (1971). The fine structure of sympathetic axons within the Pacinian corpuscle. Brain Res., 33, 279-289.

Scadding, J.W. (1981). Development of ongoing activity, mechanosensitivity and adrenaline sensitivity in severed peripheral nerve axons. Exp. Neurol., 73, 345-364.

Schiff, J.D. (1974). Role of the sympathetic innervation of the Pacinian corpuscle. J. Gen. Physiol., 63, 601-608.

Schmidt, R.F. (1973). Control of the access of afferent activity to somatosensory pathways. In Handbook of Sensory Physiology, Vol. 2. Somatosensory System. (ed. A. Iggo). Springer, Berlin-Heidelberg-New York.

Spencer, P.S. and Schaumburg, H.H. (1973). An ultrastructural study of the inner core of the Pacinian corpuscle. J. Neurocytol., 2, 217-235.

Spurling, R.G. (1930). Causalgia of the upper extremity. Treatment by dorsal sympathetic ganglionectomy. Arch. Neurol. Psychiatry, 23, 784-788.

Tapper, D.N. (1965). Stimulus-response relationships in the cutaneous slowly-adapting mechanoreceptor in hairy skin of the cat. Exp. Neurol., 13, 364-385.

Torebjörk, H.E. and Hallin, R.G. (1970). C-fibre units recorded from human sensory nerve fascicles in situ. Acta Soc. Med. Upsal., 75, 81-84.

Torebjörk, H.E., Hallin, R.G., Hongell, A. and Hagbarth, K.-E. (1970). Single unit potentials with complex waveform seen in microelectrode recordings from the human median nerve. Brain Res., 24, 443-450.

Torebjörk, H.E. and Hallin, R.G. (1974). Identification of afferent

C units in intact human skin nerves. Brain Res., 67, 387-403.

Vallbo, Å.B. and Hagbarth, K.-E. (1968). Activity from skin mechanoreceptors recorded percutaneously in awake human subjects. Exp. Neurol., 21, 270-289.

Vallbo, Å.B. (1976). Prediction of propagation block on the basis of impulse shape in single unit recordings from human nerves. Acta Physiol. Scand., 97, 66-74.

Wall, P.D., Devor, M., Inbar, R., Scadding, J.W., Schonfield, D., Seltzer, Z., and Tomkiewicz, M.M. (1979a). Autotomy following peripheral nerve lesions: experimental anaesthesia dolorosa. Pain, 7, 103-113.

Wall, P.D., and Gutnick, M. (1974). Ongoing activity in peripheral nerves: the physiology and pharmacology of impulses originating from a neuroma, Exp. Neurol., 43, 580-593.

Wall, P.D., Scadding, J.W., and Tomkiewicz, M.M. (1979b). The production and prevention of experimental anaesthesia dolorosa. Pain, 6, 175-182.

Wallin, B.G., Torebjörk, E. and Hallin, R.G. (1976). Preliminary observations on the pathophysiology of hyperalgesia in the causalgic pain syndrome. In Sensory Functions of the Skin of Primates with Special Reference to Man. (ed. Y. Zotterman). Pergamon Press, Oxford.

Wiesenfeld, Z. and Hallin, R.G. (1980). Stress-related pain behavior in rats with peripheral nerve injuries, Pain, 8, 279-284.

Wiesenfeld, Z. and Hallin, R.G. (1981). Influence of nerve lesions, strain differences and continuous cold stress on chronic pain behavior in rats, Physiol. Behav., 27, 735-740.

Wiesenfeld, Z. and Lindblom, U (1980). Behavioral and electrophysiological effects of various types of peripheral nerve lesions in the rat. A comparison of possible models for chronic pain. Pain, 8, 285-298.

PERIPHERAL CODING MECHANISMS OF TOUCH VELOCITY

*OVE FRANZÉN, **FLOYD THOMPSON, †BARRY WHITSEL, and †MICHAEL YOUNG

*Department of Psychology, University of Uppsala, Uppsala, Sweden
**Department of Neuroscience, College of Medicine, University of Florida, Gainesville, FL, USA
†Department of Physiology, University of North Carolina at Chapel Hill, NC 27514, USA

In November 1925 Adrian and Zotterman managed for the first time to record electrical impulses set up in a single fiber originating in a mechanoreceptive end-organ and thereby provide direct evidence for the basic principle that the conduction in sensory nerves is an all-or-none event. They found, furthermore, in a subsequent study that the mechanoreceptors could be dichotomized into rapidly and slowly adapting categories on the basis of the adaptive properties (Adrian and Zotterman, 1926).

The trend in recent research on cutaneous sensibility has been to merge the fields of sensory neurophysiology and perception - an approach advocated already fifty years ago as indicated in a letter written by Adrian to Zotterman in consequence of his dissertation on pain: "You are certainly right in not confining yourself to one line of attack - in fact the paper gains very greatly from the way in which the nerve impulse work fits into the work on sensation in man".

This approach has been given great impetus by the discovery that it is feasible to intercept the staccato messages in individual fibres by poking a fine tungsten electrode into the nerve of conscious human subjects and to compare the impulse traffic with the subjective sensation magnitude.

The point of departure for the present study is that movement is a stimulus attribute of great importance for the sense of vision as well as for the sense of touch. Paraphrasing Shakespeare's Ulysses we could say that 'Things in motion sooner stimulate the skin than what not stirs'. We know also from daily experience that in order to generate a distinct touch impression of the surface texture of an object we have to move our fingers over it or to gently rub the object against the skin.

The main purpose of the present study was to determine what information the peripheral mechanodetectors in hairy and glabrous skin of humans and subhuman primates can provide the central nervous system about punctuate and transversal stimulation.

The afferent activity in selected types of mechanoreceptive fibers was correlated with subjective estimates of tactile sensations enabling us to evaluate what restrictions the peripheral apparatus imposes on velocity information transmitted to the central nervous system. The response range for stimuli moving at different velocities over the skin was several orders of magnitude greater than the response range for stimuli moving perpendicular to the skin.

METHODS

Subjects

Twenty volunteers participated in these experiments and they had all normal somesthesis. Eight of them served also as subjects in a study where the technique of microneurography was applied.

Stimulation

In the psychophysical experiments two body sites were chosen for stimulation: the volar phalange of the index finger and the dorsal forearm. The punctuate stimuli delivered perpendicular to the finger pad were linearly rising pulses whose terminal amplitude was always set to 500 microns. The plastic probe mounted on a Pye-Ling vibrator had a diameter of 2 mm and could displace the skin at a maximal velocity of 6.4 cm/sec. The surface-parallel stimulation was produced by moving a brush of fine camel's hair across rectangular fields of the skin defined by plates with different apertures. A servomotor drove the brush at any velocity between 0.5 and 256 cm/sec with a precision of about 1 per cent. The electronic circuitry marked the beginning and the end of brush contact with the skin.

Scaling of subjective velocity

Three series of experiments designed to determine the stimulus-response relations for stimuli moving across the skin were carried out on seven subjects and for stimuli moving normal to the skin surface on ten subjects.

The method of free magnitude estimation was employed i.e. the subjects made his or her estimates without any designated modulus. At the beginning of each session the following instruction was read to the subject: "You will feel a series of touch stimuli that vary in velocity. They will be presented to you one at a time with an average interval of 7 seconds. Your task is to assign a number to each stimulus according to its apparent velocity, that is to say, make the numbers you use proportional to the stimulus velocity as you perceive it. I ask you also to pay attention only to the velocity aspect and neglect changes in other aspects of the percept that may occur". One session was run at each location for each subject and the stimuli were presented in random order four times in a session.

Microneurography

Single unit activity was studied in eight adults. The subjects were lying down on a bed and instructed to extend their arm laterally so as to expose the palm and the ventral surface of the arm. The nerve impulses were recorded with a tungsten electrode inserted 10 cm proximal to the elbow according to the technique introduced by Vallbo and Hagbarth (1968). The receptive fields for different receptor types were localized and determined by means of a hand held glass probe. After the identification of the receptor the mechanical stimulator was put at the center of the receptive field and the mechanical stimuli were delivered in a random order for as long as the electrode tip could be kept in a good intrafascicular position. The evoked impulses and the stimulus analogue signal were displayed on an oscilloscope and stored on tape or in a computer.

Electrophysiological experiment in monkeys

Under halothane anesthesia single nerve fibres were dissected from the medial and lateral branch of the femoral nerve supplying the skin area of the hindlimb. The conduction velocity of the sensory fibres was determined by electrically stimulating the skin in order to make an approximate classification of the fibre types under study.

RESULTS

Magnitude estimates of displacement velocity

The geometric mean of four estimates per subject assigned to a given velocity was selected by us as a measure of central tendency. In order to eliminate interobserver variance caused by differing choice of moduli or the variance due to subjects choosing to work with different number ranges the raw data were transformed according to the procedure suggested by Kalikow (1967). The group data are shown in Fig. 1, where the magnitude estimates are plotted over velocity in logarithmic coordinates.

It is evident that no simple function can describe the empirical trend of the data. Only for a limited portion of the curve a power function, $R = cV^n$ (Stevens, 1957) where R is the subjective response and V is the physical velocity could represent the trend. The slope of the log-log function which is equal to the exponent, is 0.56 for a velocity range of 0.05 - 1.6 cm/sec. In that respect the agreement with results from a previous study is quite good (Franzén and Lindblom, 1976). Zwislocki, Adams and Barlow (1969) have proposed a mathematical model for this kind of S-shaped relationships which we have applied to the human neural data to be presented in the next section.

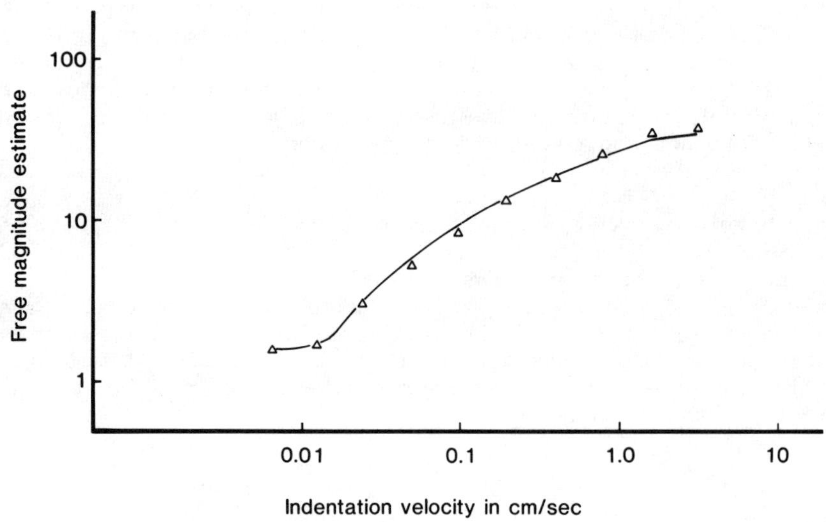

Fig. 1. Geometric mean of free magnitude estimates plotted against displacement velocity.

Impulses evoked in human mechanoreceptive A-fibres by ramp stimulation

During the course of several experiments on human subjects we have recorded activity in all types of mechanoreceptive units existing in the glabrous skin of the hand and classified them according to their adaptive properties (Johansson, 1979). Our sample consisted of 7 RA (rapidly adapting), 2 PC (Pacinian) and 8 SA (slowly adapting) units.

Since the estimated innervation density of the finger pad is highest for the RA units (Johansson, 1979), we will limit our analysis to this type of fibres. The RA units are by far the most numerous and will therefore in all likelihood 'dominate' the tactile percept, at least within the range of skin indentations we are working in the present study.

The RA units responded vigorously during ramp stimulation but generated usually no more than two spikes at velocities of 3.2 cm/sec and higher (Fig. 2).

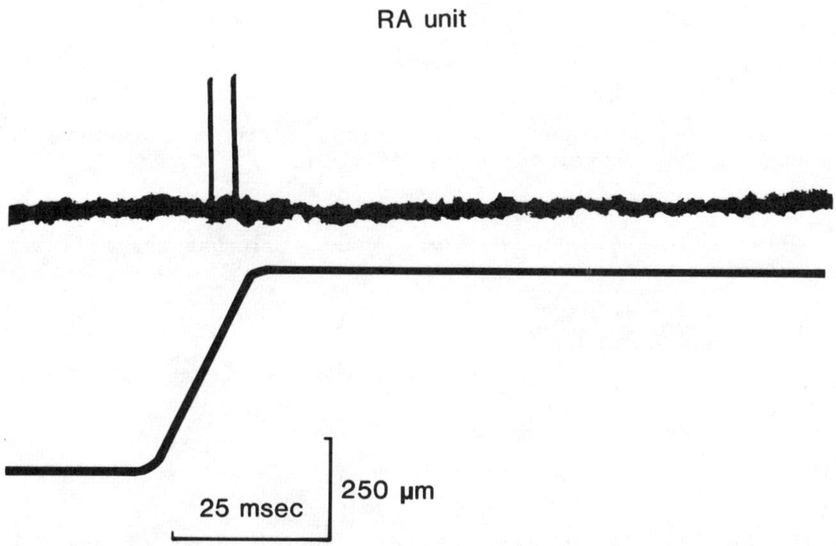

Fig. 2. Discharge of a rapidly adapting fiber terminating in the glabrous skin of the middle finger. The skin was displaced to a depth of 500 microns at a velocity of 3.2 cm/sec.

We have plotted the instantaneous firing frequency which is the reciprocal of the time interval between impulses against displacement velocity and found that the RA units saturate at about the same indentation velocity as the psychophysical function levels off (Fig. 3).

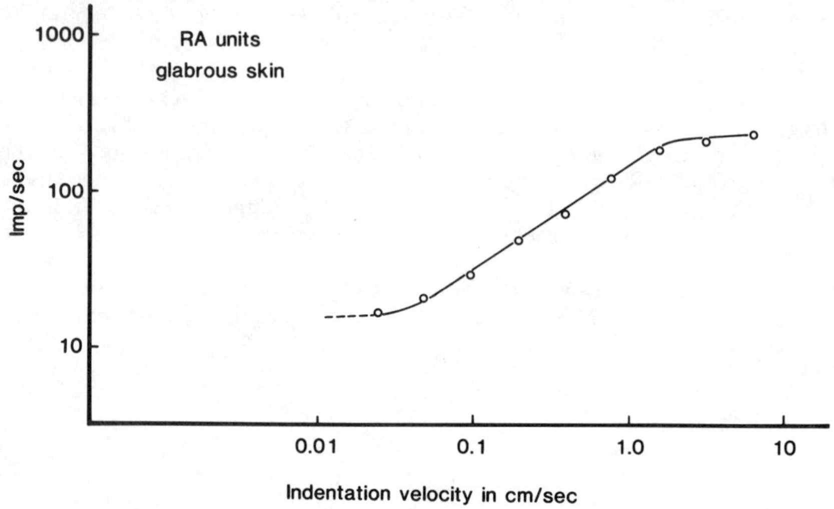

Fig. 3. Average response (instantaneous firing rate) curve for 7 RA units plotted over displacement velocity.

Zwislocki, Adams and Barlow's model that has the following mathematical form

$$R = R_m \left\{ 1 - e^{-R_o/R_m} (1 + V/N)^n \right\}$$

where R = firing rate, R_m = saturation level, R_o = threshold activity, V = velocity, N = intrinsic noise, and n = exponent

can adequately describe the average response curve of the 7 RA units if the exponent (n) is set to 0.5 which is in fact the size of the exponent 'conjectured' by Zwislocki, Adams and Barlow (1969).

Perceived velocity of tactile stimuli moving across glabrous and hairy skin

The force exerted on the skin during a stroke was close to five grams and did not alter with changing velocity (Fig. 4).

The free magnitude estimates analyzed as reported above are a monotonicly increasing function of velocity all the way up to 256 cm/sec which is the upper limit of our stimulation equipment (Fig. 5). This response range is several orders of magnitude greater than the response range for stimuli moving normal to the skin (see Fig. 1 and 3). The curvilinear trend of the data below 2 cm/sec suggests that the subjective scales may be well described by a generalized form of the power function (Ekman, 1958), $R = c\ (V - a)^n$, where 'a' is an additive constant that can be derived either by a graphic (Ekman, 1961) or an iterative method.

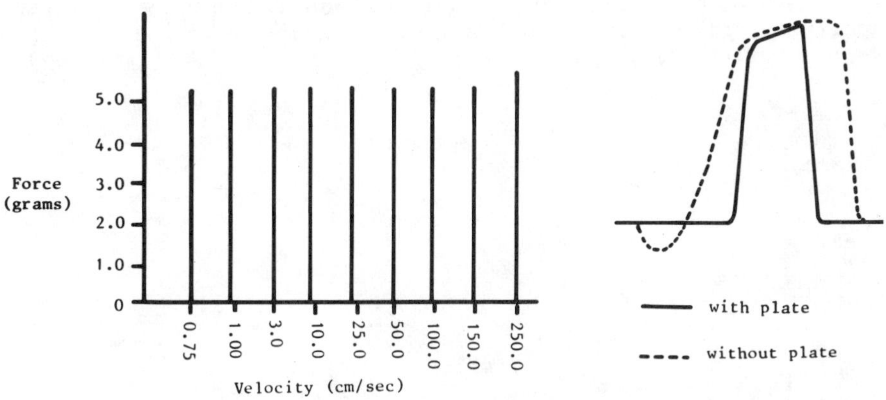

Fig. 4. Left: force exerted by the moving brush as monitored by a Sensotec transducer modified to accept a plate that, in turn, was contacted by the moving brush. The force (measured in grams) monitored in this way was virtually independent of stimulus velocity.
Right: effects of the aperture plate on the force profiles produced by the moving brush stimulus. The inner trace shows that the plate not only restricts the extent of the skin subjected to the stimulus but also leads to a more rapid rise and fall in the force profile.

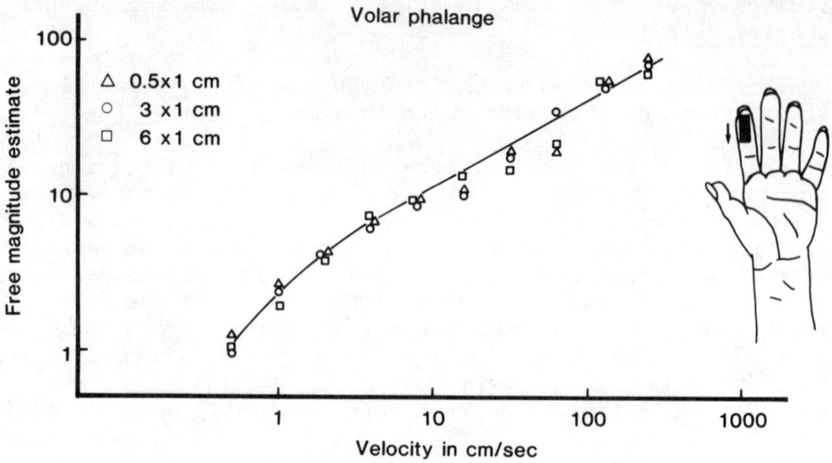

Fig. 5. Perceived velocity of brush stimuli moving over the volar phalange. The contact surface was defined by different apertures in a plate (0.5 x 1 cm; 3 x 1 cm; 6 x 1 cm) held gently against the skin surface.

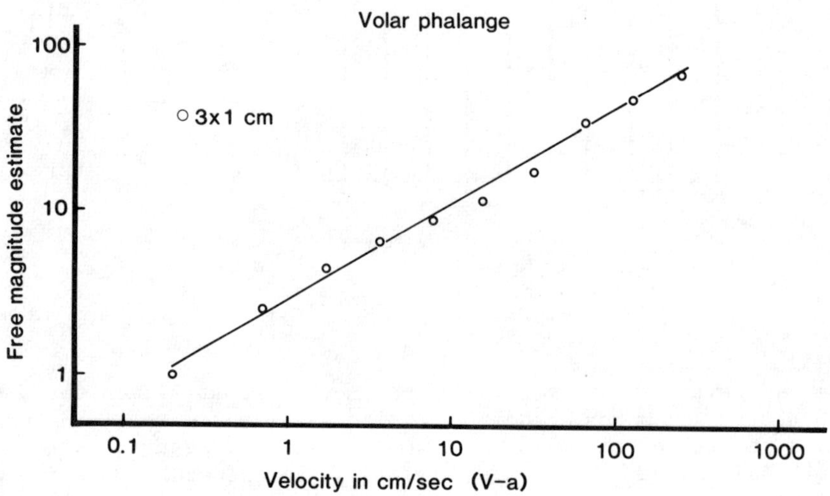

Fig. 6. Same data as in Fig. 5 (aperture (3 x 1 cm) except for a constant subtracted from each velocity. The straight line with a slope of 0.6 was determined by the method of least squares.

A plot of log R against log (V - 0.3) is linear over the full range of velocities (Fig. 6) indicating that a power function describes in a satisfactory fashion the psychophysical relation. The straight line has a slope of 0.6 (the exponent of the power function).

It may be worth noticing that a dip in the function occurs in the range of velocities in which the effect of changes in velocity on perceived traverse length on the skin is relatively minor (unpublished observations). When the brush swept in a distal-to-proximal direction on the dorsal forearm almost identical functions (Fig. 7) were obtained as in Fig. 5.

Neither innervation density nor aperture size seem to influence the over-all shape of the psychophysical velocity functions. We have found, however, that the greater the density of cutaneous innervation the greater the sensitivity to direction of stimulus motion.

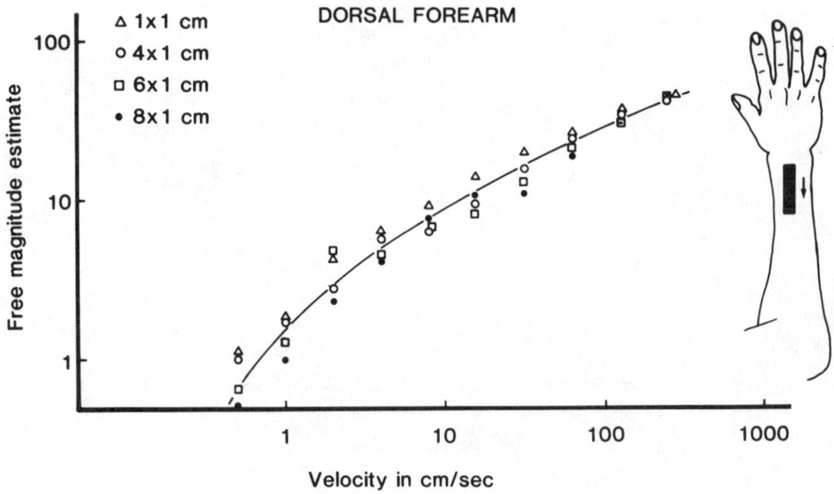

Fig. 7. Subjective velocity of brushing stimuli moving over the dorsal forearm. The size of the different apertures is indicated in the graph.

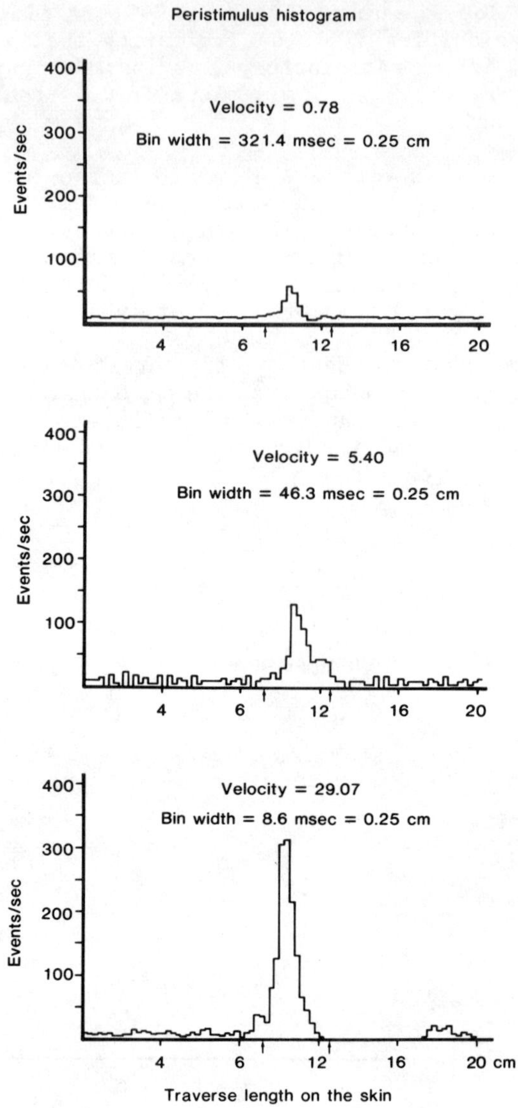

Fig. 8. Response profiles resulting from five stimulus replications of brush movements from proximal to distal across the receptive field of an SA II unit located on the preaxial ventral calf of a monkey. Bin width for each velocity was adjusted to represent the same distance traversed on the skin.

Responses of first order afferents in primate hairy skin to brush stimuli

It is generally agreed upon that the mechanoreceptors of the human skin are very similar to those found in the integument of monkeys (Stopford, 1918; Werner & Mountcastle, 1965). We have identified SA and hair fibers in the medial and lateral branch of the femoral nerve and exposed their receptive fields to the same moving-brush stimuli which we have employed in the psychophysical experiments. The peristimulus histograms for an SA II unit whose receptive field was located on the preaxial ventral calf are displayed in Fig. 8. Note the cessation of spontaneous activity after the traverse of the skin by the brush at a velocity of 29 cm/sec. The bin width for each velocity was always adjusted to correspond to a standard unit of length on the skin.

There are several response measures that could characterize the discharge rate of the units elicited by brush movements. It was consistently observed, however, that mean firing rate produced functions that were monotonically related to the velocity and that peak firing rate tended to produce functions with high variability that sometimes saturated or even declined at extreme velocities. We have chosen a segment of the record where the discharge significantly exceeded the spontaneous firing level, and calculated the average firing frequency within that time interval.

The straight lines in log-log coordinates (Fig. 9) represent power functions with a slope (exponent) of 0.42 for the SA II unit and 0.38 for the hair fiber.

In striking contrast to the velocity response functions obtained using punctuate stimuli (Fig. 1 and Fig. 3), no sign of saturation was observed for transversal stimuli even at 200 cm/sec (Fig. 9). These units do signal changes in motion velocity over the same range as established in the psychophysical studies and, therefore, it appears that either the SA or the RA fibers could satisfactorily account for the capacity of humans to estimate the physical velocity of an object moving parallel to the skin surface. None of the peripheral units exhibited any preference with respect to stimulus direction.

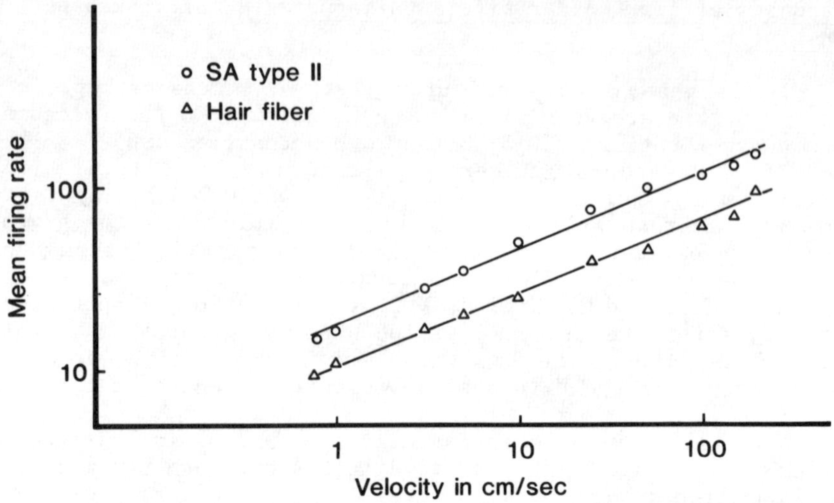

Fig. 9. Mean firing rate as a function of brush velocity for primary afferent fibers in the hairy skin. The slope (exponent) of the power function for the SA II fiber was 0.42 and for the hair fiber 0.38.

DISCUSSION

Several of the psychophysical and neurophysiological observations obtained in this study have implications which go beyond the establishment of quantitative relations existing between a stimulus parameter and single unit activity recorded at peripheral and cortical levels of the somatosensory projection pathways. First, the data indicate that while mechanoreceptors encode a limited range of velocities when vertical displacement stimuli were used, a considerably larger range of velocities is signalled when a tactile stimulus moves across the skin. In general, this dependency of single neuron velocity response on tactile stimulation indicates that the capacity of mechanoreceptive afferents or cerebral cortical neurons to encode velocity is, as the receptive field, a functional concept which cannot be considered in isolation from the stimulus conditions employed to study it (Johansson, 1978; Duncan, Dreyer, McKenna and Whitsel,

1982). The mechanisms underlying the different velocity-response relations observed for the same neuron under different modes of stimulation were not the aim of our experiments. However, they undoubtedly involve different interactions between neural activity set up in each of the terminal branches of a single afferent fiber by the two modes of tactile stimulation we employed in the present study. The close agreement between the velocity response (neural and perceptual) relations exhibited by S I neurons and mechanoreceptive fibers, taken together with the large differences in receptive field size at the two levels, suggests that the pattern of convergence between cutaneous periphery and S I cortical neurons is highly structured.

The second observation having important functional implications is that stimulus traverse length, direction and locus did not exert any noticeable influence on a subject's estimate of the velocity of surface-parallel moving stimuli. These findings imply that the percept of stimulus velocity, unlike that of stimulus direction, is relatively uninfluenced by factors which fluctuate unpredictably over a wide range of values when a stimulus object is tactually scanned.

In future experiments we plan to evaluate in greater detail the capacity of individual mechanoreceptive fibers and cortical neurons to discriminate between tactile moving stimuli differing only in velocity. These studies could also potentially prove valuable for the assessment of disorders affecting the somatosensory nervous system.

ACKNOWLEDGEMENTS

This study was supported by a grant from the Swedish Council for Research in the Humanities and Social Sciences (OF) and by NIH grant NS 10865 (BLW).

REFERENCES

Adrian, E.D. and Y. Zotterman (1926). The impulses produced by sensory nerve endings. Part 3. Impulses set up by Touch and Pressure. J. Physiol., 61, 465-483.

Duncan, G.H., D.A. Dreyer, T.M. McKenna and B.L. Whitsel (1982). Dose- and time-dependent effects of ketamine on S I neurons with cutaneous receptive fields. J. Neurophysiol. 47, 677-699.

Ekman, G. (1958). Two generalized ratio scaling methods. J. Psychol., 45, 287-295.

Ekman, G. (1961). A simple method for fitting psychophysical power functions. J. Psychol., 51, 343-350.

Franzén, O. and U. Lindblom (1976). Coding of velocity of skin indentation in man and monkey. A perceptual-neurophysiological correlation. In Sensory Functions of the Skin (ed. Y. Zotterman). Pergamon Press, Oxford and New York, pp. 55-65.

Johansson, R.S. (1978). Tactile sensibility in the human hand: receptive field characteristics of mechanoreceptive units in the glabrous skin. J. Physiol., 281, 101-123.

Johansson, R.S. (1979). Tactile afferent units with small and well demarcated receptive fields in the glabrous skin area of the human hand. In Sensory Functions of the Skin of Humans (ed. D.R. Kenshalo). Plenum Publishing Corporation, pp. 129-145.

Kalikow, D.N. (1967). Psychofit. Unpublished computer program (Fortran) for the analysis of magnitude estimates. Brown University, RI, USA.

Stevens, S.S. (1956). The direct estimation of sensory magnitudes - loudness. Am. J. Psychol., 69, 1-25.

Stevens, S.S. (1957). On the psychophysical law. Psychol. Rev., 64, 153-215.

Stopford, J.S.B. (1918). The variation in distribution of the cutaneous nerves of the hand and digits. J. Anat. (Lond.), 53, 14-25.

Vallbo, Å.B. and Hagbarth, K.-E. (1968). Activity from skin mechanoreceptors recorded percutaneously in awake human subjects. Expl. Neurol., 21, 270-289.

Werner, G. and V.B. Mountcastle (1965). Neural activity in mechanoreceptive cutaneous afferents: Stimulus-response relations, Weber functions, and information transmission. J. Neurophysiol. 28, 359-397.

Zwislocki, J.J., W.B. Adams & R.B. Barlow (1969). Intensity characteristics of sensory receptors. Paper presented at the Third International Biophysics Congress of the International Union for Pure and Applied Biophysics, Cambridge, Mass., August 1969.

VELOCITY OF INDENTATION AS A VARIABLE IN TACTILE SENSATION

DAN R. KENSHALO, Sr., *JOEL D. GREENSPAN, and ROSS HENDERSON

Department of Psychology and the Psychobiology Research Center, Florida State University, Tallahassee, FL 32306, USA
**Department of Physiology, University of North Carolina, School of Medicine, Chapel Hill, NC 27514, USA*

Most investigators of tactile sensitivity have assumed that one of the static aspects of the tactile stimulating event, pressure (Weber, 1846) or tension (Meissner, 1859; von Frey and Kiesow, (1899) was the relevant metric that best described the relationship between the stimulating event and the resulting sensation. Recently other investigators (e.g., Harrington and Merzenich, 1970; Knibestöl and Vallbo, 1980; Mountcastle and Darian-Smith, 1968) have used skin indentation depth as the appropriate metric to describe tactile stimuli.

Velocity, a dynamic component of tactile stimulation, was implicated as an important part of the stimulating event when Grindley (1936) noted that slower rates of indentation required deeper depths of indentation to produce threshold tactile sensations. Velocity of indentation was also found to be a critical variable in the adaptation of tactile sensations (Nafe and Wagoner, 1941a, b).

The effects of velocity and depth of indentation on the intensity of suprathreshold tactile sensations have been investigated by both Harrington and Merzenich (1970) and Jones (1960). In both, variations in velocity had little, if any, effect on the estimated magnitudes of the tactile sensations. However, the range of velocities and indentation depths used were limited.

We report here measurements of the absolute tactile threshold in terms of indentation depth as a function of the velocity of indentation. These measurements were used to establish the lower limits of 51 suprathreshold combinations of indentation depth and velocity to be used as stimuli for which subjects were asked to estimate the magnitude (intensity) of the resulting tactile sensations. The principal aim of the study is to determine the effect of skin indentation and velocity on the judged intensity of the tactile sensations.

METHODS

Seven paid students, 2 male and 5 female, between 20 and 27 years old served as subjects in the measurement of absolute tactile thresholds and estimating the sensation intensity of suprathreshold stimuli. Stimuli were delivered to the thenar eminence by a Goodman shaker driven by an electrical function generator. Negative feedback from a position sensor between the shaker output and the skin contactor ensured that the mechanical analog conformed to the electrical function. The contactor was 1 cm^2 in area and was surrounded by a static ring with 1 mm clearance between the contactor and the ring.

Absolute tactile thresholds were measured in terms of indentation depth by ramp-and-hold stimuli arranged in 0.1 log steps between 0.006 and 1.2 mm. Thresholds were measured at each of seven ramp velocities (0.1, 0.15, 0.2, 0.3, 0.5, 1.0 and 10 mm/sec).

Magnitude estimates of tactile sensation intensity were made in response to 51 suprathreshold combinations of indentation depth and velocity. Each combination was presented during each of the 8 measurement sessions. No modulus was used.

In both types of measurements, threshold and magnitude estimate, skin indentation was maintained until the subject made a report.

RESULTS

Absolute Thresholds

It is clear from Fig. 1 that velocities of indentation faster than 0.2 mm/sec were without appreciable effect on indentation depth at absolute threshold. A 33-fold change in velocity produced only a 2-fold change in threshold. However, at velocities of 0.2 mm/sec and slower a 2-fold change in velocity produced a 10-fold change in threshold.

Magnitude Estimates of Tactile Sensations

As shown in Fig. 2, variations in the velocity of indentation produced a family of magnitude estimate curves, one for each velocity. The main feature of the functions is that tactile sensations do not grow as single functions of indentation depth, at least at velocities of 0.4 mm/sec and faster. Rather, two exponants appear to be required to account for tactile sensation growth for small indentation depths. At the three fastest velocities, sensations increased in intensity with slopes of from 0.37 to 0.49. At deeper indentation depths tactile sensations grew at much faster rates. Here the slopes of the functions ranged from 1.07 to 1.43.

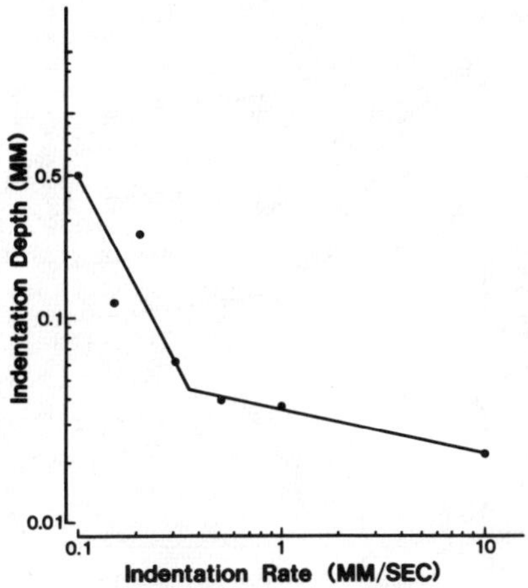

Fig. 1. Absolute thresholds of tactile sensations measured at several ramp velocities of ramp-and-hold stimuli.

Figure 2 contains the information necessary to derive the combinations of velocity and indentation depth that will produce equally intense tactile sensations. The contours of Fig. 3 were generated by taking horizontal cuts at various sensation levels so the plotted points of Fig. 3 represent the intersections of the cuts with the psychophysical functions. The absolute threshold curve of Fig. 1 is reproduced for comparison.

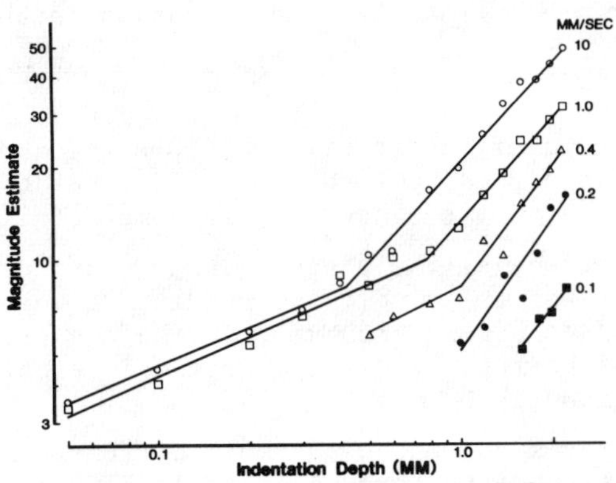

Fig. 2. Growth in intensity of tactile sensations as a function of depth of indentation. The velocity of indentation is the parameter.

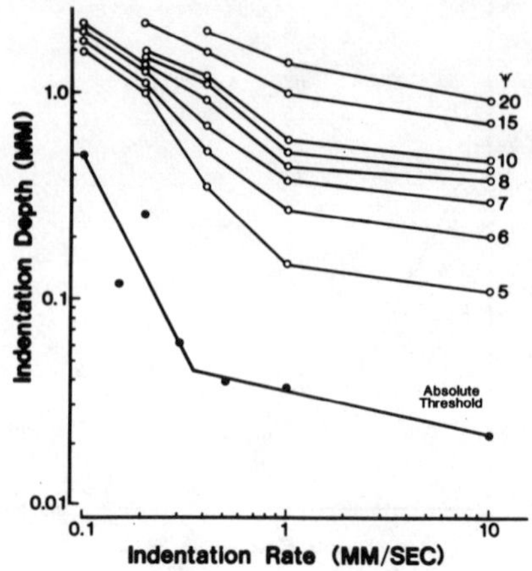

Fig. 3. Equal intensity contours for tactile sensations with velocity and depth of indentation as co-variables. The values on the right (ψ) show the estimated magnitudes at which equal intensities were determined. The text describes the procedure followed to derive this family.

At all suprathreshold intensities the depth required to produce equally intense tactile sensations increased little even though velocity decreased from 10 to 1 mm/sec. Velocities slower than 1 mm/sec had greater interaction with depth as shown by the steeper slopes of the iso-sensation intensity contours. Furthermore, the effect of velocity on depth appears somewhat greater at low than at high sensation intensities.

DISCUSSION

The general form of the tactile absolute threshold curve is similar to those reported by Lindblom (1974) and Johansson and Vallbo (1979). They differ, however, in some details.

Lindblom reported, as the present study shows as well, that indentation depth was relatively independent of velocity down to 0.2 to 0.3 mm/sec below which velocity became critical. Johansson and Vallbo report that the critical velocity point on their curve was about 1.0 mm/sec.

While the three studies showed small differences in the absolute indentation depth thresholds these differences agreed with regional differences in sensitivity of the hand at which the measurements were made (Johansson and Vallbo, 1979).

Magnitude Estimates. The effect of velocity was largely absent at the absolute threshold, but its effect was much more pronounced at suprathreshold intensities of stimulation as shown in Fig. 2.

Harrington and Merzenich (1970) and Jones (1960) failed to find any appreciable or systematic effect of velocity of indentation on the estimated intensity of suprathreshold stimuli. Harrington and Merzenich used indentation depths of .064 to .832 mm presented at velocities of 3.4, 9.6 and 33 mm/sec. They do not specify but it appears likely that depth was varied within sessions and rate across sessions, a procedure not conducive to finding an effect of velocity on tactile sensation intensities. Also most of the indentation depths used by them come within the range where the effect of velocity on sensation intensity was minimal in the present study. The slope of their function (velocities combined) was 0.54 compared to slopes between 0.38 to 0.49 for the slow growth segments of the functions shown in Fig. 2 of this study.

Jones (1960) asked his subjects to make magnitude estimates of a single indentation depth presented at velocities ranging from 0.27 to 120 mm/sec. Details were omitted but if an indentation depth of 1.2 mm were used, a function constructed from Fig.2 would have a slope of about 0.3 compared to the 0.11 slope he reported.

Receptor Model

A descriptive receptor model is suggested by the results of this study. It is based on the assumption that tactile sensations are functions of the total impulse density in the primary afferents and that the two segment psychometric functions result from engagement of two different mechanoreceptor types.

At absolute threshold, indentation velocity was largely without effect down to about a 0.3 mm/sec velocity. At slower velocities, threshold increased rapidly. This suggests a receptor with response characteristics like the rapidly adapting (RA) mechanoreceptors found in volar hand skin (Knibestöl, 1973). Many of these RA units respond during indentation with only a few (2 to 3) impulses. The number of impulses does not change with changes in indentation velocity. Only the interspike intervals shorten as indentation velocity is increased. Others respond (1 impulse) at slow indentation velocities (about 0.4 mm/sec). As indentation velocity is increased the number of impulses increase to reach a maximum (10 to 12 impulses) at an intermediate indentation velocity of 7 to 10 mm/sec, then decrease at faster (up to 72 mm/sec) indentation velocities. RA units exhibit a critical velocity (0.4 to 39.3 mm/sec, Knibestöl's sample) below which they will not discharge unless indentation depth is increased. The sharp inflection point on the absolute threshold curve near the 0.3 mm/sec velocity (Fig. 1) may represent the lower limit of critical velocity of the RA mechanoreceptors engaged by the tactile stimuli.

Each magnitude estimate function (velocities 0.4, 1.0 and 10 mm/sec) required regression lines of two different slopes to describe the growth of tactile sensation intensity. This suggests that two different receptor populations were engaged. During the slow segment of the functions (from 0.05 to about 0.8 mm indentation depth at 1.0 and 10 mm/sec velocities sensation magnitude was little effected by the 10-fold change in indentation velocity. This is shown by their close vertical proximity. Indentation depth appeared to be more important. This suggests engagement of RA receptors. The wide horizontal (due to indentation depth) and vertical (due to indentation rate) separation, of the slow phase of the 0.4 mm/sec velocity function from those of 1 and 10 mm/sec functions, may have occurred because the 0.4 mm/sec velocity function is close to the critical rate of RA receptors (Knibestöl, 1973). Functions for the 0.2 and 0.1 mm/sec velocities did not reveal a slow segment. This may be because these rates are slower than the critical velocities of RA receptors.

During the fast segment, a second receptor type, with response characteristics different from those engaged during the slow phase, may have become engaged. The greater magnitude estimates to increases in both indentation velocity and depth suggest a receptor more sensitive to these stimulus parameters than the RA receptors. A likely candidate is the slowly adapting type I receptor (SAI) in the skin of the volar human hand (Knibestöl, 1975). While SAI receptors have higher thresholds to skin indentation than RA's, SAI's show considerably larger changes in their responses to changes in the velocity and depth of indentation, at least in subhuman mammals (Iggo and Muir, 1979; Pubols et al, 1971; Pubols and Pubols, 1976). It should be noted here that contrary to this hypothesis Knibestöl and Vallbo (1980) were unable to demonstrate close agreement between stimulus-response functions of SAI mechanoreceptors and magnitude estimates of tactile sensation intensity in the human hand.

There are, of course, other candidate receptors in the volar human hand skin--Pacinian corpuscles (PC) and slowly adapting type II (SAII) receptors. It is not possible, at this time, to absolutely exclude the potential contribution of PC units to the psychophysical observations presented here. However, several considerations lessen that potential. First, the ramp-and-hold stimuli were electronically filtered to eliminate frequencies above 60 Hz. Second, 50 percent of PC units have critical slopes (velocity of indentation required to produce one impulse on 50 percent of the stimulations) greater than 5 mm/sec. On the other hand, only 5 percent of RA units have critical velocities greater than 5 mm/sec (Johansson and Vallbo, 1979). Third, in observations where simultaneous measurements of neural activity and sensations were made there was good agreement between the presence of an impulse from a RA unit and a tactile sensation

(Vallbo and Johansson, 1976). In several instances where impulses were identified as originating from PC units, the subject failed to detect any sensation whatsoever (Johansson and Vallbo, 1979). These considerations lead us to conclude that the RA units are almost totally responsible for the absolute tactile threshold, likely mediate the slow segment of the psychophysical function, and also may make a contribution to the fast segment.

As with PC receptors, the data are insufficient to exclude with certainty the contribution of SAII receptors to the psychophysical functions of Fig. 2. Like the SAI, SAII receptors give responses conditioned by both indentation depth and velocity in cat (Chambers et al, 1972); those found in the human hand also have their responses conditioned by indentation depth. However, when a single SAII fiber was electrically stimulated in awake humans, no sensation was ever elicited, even when the electrical pulse rate exceeded 100 Hz (Törebjork and Ochoa, 1980). When individual SAI units were similarly stimulated, subjects reported "sensations of light pressure". Similar stimulations of RA units elicited varying qualities of sensation, i.e., touch, tickle, and vibration (Vallbo, 1981).

ACKNOWLEDGEMENTS

This paper is a condensed version of manuscript submitted to Somatosensory Research (in press) this year.

REFERENCES

Chambers, M.R., Andres, K.H., von Düring, M., and Iggo, A. (1972). The structure and function of the slowly adapting type II mechanoreceptor in hairy skin. Quart. J. Exp. Physiol., 57, 417-445.

Frey, M. von and Kiesow, F. (1899). Über die Function der Tastkorperchen. Z. Psychol. Physiol. Sin., 20, 126-163.

Grindley, G.C. (1936). The variation of sensory thresholds with rate of application of the stimulus. I. The differential threshold for pressure. Brit. J. Psychol. 27, 86-95.

Harrington, T., and Merzenich, M.M. (1970). Neural coding in the sense of touch: Human sensations of skin indentation compare with the responses of slowly adapting mechanoreceptive afferents innervating the hairy skin of monkeys. Exp. Brain Res. 10, 251-264.

Iggo, A., and Muir, A.R. (1969). The structure and function of a slowly adapting touch corpuscle in hairy skin. J. Physiol. (Lond). 200, 763-796.

Johansson, R.S., & Vallbo, A.B. (1979). Detection of tactile stimuli. Thresholds of afferent units related to psychophysical thresholds in the human hand. J. Physiol. (Lond). 29, 405-422.

Jones, F.N. (1960). Subjective intensity functions in somesthesis. In Symposium on Cutaneous Sensitivity (ed. G.R. Hawkes). U.S. Army Med. Res. Lab. Ft. Knox, KY.

Knibestöl, M. (1973). Stimulus-response functions of rapidly adapting mechanoreceptors in the human glabrous skin area. J. Physiol.(Lond.) 232, 427-452.

Knibestöl, M. (1975). Stimulus-response functions of slowly adapting mechanoreceptors in the human glabrous skin area. J. Physiol.(Lond.) 245, 63-80.

Knibestöl, M., and Vallbo, Å.B. (1980). Intensity of sensation related to activity of slowly adapting mechanoreceptive units in the human hand. J. Physiol. (Lond.) 300, 251-268.

Lindblom, U. (1974). Touch perception threshold in human glabrous skin in terms of displacement amplitude on stimulation with single mechanical pulses. Brain Res. 82, 205-210.

Meissner, G. (1859). Untersuchungen über den Tastsinn. Z. Nat. Med. Grundlagenforschun. 7, 92-119.

Mountcastle, V.B., and Darian-Smith, I. (1968). Neural mechanisms in somesthesia. In Medical Physiology (ed. V.B. Mountcastle) C.B. Mosby, St. Louis, Mo.

Nafe, J.P., and Wagoner, K.S. (1941a). The nature of sensory adaptation. J. Gen.Psychol. 25, 295-321.

Nafe, J.P., and Wagoner, K.S. (1941b). The nature of pressure adaptation. 25, 323-351.

Pubols, B.H., Jr. and Pubols, L.M. (1976). Coding of mechanical stimulus velocity and indentation depth by squirrel monkey and raccoon glabrous skin mechanoreceptors. J. Neurophysiol. 39, 773-787.

Pubols, L.M., Pubols, B.H., Jr., and Munger, B.L. (1971). Functional properties of mechanoreceptors in glabrous skin of the raccoon's forepaw. Exp. Neurol., 31, 165-182.

Törebjork, H.E., and Ochoa, J. (1980). Specific sensations evoked by activity in single identified sensory units in man. Acta Physiol. Scand. 110, 445-447.

Vallbo, Å.B., (1981). Sensations evoked from the glabrous skin of the human hand by electrical stimulation of unitary mechanosensitive afferents. Brain Res., 215, 359-363.

Vallbo, Å.B., and Johansson, R. (1976). Skin mechanoreceptors in the human hand: Neuronal and psychophysical thresholds: An inference of some population properties. In Sensory Functions of the Skin of Primates. (ed. Y.Zotterman). Pergamon, New York.

Weber, E.H. (1846). Der Tastsinn und das Gemeingefühl. Wagner's Handwörterbuch der Physiologie. 3, 481-588.

SPATIAL AND NONSPATIAL NEURAL MECHANISMS UNDERLYING TACTILE SPATIAL DISCRIMINATION

KENNETH O. JOHNSON and JOHN R. PHILLIPS

Bard Laboratories of Neurophysiology, Department of Neuroscience, The Johns Hopkins School of Medicine, Baltimore, MD 21205, USA

This paper is concerned with a series of psychophysical and neurophysiological investigations aimed at understanding tactile spatial discrimination and its underlying neural mechanisms. Two aspects of those studies are discussed here. They are (i) the differentiation of surface feature discrimination into two categories, one based on spatial neural coding mechanisms and one based on nonspatial mechanisms, and (ii) the identity of the afferent population(s) that carry the critical spatial information. Because our objective is the study of parallel processing within the somatosensory system, emphasis is placed on identifying and studying those aspects of feature discrimination that are based on spatial (parallel) patterning in the afferent discharge. For this reason, particular attention is given to the interpretation of spatial discrimination as measured in psychophysical experiments and to the criteria that must be satisfied by the neuronal populations that carry the spatially distributed information. The populations considered are the Pacinian (PC), slowly adapting (SA), and quickly adapting (QA) cutaneous mechanoreceptive afferents.

The experimental approach taken in these studies is based on the diagram in Figure 1. The spatial parameters of a set of tactile stimuli are varied in a psychophysical study of discrimination behavior. Then, exactly the same stimuli are used in neurophysiological studies in order to reconstruct the neural representations of the set of stimuli. The aim is to infer the bases on which discrimination between neural representations, and hence stimuli, is made. However, care must be taken when comparing psychophysical performance with some measure of neural activity. The fact that a subject easily discriminates two surfaces that differ only in their spatial topography does not imply that central mechanisms have used only spatial structure within the afferent discharge as a basis for discrimination. As illustrated in Figure 1, changes in the spatial structure of the stimulus may also be represented as changes in the intensive, temporal or modal structure of the afferent response. For example, when a subject palpates two surfaces he may recognize them as different because one surface evokes a higher total impulse rate (intensive), or it evokes impulse patterns in single afferents that are more regular (temporal), or it preferentially activates the Pacinian population (modal).

In order to clarify the following discussion, several terms and phrases are defined in the footnote below.

Human Capacity for Spatial Discrimination

Because of its relative simplicity, the problem of identifying the neural mechanisms underlying surface feature discrimination for stationary stimuli is addressed first. When spatially different surfaces are applied to the skin without lateral movement there are relatively few neural bases for discrimination. Differences in spatial structure between surfaces may be reflected as differences in the spatial patterning of the afferent representations or in some intensive facet of the responses (e.g. total impulse rate or total number of active fibers) but not in the temporal or modal dimensions of the neural representations. When the stimulus is stationary the rapidly adapting afferents (PC's and QA's) are silent and the slowly adapting afferents provide responses whose mean impulse rates are related to the local tissue deformation. Thus the modal composition of the afferent discharge is unrelated to the spatial structure of the stimulus; the lack of rapidly adapting input only signals that the stimulus is stationary. In the temporal dimension, there is no information in the SA responses beyond the measure of mean rate, which is independent of the exact timing of individual impulses. Thus, when stationary stimuli are used the problem is reduced to ensuring that discrimination is not based on intensive cues.

There are two approaches to this problem, each having advantages and disadvantages. First, one can use a classical experimental

---

Surface feature discrimination - discrimination based on structural differences between surfaces.
Spatial discrimination - Surface feature discrimination based on preservation of the surface geometry in the spatial patterning of the afferent discharge.
Nonspatial discrimination - Discrimination based on nonspatial (intensive, temporal or modal) facets of the afferent discharge.
Texture discrimination - surface feature discrimination based on distributed rather than localized structural properties.
Form discrimination - surface feature discrimination based on localized geometric relationships.

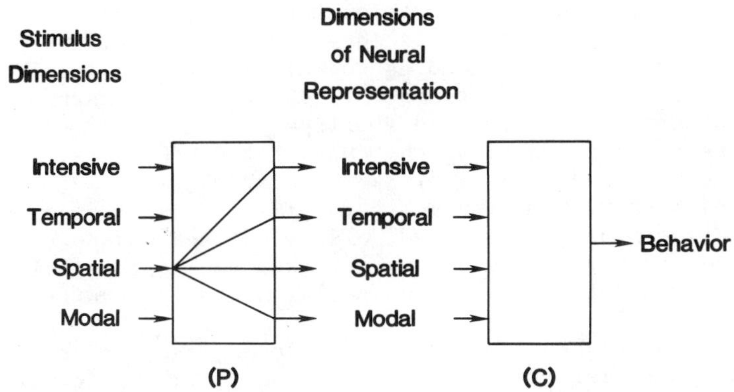

Figure 1. A schematic diagram of a sensory pathway, showing that stimulus dimensions are not necessarily mapped 1:1 into equivalent dimensions in the neural represenation. Surface stimuli differing only in the spatial domain, may evoke neural representations that differ in intensity, temporal structure, spatial structure, and modal composition. In the diagram, P represents transduction and other processes peripheral to the recording level that affect the neural representation and C represents central processes leading to psychophysical behavior. The dimensions refer to the classes of variables that must be included in any thorough specification of the stimulus or its neural representation. For the stimulus, the modal dimension refers to its physical category (mechanical, thermal, etc.). For the neural representation, it refers to the neuronal subpopulations activated by the stimulus (SA, QA, PC, etc.).

design involving discrimination between a small number of stimuli and attempt to minimize the available intensive cues. This has the advantage that well developed theoretical bases for relating the psychophysical and neurophysiological results are available (Johnson, 1980). A disadvantage of such designs is that it is dangerous to infer that subjects have not used subtle intensive cues when making their responses. An alternative is to accept that intensive cues are inevitable and to eliminate their effects on behavior by making the stimuli so complex and multidimensional that a unidimensional intensive cue becomes useless. For example, it seems certain that when subjects achieve high recognition rates in letter or Braille character recognition, their performance is based on spatial neural coding. These complex stimuli have been used in our studies for this reason. The disadvantage of this kind of design is that the results are complex and difficult to interpret. Consequently we have used both approaches.

The results of three psychophysical tasks aimed at measuring the human capacity for stationary spatial discrimination are illustrated in Figure 2. Their consistency and the precautions taken to eliminate nonspatial cues lead us to believe that they all reflect discrimination performance based only on spatial patterning in the afferent discharge (Johnson and Phillips, 1981). The consistency of the results is illustrated by the fact that 0.9 mm gaps, gratings with 0.9 mm gaps and bars (i.e. 1.8 mm grating periods), and letters composed of 0.9 mm gaps and bars (i.e. 4.5 mm letter heights) were all discriminated at a level midway between chance and perfect performance, and therefore 0.9 mm is taken as the threshold for spatial discrimination (based on spatial neural mechanisms).

When movement across the skin is allowed, the number of neural coding possibilities is increased enormously, making it more difficult to eliminate nonspatial cues. The grating, which proved to be an effective stationary stimulus in the psychophysical experiments described above, will not serve to investigate spatial discrimination when scanning is allowed. During scanning, the repetitive mechanical stucture is converted to vibratory cutaneous deformation and the resulting temporal responses (Darian-Smith and Oke, 1981) in all of the afferents provide a multitude of temporal and intensive cues for differentiating the gratings. This was demonstrated by Lamb (1983a&b) using scanned, regular dot arrays in psychophysical and neurophysiological experiments. He measured extremely fine discrimination performance (DL = 20 and 40 microns for arrays with spacings of 1.0 and 2.0 mm respectively) and identified a simple rate code as the basis for this performance. Consequently, we avoided using gratings or other periodic stimuli in the investigation of spatial discrimination with scanned stimuli. In contrast to gratings, letters retain their advantages when employed as moving stimuli in a pattern recognition task, because accurate pattern identification from a large number of alternatives requires accurate spatial representation in the afferent discharge.

In order to assess the effect of scanning on spatial discrimination the letter recognition task (task IV in Figure 2) was repeated, allowing the subjects to scan the stimuli in whatever way seemed most appropriate. The result (Phillips et al., 1983) was a modest 25% improvement in performance in all subjects (i.e., when matched for identical performance levels, the letter sizes were 25% smaller when scanning was allowed). This observation suggests that when scanning is allowed the resolution limit for spatial discrimination (based on spatial neural imagery) drops to about 0.7 mm (75% of 0.9). This modest increase in performance does not account for the major improvement in surface feature discrimination commonly experienced for scanned touch. Later in this paper we cite evidence that implies that it is the nonspatial component of surface feature discrimination that is markedly affected by scanning.

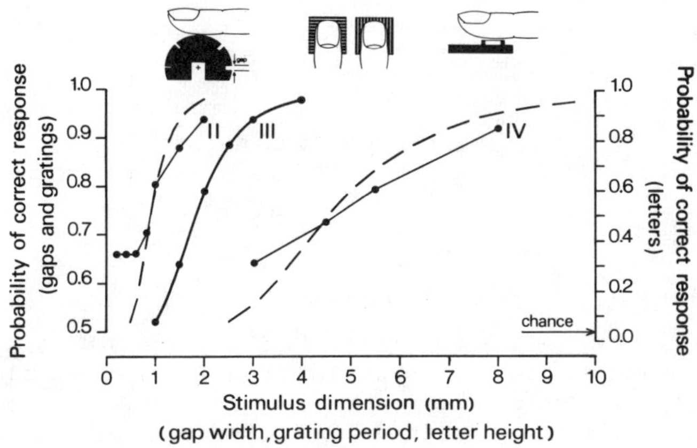

Figure 2. Human performance in gap detection (II), grating orientation (III) and letter recognition (IV) tasks. The left and right ordinates have been scaled to match chance behavior and perfect discrimination. Black dots and solid lines represent observed performance. Adapted from Johnson and Phillips (1981).

Which Afferent Populations Convey the Spatial Information?

Factors that limit a population's capacity to transmit information based on spatial patterning, when the neural imagery is stationary, are (i) the afferent sampling density, (ii) the spatial response properties of single afferents, and (iii) variability in their response properties. Two more factors, (iv) their temporal response properties and (v) the conduction delay dispersion between fibers, apply only when the spatial imagery is time-varying. Whichever afferent population conveys the information underlying the observed psychophysical behavior, it must meet certain minimum requirements with respect to these factors.

Sampling density. Estimates of the innervation densities of SA, QA, and PC fibers in the human (Johansson and Vallbo, 1979) and monkey (Darian-Smith and Kenins, 1980) finger pad favor QA afferents as the critical population setting the limit of spatial resolution. For both man and monkey the estimated limits for QA innervation density at the fingertip is approximately 150 afferents/sq.cm which corresponds to an average spacing of 0.8 mm. The SA estimates for man and monkey are less similar, being approximately 70 afferents/sq.cm for the human (average spacing 1.2 mm) and 130 afferents/sq.cm for the monkey (average spacing 0.9 mm). Although there is no well developed theoretical framework for analyzing the relationships between spatial parameters of the

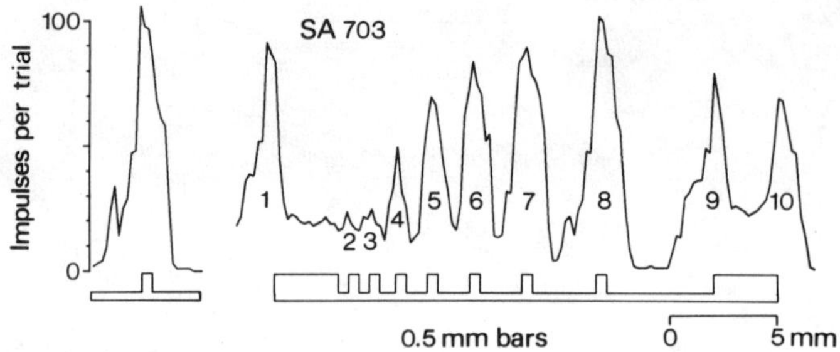

Figure 3. SA spatial response profile. The ordinate represents the number of impulses obtained in a 1.0 sec period (trial) following application of the grating illustrated at the bottom of the figure. The grating, with 0.5 mm bars and gaps of varying widths, depressed the skin by 1000 microns during each trial and was moved laterally by 200 mu between trials.

neural activity and psychophysical behavior (and therefore to establish an absolute requirement for innervation density) the estimated innervation densities for SA's in humans appear too low to account for observed psychophysical behavior. The Shannon sampling theorem in two dimensions (Rosenfeld and Kak, 1976) in conjunction with the threshold for spatial discrimination (0.9 mm, see above) suggests that the afferent population conveying the relevant information has a mean afferent spacing of 0.9 mm or less. The PC afferent population (approximately 10 afferents/sq.cm) fails this criterion and most of the remaining criteria and will not be considered further in relation to spatial mechanisms.

Spatial response properties. If the QA or SA afferents are presumed to account for human behavior then they must resolve the gaps, gratings and letters that were resolved by humans in the psychophysical experiments. In the following discussion it must be borne in mind that the neurophysiological data are derived from monkeys and that comparison with human psychophysics involves the usual cross species assumptions. Neurophysiological experiments, using the same gratings and stimulus conditions as in the psychophysical experiments, showed that only the SA afferents provided information that could account for the psychophysical results (Phillips and Johnson, 1981a). Using indentation without lateral movement, as in the psychophysical experiments, the QA afferents provided no information regarding the structure of the gratings for gaps and bars smaller than 3.0 mm. The SA afferents resolved 0.5 mm gaps and bars, the finest that affected human behavior in the psychophysical experiments. The spatial resolution exhibited by a typical SA afferent is shown in Figure 3 for an aperiodic grating.

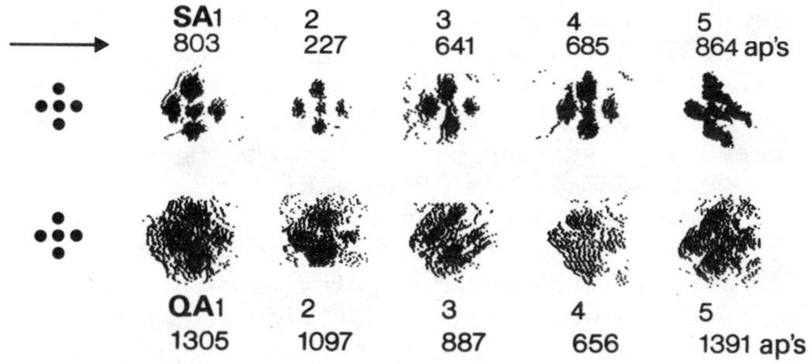

Figure 4. SA and QA responses to Braille-like dot patterns scanned across the skin. The large dots represent Braille dots embossed on plastic (1.2 mm diameter, 2.0 mm center-to-center spacing). Each cluster of small dots represents the pattern of action potentials evoked from a single afferent on many successive scans. The braille dot patterns were shifted by 125 microns on each scan. The number below each fiber designation is the number of action potentials in the cluster, which illustrates that there is no simple relationship between resolution and sensitivity.

The data of Figure 4 are drawn from a study (Johnson and Lamb, 1981) aimed at assessing the spatial response properties of the afferents for scanned stimuli. Braille dots and Braille-like dot patterns were scanned across the receptive fields of SA and QA afferents. As can be seen, the SA responses represent Braille dot patterns much more effectively than do the QA's. These results were derived from scanning trials with a relatively low scanning rate (4 cm/sec). At 16 cm/sec, near the upper limits for Braille reading, the spatial resolution of the SA's was affected less than the QA's, which failed entirely to resolve the dot patterns.

These data appear to implicate the SA afferents as the critical population in spatial discrimination near the resolution limits. The SA's also appear specialized for conveying spatial information in other ways. Their responses exhibit marked edge sensitivity, surround suppression (see Figure 3), and complex receptive field properties that result in increased spatial resolution when the stimulus structure is complex (Phillips and Johnson, 1981a&b).

Variability. There are two kinds of variability that may affect the spatiotemporal information transmitted by a single population: variability of single afferent responses and variability in response properties between afferents. It is evident from examination of repetitive presentations of spatiotemporal stimuli within and between fibers of the same class (e.g. Johnson and Lamb, 1981,

Phillips and Johnson, 1981a) that the variability within fibers is very small while the variability between fibers is large; for example, the responses of different SA afferents to identical Braille-like dot patterns vary almost 4:1 in mean rate (cf. SA1 and SA2 in Figure 4). The comparable QA variability appears to be somewhat smaller, the mean rates varying by 2:1 for the same stimuli. The significance of this second form of variability is difficult to assess as there is no well developed theory for analyzing information relayed via spatiotemporal patterning in neuronal populations. The question is whether this inhomogeneity has a significant effect on the quality of the neural imagery distributed across the population. The available evidence concerning this variability does not provide a basis for inferring that either the SA or QA populations are incapable of accounting for human spatial discrimination.

Temporal response properties. When a stimulus is stationary, the temporal response properties of the afferents are largely irrelevant. However, when a neural image sweeps across a population the single units in the population are subjected to a dynamic stimulus whose upper temporal frequency is the product of the highest spatial frequency and the sweep rate. An estimate of this upper temporal frequency comes from Braille where the rates reach 80 dots/sec across a single receptive field: a rate well within the temporal range of all of the afferents considered as candidates for spatial coding. However, an observation by Ochoa and Torebjork (1983) casts some doubt on the SA temporal resolution. They report that when single SAI afferents (their nomenclature) are excited electrically at rates in excess of 6 impulses/sec, humans fail to perceive the temporal cadence of the stimulus. If this failure is due to poor temporal resolution in the central pathways it would make it unlikely that the SA's are responsible for spatial discrimination during scanning.

Conduction delay dispersion between fibers. There is a significant conduction delay from the finger to the brain (approx. 20 msec at 50 m/sec) which is not important in itself but the variability in conduction velocities between fibers imposes a potential limit on the coherent spatiotemporal information that can be transmitted. Conduction delay dispersion results in disorganization in rapidly changing neural imagery whose effect is readily calculated. Using data from the monkey (Darian-Smith and Kenins, 1980), the conduction dispersions for SA's and QA's are approximately 10 and 16 msec (Johnson and Lamb, 1981). At a scanning rate of 16 cm/sec, a rate near the upper limit for Braille reading, this corresponds to spatial confusions of 1.5 mm in the SA imagery and 2.5 mm in the QA imagery. If the figures for humans are comparable they suggest that the QA properties are less suited for conveying rapidly changing spatiotemporal imagery.

Nonspatial Feature Discrimination

All of the psychophysical and neurophysiological data indicate that the neural mechanisms underlying spatial discrimination fail to resolve surface features finer than about 0.9 mm (i.e., spatial frequencies finer than 0.6 cycles/mm). This implies that the discrimination of surfaces with spatial frequencies wholly finer than 0.6 cycles/mm must be based on nonspatial neural coding mechanisms, i.e. information relayed by virtue of the temporal, intensive, or modal structure of the afferent discharge. Many common surfaces that are readily recognized and easily discriminated (e.g. fine grades of sandpaper, grades of paper, frosted glass, many fabrics) fall in this category. Experiments were conducted to investigate possible nonspatial coding mechanisms using surfaces composed of embossed dots. Nine surfaces with center-to-center dot spacings ranging from 0.5 to 3.2 mm were employed; four had dimensions wholly finer than the threshold for spatial discrimination. These surfaces were applied to the receptive fields of SA, QA, and PC afferents with horizontal scanning movement (Darian-Smith, Davidson and Johnson, 1980) yielding plots of the type illustrated in Figure 4 and neurophysiological data for testing hypotheses concerning the mechanisms responsible for discrimination. Of the many possible nonspatial mechanisms, two are particularly attractive since they provide a potential basis for a wide range of fine textural discriminations. They are temporal mechanisms that use temporal structure in the afferent discharge and modal mechanisms that combine the rate information from the mechanoreceptive submodalities.

In the neurophysiological data derived from the dot patterns, we could find no simple, invariant relationship between the temporal structure of the afferent discharge and the surface, nor could Darian-Smith and Oke (1980) using grating stimuli. However, that does not rule out temporal codes entirely since they may involve integration between neurons or mechanisms that are more complex than those we examined. A basis for an alternative mechanism is illustrated in Figure 5. The relative engagement of the three submodalities provides a consistent basis for differentiating the surfaces with spacings finer than 2.0 mm, since the relative response rates were largely unaffected by changes in contact force and scanning velocity. The coarser surfaces, not discriminated by this mechanism, have dimensions above the spatial threshold and were resolved by SA afferents, suggesting a dual mechanism for the discrimination of these surfaces.

In addition to its capacity for differentiating these dot patterns, the relative engagement mechanism represents an attractive hypothesis because it is a proven, powerful mechanism in other sensory systems (Richards, 1979) and also because it accounts for the multidimensional nature of fine texture perception.

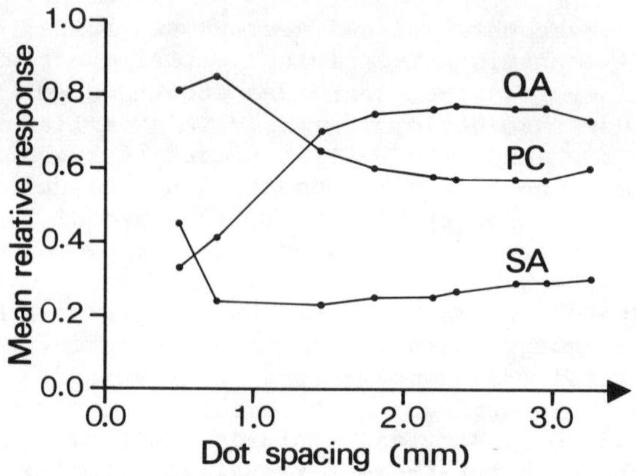

Figure 5. Relative magnitudes of average response rates in SA, QA and PC populations responding to regular dot arrays scanned across the skin surface. Changes in scanning velocity and contact force had a major effect on absolute but not relative rates between fiber classes. Relative rates (averaged across all scanning velocities and contact forces) are illustrated here. (Adapted from Johnson, 1983)

Form and Texture Discrimination.

Earlier in the paper, surface feature discrimintion was classified two ways - one in terms of the surface characteristics being discriminated (texture and form discrimination) and the other in terms of the underlying neural mechanisms (spatial and nonspatial discrimination). The classifications are similar but not entirely equivalent. Spatial neural mechanisms provide the basis for form discrimination ranging from complex pattern discrimination (e.g. Braille) to the appreciation of shape in the form of edge and curvature relationships. Similarly, nonspatial mechanisms provide the basis for texture discrimination for the wide range of surfaces whose distributed geometry is characterized by spatial frequencies wholly finer than 0.6 cycles/mm (e.g. medium to fine sandpaper, frosted glass, grades of paper, most fabrics). However, there are enough exceptions to the rule to make it clear that there is no strict relationship between spatial and form discrimination on the one hand and nonspatial and texture discrimination on the other. For example, gaps of 0.2, 0.4 and 0.6 mm are detected at a level well above chance, as illustrated in Figure 2, but this performance is more easily explained on the basis of an intensive cue resulting from the SA's sensitivity to edges than to any spatial neural code. A more common example of the use of nonspatial neural mechanisms for extremely fine form discrimination is the use of a fingernail or other sharp object to detect a scratch or localized irregularity in a surface. Similarly, texture discrimination may depend on spatial neural mechanisms. The dot patterns discussed above provide one example. An example of a common surface that probably requires both spatial and nonspatial discrimination is corduroy, the appreciation of macroscopic periodicity in the surface structure of corduroy is probably an essential part of its differentiation from other fabrics with similar microscopic structures.

Summary

In conclusion, we believe that there are two forms of surface feature discrimination based on distinctly different neural mechanisms. The first form is based on neural mechanisms that preserve the spatial detail of the stimulus in the spatial patterning of the neural response. This form is engaged during tactual pattern recognition (e.g. Braille and letter recognition). The second form is based on neural mechanisms that encode spatial details of the stimulus in other (nonspatial) dimensions of the neural representation. The properties of the spatial and nonspatial forms of surface discrimination are different in almost every respect. The nonspatial mechanisms encode microscopic dimensions of a surface, depend on movement between the skin and the surface, and probably depend on information relayed by the rapidly adapting afferents (QA's and/or PC's). The spatial mechanisms preserve the macroscopic surface structure, they do not depend on relative movement between the skin and the surface, and they appear to rely on information relayed by the slowly adapting afferents. While some tactual perception may rely exclusively on spatial or nonspatial mechanisms (e.g. letter recognition on the spatial mechanisms and fine texture perception on the nonspatial mechanisms) most surface pattern perception is probably based on information from both mechansisms.

References

Darian-Smith, I., Davidson, I., and Johnson, K.O. (1980). Peripheral neural representation of spatial dimensions of a textured surface moving across the monkey's finger pad. J. Physiol. (London), 309, 135-146.

Darian-Smith, I. and Kenins, P. (1980). Innervation density of mechanoreceptive fibers supplying glabrous skin of the monkey's index finger. J. Physiol. (London), 309, 147-155.

Darian-Smith, I. and Oke, L.A. (1980). Peripheral neural representation of the spatial frequency of a grating moving across the monkey's finger pad. J. Physiol. (London), 309, 117-133.

Johansson, R.S. and Vallbo, A.B. (1979). Tactile sensibility in the human hand: relative and absolute densities of four types of mechanoreceptive units in glabrous skin. J. Physiol. (London), 286, 283-300.

Johansson, R.S. and Vallbo, A.B. (1983). Tactile sensory coding in the glabrous skin of the human hand. Trends Neurosci., 6, 27-32.

Johnson, K.O. (1980). Sensory discrimination: decision process. J. Neurophysiol., 43, 1771-1792.

Johnson, K.O. (1983). Neural mechanisms of tactual form and texture discrimination. Fed. Proc., 42, 2542-2547.

Johnson, K.O. and Lamb, G.D. (1981). Neural mechanisms of spatial tactile discrimination: Neural patterns evoked by Braille-like dot patterns in the monkey. J. Physiol. (London), 310, 117-144.

Johnson, K.O. and Phillips, J.R. (1981). Tactile spatial resolution: I. Two-point discrimination, gap detection, grating resolution, and letter recognition. J. Neurophysiol., 46, 1177-1191.

Krueger, L.E. (1970). David Katz's Der Aufbau der Tastwelt (The World of Touch): a synopsis. Percept. Psychophys., 7, 337-341.

Lamb, G.D. (1983a). Tactile discrimination of textured surfaces: Psychophysical performance measurements in humans. J. Physiol. (London), 338, 551-565.

Lamb, G.D. (1983b). Tactile discrimination of textured surfaces: Peripheral neural coding in the monkey. J. Physiol. (London), 338, 567-587.

Lederman, S.J. (1982). The perception of texture by touch. In Tactual Perception: A Sourcebook. (eds. W. Schiff and E. Foulke). Cambridge Univ. Press, Cambridge.

Ochoa J. and Torebjork E. (1983) Sensations evoked by intraneural microstimulation of single mechanoreceptor units innervating the human hand. J. Physiol. (London), in press.

Phillips, J.R. and Johnson, K.O. (1981a). Tactile spatial resolution: II. Neural representation of bars, edges, and gratings in monkey afferents. J. Neurophysiol., 46, 1192-1203.

Phillips, J.R. and Johnson, K.O. (1981b). Tactile spatial resolution: III. A continuum mechanics model of skin predicting mechanoreceptor responses to bars, edges, and gratings. J. Neurophysiol., 46, 1204-1225.

Phillips, J.R., Johnson, K.O., and Browne, H. (1983). The equivalence of visual and tactile letter recognition. Percept. Psychophys. in press.

Richards, W. (1979) Quantifying sensory channels: generalizing colorimetry to orientation and texture, touch, and tones. Sensory Processes 3: 207-229.

Rosenfeld, A. and Kak, A.C. (1976). Digital Picture Processing. Academic Press, New York.

INFLUENCES OF CUTANEOUS SENSORY INPUT ON THE MOTOR COORDINATION DURING PRECISION MANIPULATION

R.S. JOHANSSON and G. WESTLING

Department of Physiology, University of Umeå, S-901 87 Umeå, Sweden

Despite that tactile skin sensation has been regarded as essential for refined motor acts since long ago (e.g. Mott & Sherrington, 1895), little is known about the functional role of tactile afferent input in the control of hand movements and posture. In various lesions of sensory nerves supplying the hand, patients often exhibit motor defects although the innervation of the muscles may be intact. Indeed, rather than emphasizing the loss of sensibility, such patients make complaints about their motor deficiences (e.g. Moberg, 1962). Typically, they show difficulties with gripping and holding objects and clumsiness during fine manipulatory movements. Tasks involving the precision grip (Napier, 1956) between the tips of fingers and thumb appear to be particularly affected. These difficulties indicate an impaired balance, or coordination, between the grip forces and the load forces within the grip. Although the precision grip has frequently been examined with regard to a variety of motor control problems (e.g. Lawrence & Kuypers, 1968; Long et al., 1970; Brinkman & Kuypers 1973; Smith et al., 1975; Roland, 1978; Passingham et al., 1978; Lemon, 1981; Smith & Bourbonnais, 1981), there have been no quantitative studies concerning this force coordination. Neither has the possible role of the tactile input been considered in this context.

In order to elucidate these problems we designed a motor task which required a multidimensioinal force output, but still simple enough to be amenable to quantitative analysis. Our approach was to analyse the coordination between the grip force and the vertical lifting force, i.e. the load force, while subjects, naive to the purpose of the experiments, picked up small objects from a table, using the index finger and thumb. In this task the balance between the two forces may be critical. Too weak a grip force, the object will slip. The minimal grip force required to avoid slips (GF_s) is determined by the magnitude of the load force (LF), but also by the coefficient of friction (u) between the skin and the object, according to the following eqn.: $GF_s = (1/u) \times (LF/2)$, (the explanation for the denominator 2 is that the load is distributed on two fingers). Too strong a grip force, on the other hand, may cause damage to the object or the hand as well as unnecessary muscle fatigue. Moreover, it may render difficult further manipulation superimposed on the grip. For an optimal performance, all these factors should be taken into account.

The present contribution is mainly concerned with the frictional factor. Three main questions will be considered. The first question is whether the variation in friction between the glabrous skin and objects of different surface structures is so great, that one would suspect an adaptation of the force coordination. If so, does the coordination between the two forces adapt to the different frictional conditions? Finally, does tactile afferent input play a role for such an adaptation?

The lifted <u>object</u> was equipped with strain gauge transducers allowing us to measure continuously the <u>grip force</u> and the <u>load force</u>. Its vertical <u>position</u> was recorded with an ultrasonic position transducer. In addition, to detect <u>vibrations</u> in the object caused by slips, an accelerometer was attached. The touched surfaces were two discs (diameter: 30 mm) symetrically mounted on each side of the object in two parallel planes (30 mm apart). The surface structure could be varied between consecutive lifting trials by exchanging these discs. Likewise, the weight of the object could be altered between trials. The subject could not discriminate visually between the different surface structures and the different weights used. (For further details see Westling & Johansson, 1983.)

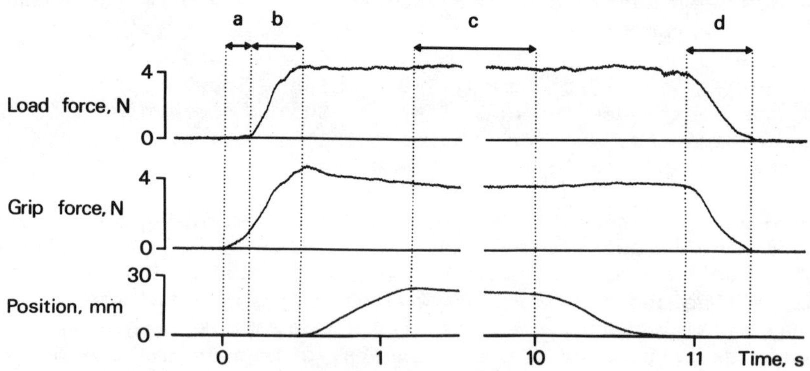

Fig. 1. Load force, grip force and vertical position as a function of time for a lifting trial. (a - d see text.)

The Structure of the Lifting Trials

Figure 1 shows a typical lifting trial, in which the subject was asked to lift the object, hold it in the air for 10 or 15 sec and then replace it. After the index finger and thumb were brought in contact with the object the grip force increased initially (a in Fig. 1), while there were only small load force changes. Thereafter, the load force and the grip force increased in parallel (b in Fig. 1). Soon the load force overcame the force of gravity, and the object started to move. The grip force showed a peak value while the object accelerated, followed by a decay. About 1 s later, a static level was reached (c in Fig. 1). During the replacement, when the object contacts the table, there was a short delay, after which the two forces decreased in parallel (d).

Friction Between the Skin and the Object

To determine the friction between the object and the skin, the minimal grip force required to avoid slips (GF_s), denoted as the slip force, was measured at a given load force, as illustrated in Fig. 2. The inverse of the coefficient of friction was then calculated. The reason for using the inverse of the coefficient was the easily comprehended proportional relationship to the slip force (see eqn. above).

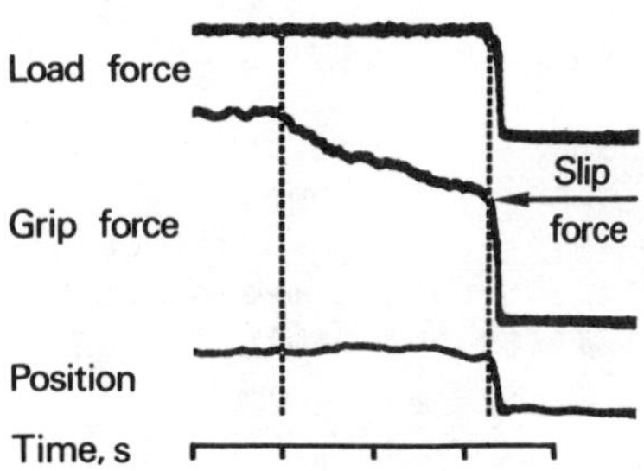

Fig. 2. Determination of the slip force (arrow). Instead of replacing the object in an ordinary fashion, the subject was asked to slowly separate the fingers until the object slipped (right vertical line). Left vertical line indicates the start of this separation.

A number of different surface materials were tested (weight of the object: 400 g). Three of these were selected as representative for the frictional range of common materials: finely textured silk, suede and sandpaper (no. 320). As can bee seen in Fig. 3 A, presenting data from 10 subjects, these materials clearly exhibited different friction in relation to the skin. Silk was most slippery and sandpaper the least slippery material. It is likely that some of the pronounced interindividual variation may be explained on the basis of differences in sweating rate between individuals (see below). Indeed, the two extreme subjects showing the lowest friction were postmenopausal women (subject I and J in Fig. 3 A). To examine whether the friction was influenced by the magnitudes of the forces within the grip (cf. Comaish & Bottoms,

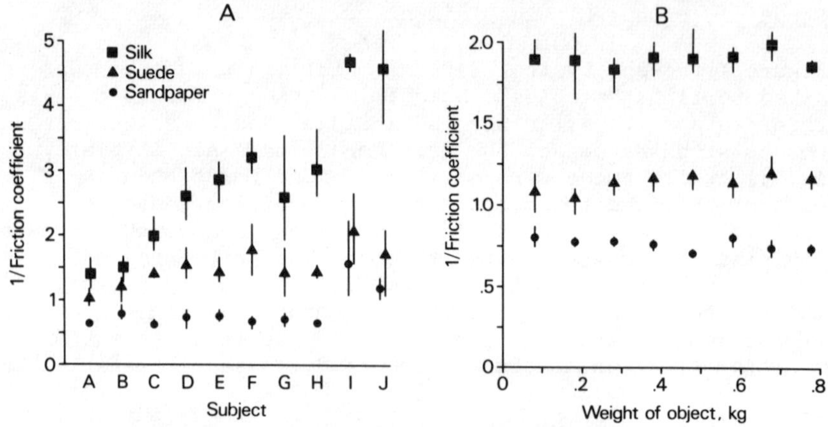

Fig. 3. A: Friction between skin and three surface structures. Ten subjects. B: Relationship between friction and load force expressed in weight of the lifted object (single subject). A and B: Symbols represent median value of 5 trial and bars corresponding ranges.

1971), slip forces were measured at different load forces and the inverse of the coefficient of friction was calculated. It turned out that the friction was fairly constant for each of the three materials, thus obeying the Amonton´s law of friction (Bowden & Tabor 1973) within the load force range studied (Fig. 3 B).

It was concluded that the friction between common surface structures and the skin of the fingers might vary considerably between different surface materials as well as between different subjects in contact with a given material. Hence, a mechanism compensating for this variation while lifting and manipulating objects might be desirable.

Adaptation to the Frictional Condition

Force coordination during static conditions. To find out whether there was an adjustment of the force coordination during the static phase of the lifting trials (c in Fig. 1), the relationship between the static grip forces and the corresponding static load forces were analysed for trials carried out with the three surface structures. This relationship is illustrated in Fig. 4 A (black dots) in which the static load force is expressed as the weight of the object. It can be seen that the relationship between the static grip force and the weight was close to linearity, indicating an adjustment of the grip force to the weight. The slope, however, varied with the surface structure. A comparison with the corresponding slip force data (triangles in

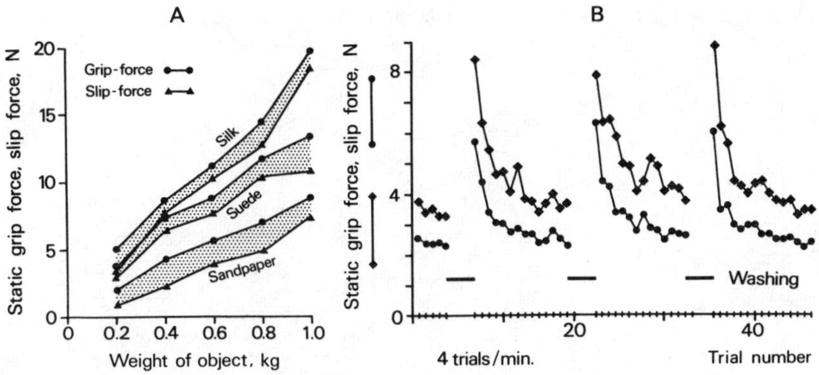

Fig. 4. A: Static grip force and corresponding slip force as a function of the object's weight and surface structure for 15 trials by single subject. Dashed areas illustrate the safety margins to prevent slips. B: Static grip forces (diamonds) and corresponding slip forces (dots) before and after washing and drying the subject's hand. 48 consecutive trials by a single subject. Suede was used in all trials. A and B: Each trial was terminated by slip force measurement, as illustrated in Fig. 2.

Fig. 4 A) revealed that this slope was adjusted to the frictional conditions, i.e. it was about proportional to the inverse of the coefficient of friction. Thus, the coordination between the static grip and load forces was adapted to the friction between the skin and the object. The dashed areas illustrate the relatively small <u>safety margin</u> to prevent slips. This safety margin could vary between subjects. (For further data see Westling & Johansson, 1973).

It may be argued that this adjustment of the force coordination with surface structure could have been made on the basis of the different texture properties of the touched materials rather than the friction <u>per se</u>. To explore this alternative, series of lifting trials were carried out immediately before and after washing (soap and water) and drying the hands of the subjects. During the washing procedure sweat was removed from the skin, i.e. the skin was made less adhesive and the friction in relation to the object was temporarily decreased (cf. slip force curves in Fig. 4 B). By keeping the surface structure (i.e. the texture) and the weight of the object constant in these trials, the effects of pure frictional changes could be studied. As can be seen in Fig. 4 B, the grip force was clearly adapted to these changes. Thus, these results strongly suggested that the static force coordination was adapted to the friction rather than to the texture.

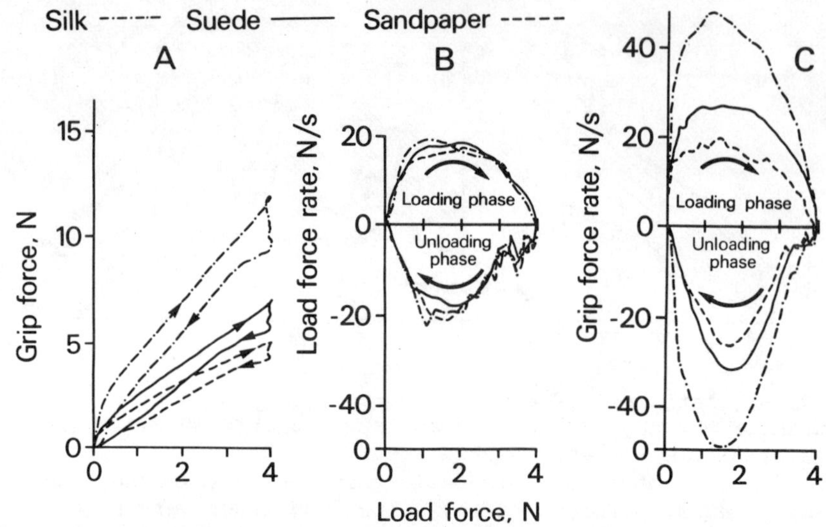

Fig. 5. A: Relationship (coordination) between grip force and load force during trials with silk, suede and sandpaper. Weight of object constant at 400 g. Averages of 120 trials by 9 subjects. The initial period of grip force increase appears as the small, almost vertical, part of the curves close to the origin, whereas the following, approximately linear and slanted, part of the curves show the force coordination during the phase of paralell force increase. The nearly vertical part at the top of the curves was accounted for by the grip force decay following its peak. Finally the phase of parallel force decrease appears as the approximately linear part of the curves intersecting the origin. B and C: grip and load force rates, respectively, as a function of load force and surface structure (same trials as in A). Arrows indicate the course of load force changes.

Force coordination during force changes. To analyse whether the force coordination was adapted to meet different frictional demands also during the dynamic phases, i.e. the phases of parallel grip and load force change (b and d in Fig. 1), the two forces were displayed against each other as shown in Fig. 5 A. The three curves refer to the three different surface structures, respectively. The two approximately linear and slanted parts of each curve refer to the phases of parallel force increase (arrow towards higher forces) and parallel force decrease (arrow towards lower forces). Note that the slopes of the curves were different: the more slippery the structure, the higher the grip force at any given load force. This adjustment with friction was mainly accounted for by differences in the rate of grip force change; the grip force rate was higher the more slippery the surface structure. In contrast, the load force rate appeared to be independent of the surface structure. These findings are illustrated in Figs. 5 B and C, which show the rates of the two forces, respectively, as a function of the load force level. Since there was a parallel force increase until the object started to move, the variation in grip force rate with friction determined the grip force magnitude and thus the force balance during the static phase. Changes in the weight of the object principally influenced the duration of the phases of parallel force change, but not the balance between the forces, i.e. the heavier the object the longer the periods of parallel force change.

Dependence on signals in cutaneous afferents. It would seem reasonable that signals from receptors in the skin of the fingers would play a role in the adaptation of the force coordination to different frictional conditions. To test this hypothesis, trials with different surface structures were carried out by subjects in whome the tips of the index finger and thumb were anesthetized by local intradermal infiltration of Marcain. This anesthesia prohibited the cutaneous afferent signals from reaching the C.N.S., but left the motor and proprioceptive innervation of the hand intact. It turned out that the grip and load forces still changed in parallel but the appropriate adaptation to the frictional conditions was lost. This implied that slips now occurred when slippery surface structures were used, and that the grip force could be unnecessarily high using less slippery materials.

Adjustment to Frictional Changes and Tactile Afferent Input

To elucidate when and how the adjustment of the force coordination to the frictional demands took place, this coordination was analysed in lifting trial series during which the surface structure was changed between consecutive trials. It was found that a new structure influenced the rate of grip force increase already about 0.1 sec after the object was initially gripped, i.e. approximately at the moment the grip and load forces began to increase in parallel. This indicated that afferent information related to the friction must have excerted their influences already during the initial period of grip force increase (a in Fig. 1). However, for some trials, particulary those carried out with the most slippery material, this initial adjustment seemed not to be sufficient since a further, secondary adjustment of the coordination occurred later. Such adjustments took place in response to small short-lasting slips, revealed as vibrations of the object recorded by the acceleration transducer. The latencies between the onset of the vibratory events and the appearance of the adjustments were 0.06 - 0.08 sec. These adjustments, which were most frequently observed during the phase of parallel force increase, took place as changes in the force rates which resulted in higher grip/load force ratios that were maintained throughout the trials, i.e. the safety margin increased. Sometimes late adjustments were also observed during the static phase, particulary for subjects with very small safety margins. In these cases the coordination changes appeared as a grip force increase to a new, higher stable value.

The slips eliciting the described coordination changes were generally very small and often they were not even noticed by the subjects. Most likely they were limited to only one of the two skin areas in contact with the object. If occurring during the static phase they rarely gave rise to much position change of the object. Nevertheless, they were large enough to elicit a short lasting burst of action potentials in tactile units with high dynamic sensitivity, i.e. the FA I (RA), FA II (PC) and SA I units (for refs. see Johansson & Vallbo, 1983). The responses of SA II units, on the other hand, appeared to be more related to the three dimensional force distribution within the grip. These observations were made on the basis of single unit nerve recordings obtained from the median nerve using the micro-neurographic technique of Vallbo and Hagbarth (1968). It was concluded that signals in tactile afferents elicited by small slips most likely <u>triggered</u> the observed coordination shifts.

Of particular interest is that the resultant new force coordinations following slips in each instance were maintained throughout the lifting trial, suggesting that the relationship between the two forces was controlled on the basis of a <u>memory trace</u>. The updating of this trace was most likely accounted for by tactile information entering <u>intermittently</u> at inappropriate force coordination, as during slips. Further evidence for a memory influ encing the force coordination is the fact that the frictional condition in the previous trial could excert certain influences on the force coordination remaining throughout the current trial (Westling & Johansson, 1983). Moreover, anesthesia of the fingers appeared not to principally alter the motor behaviour, except for the lack of the adjustment to friction. This indicated that, once set, the coordination could be maintained without requiring cutaneous afferent information. A memory, setting the coordination, would, if adequately updated, allow the C.N.S. to <u>simultaneously</u> change the grip and load forces in a manner appropriate for the current friction. Thus, during force changes there would be no time lag between the two forces disturbing their balance. Such a lag would occur if, for instance, the grip force was regulated solely on the basis of a continuous feed-back loop involving tactile receptors. The proposed memory based design would be of particular value during quick handling of fragile objects, e.g. while picking berries.

There are several studies indicating that mechanoreceptors in the fingers provide an important afferent source for control of finger movements. The prevalent view seems to be that function of the input from skin and joint afferents is to provide a general tonic facilitatory effect on motor commands accounting for certain movements (e.g. Rood, 1860; Marsden et al., 1977; McCloskey & Gandevia, 1978; Torebjörk et al., 1978; Garnett & Stephens 1981). It has also been demonstrated that modest electrical or mechanical stimulation of fingers may elicit multiphasic reflex modulation of ongoing EMG in hand muscles (Caccia et al., 1973; Garnett & Stephens, 1980). In the first human dorsal interosseus muscle, i.e. one of the muscles which significantly contributed to the grip force in the present experiments, the late and most pronounced exitatory component appears at latencies comparable to those observed between the slips and the evoked grip force increases (cf. Garnett & Stephens, 1980). This reflex component is considerd to be of supraspinal origin, involving transmission of impulses through the dorsal columns and the corticospinal tract (Jenner & Stephens, 1982). The participation of supraspinal mechanisms would agree with the abundant evidence that the motor cortex and the pyramidal tract are of fundamental importance for the performence of fine finger and hand movements, particulary those involved in the the precision grip (Lawrence & Kuypers, 1968; Brinkman & Kuypers, 1973; Phillips & Porter, 1977; Passingham et al., 1978; Evarts, 1980). Likewise, transmission through dorsal columns appear to be a prerequsite for an adequate opposition of the thumb and forefinger as when picking up small objects (Vierck, 1978). On the basis of these considerations it seems resaonable to assume that the controlling processes underlying the coordinated motor output as considered in the present study was, at least, partly dependent on supraspinal neuronal mechanisms.

There are several situations described when light mechanical stimuli of the animal´s skin trigger intense motor reactions. One classical example is the extensor thrust (Sherrington, 1947). More recently, a variety of such reactions related to locomotion have been described (see Forssberg, 1979). One finding of particular interest is their adaptive nature, i.e. the responses triggered by a given type of stimuli may differ considerably, dependent on the phase of movement or postural situation at which they are elicited. As to the hand, there is considerable evidence that voluntary movements may depend on basic reactions triggered by peripheral tactile stimuli. Such reactions appear to be most apparent in goal directed manipulative movements. For instance, on the basis of studies on clinical and experimental material, Denny-Brown (1966) could distinguish between the "grasp reflex" and the more complex "instinctive grasp reaction" as well as tactile avoiding reactions. The pathological feature of these reactions was considered to be an inability to adequately supress the first phase, i.e. they appeared to be inappropriately triggered. The neuronal mechanisms accounting for the slip-triggered coordination shifts in the present study as well as the type of reactions described by Denny-Brown may be explained in terms of the "two-stage model of adaptive motor control" recently proposed by Houk (1978) (see also Houk & Rymers, 1981). According to this model, a particular pattern of afferent information triggers the release of a particular set of preprogrammed motor commands in a discontinuous fashion. The stimulus-response relations, determined by a supraspinal "stimulus-response processor" operating in a quasilogical manner, are dependent on the task or the postural goal of the subject, i.e. on a supervisory process. The minimal latencies at which this adaptive mechanism would influence the motor output during kinaesthetic triggering of "reaction-time" or "intended" arm movements are similar to the latencies between the slips and the elicited coordination changes (cf. Crago et al., 1976; Evarts & Vaughn, 1978). The latter latencies (0.06 - 0.08 s) clearly indicate that the controlling processes were too fast to involve direct voluntary control. To avoid interferences with the overall goal of the manipulative tasks, it seems reasonable that these processes should proceed without requiring much conscious attention, i.e. as automatic subroutines.

Conclusion

The friction between a gripped object and the skin might vary considerably with the surface structure of the object and for any given surface material between subjects. Likewise, the amount of sweat on the skin surface seemed to heavily influence the friction.

While lifting objects, the coordination between the grip and load forces was adapted to the actual frictional condition: the more slippery the surface structure, the stronger the grip force at any given load force. The control of this coordination may be described in terms of two automatically operating functionally different mechanisms. First, there was a parallel change of grip and load forces, which may be regarded as generated by a motor program executed after the object was initially gripped. Second, the balance between the two forces was adapted, via a memory, to the frictional demands; i.e. changes of load forces were accompanied by changes of the grip forces providing a relatively small safety margin to prevent slips. This adaptation was most likely dependent on tactile afferent input entering intermittently at inappropriate grip forces, such as occurs during slips. Thus, impulses in tactile afferent units apperaed to automatically influence the motor coordination during this goal directed manipulative task. These principles may operate in most types of prehensile movements, so that actively generated load forces will always be accompanied by adequate grip forces.

This work was supported by grants from the Swedish Medical Research Council, the Gunvor och Josef Aner´s Stiftelse and the University of Umeå, which are gratefully acknowledged.

References

Bowden, F.P. and Tabor, D. (1973). Friction - An Introduction to Tribology, Garden City, N.Y.: Anchor Press.

Brinkman, J. and Kuypers, H.G.J.M. (1973). Cerebral control of contralateral and ipsilateral arm, hand and finger movements in the split-brain Rhesus monkey. Brain, 96, 653-674.

Caccia, M.R., McComas, A.J., Upton, A.R.M. and Blogg, T. (1973) Cutaneous reflexes in small muscles of the hand. J. Neurol. Neurosurg. Psychiat., 36, 960-977.

Crago, P.E., Houk, J.C. and Hasan, Z. (1976). Regulatory actions of human stretch reflex. J. Neurophysiol., 39, 925-935.

Comaish, S. and Bottoms, E. (1971). The skin and friction: deviations from Amonton´s law, and the effects of hydration and lubrication. Br.J. Derm., 84, 37-43.

Denny-Brown, D. (1966). The cerebral control of movement. Liverpool: The Liverpool Univ. Press.

Evarts, E.V. and Vaughn, W.J. (1978). Intended arm movements in response to externally produced arm displacements in man. In Cerebral Motor Control in Man: Long loop Mechanisms, ed. J. E. Desmedt, pp. 178-192, Basel:Karger.

Evarts, E.V. (1980). Brain mechanisms in voluntary movement. In Neural Mechanisms in Behavior, ed. D. McFadden, pp. 223-259. New York:Springer.

Forssberg, H. (1979). Stumbling corrective reaction: A phase-dependent compensatoey reaction during locomotion. J. Neurophysiol., 42, 936-953.

Garnett, R. and Stephens, J.A. (1980). The reflex responses of single motor units in human first dorsal interosseus muscle following cutaneous afferent stimulation. J. Physiol. (Lond.), 303, 351-364.

Garnett, R. and Stephens, J.A. (1981). Changes in the recruitment threshold of motor units produced by cutaneous stimulation in man. J. Physiol. (Lond.), 311, 463-473.

Houk, J.C. (1978). Participation of reflex mechanisms and reaction time processes in the compensatory adjustments to mechanical disturbances. In Cerebral Motor Control in Man: Long Loop Mechanisms, ed. J. E. Desmedt, pp. 193-215, Basel: Karger.

Houk, J.C. and Rymers, W.Z. (1981). Neural control of muscle lenght and tension. In Handbook of Physiology, The Nervous System, Vol. 2, ed. V.B. Brooks, pp. 257-323, Bethesda, Maryland: Am. Physiol. Soc.

Jenner, J.R. and Stephens, J.A. (1982). Cutaneous reflex responses and their central nervous pathways studied in man. J. Physiol. (Lond.), 333, 405-419.

Johansson, R.S. and Vallbo, Å.B. (1983) Tactile sensory coding in the glabrous skin of the human hand. Trends in Neuroscience, 6, 27-31.

Lawrence, D.G. and Kuypers, H.G.J.M. (1968). The functional organization of the motor system in the monkey. I. The effects of bilateral pyramidal lesions. Brain, 91, 1-14.

Lemon, R.N. (1981). Functional properties of monkey motor cortex neurones receiving afferent input from the hand and fingers. J. Physiol. (Lond.), 311, 497-311.

Long, C.H., Conrad, P.W., Hall, E.A. and Furler, S.L. (1970). Intrinsic and extrinsic muscle control of the hand in power grip and precision handling. J. Bone Jt. Surg., 52A, 853-867.

Marsden, C.D., Merton, P.A. and Morton, H.B. (1977). The sensory mechanism of servoaction in human muscle. J. Physiol. (Lond.), 265, 521-535.

McCloskey, D.I. and Gandevia, S.G. (1978). Role of skin, joints and muscles and of corollary discharges, in human discrimination tasks. In Active Touch, ed. Gordon, G., pp 177-187. Oxford: Pergamon.

Moberg, E. (1962). Criticism and study of methods for examining sensibility in the hand. Neurology, 12, 8-19.

Mott, F.W and Sherrington, C.S. (1895). Experiments upon the influence of sensory nerves upon movement and nutrition of the limbs. Proc. R. Soc., B., 57, 481-488.

Napier, J.R. (1956). The prehensile movements of the human hand. J. Bone Jt. Surg., 38B, 902-913.

Phillips, C. G. and Porter, R. (1977). Corticospinal Neurones. London: Academic Press.

Passingham. R., Perry, H. and Wilkinsson, F. (1978). Failure to develop a precision grip in monkeys with unilateral neocortical lesions made in infancy. Brain Res., 145, 410-415.

Roland, P.E. (1978). Sensory feedback to the cerebral cortex during voluntary movement in man. Behav. Brain Sci. 1, 129-171.

Rood, O.N. (1860). On contraction of the muscles induced by contact with bodies in vibration. Am. J. Sci. Arts, 24, 449

Sherrington, C.S. (1947). The integrative action of the nervous system (2nd ed.). New Haven CT: Yale Univ. Press.

Smith, A.M. and Bourbonnais, D. (1981). Neuronal activity in cerebellar cortex related to control of prehensile force. J. Neurophysiol., 45, 286-303.

Smith, A.M., Hepp-Reymond, M.C. and Wyss, U.R. (1975). Relation of activity in precentral cortical neurons to force and rate of force change during isometric contractions of finger muscles. Exp. Brain Res., 23, 315-332.

Torebjörk, H.E., Hagbarth, K.-E. and Eklund, G. (1978). Tonic finger flexion reflex induced by vibratory activation of digital mechanoreceptors. In Active Touch, ed Gordon, G., pp. 197-203. Oxford: Pergamon.

Vallbo, Å.B. and Hagbarth, K.-E. (1968). Activity from skin mechanoreceptors recorded percutaneously in avake human subjects. Expl. Neurol., 21, 270-289.

Vierck, C.J. (1978). Interpretations of the sensory and motor consequences of dorsal column lesions. In Active Touch, ed. Gordong, G., pp. 139-159, Oxford: Pergamon.

Westling, G., and Johansson, R.S. (1983). Factors influencing the force control during precision grip. Exp. Brain Res., In press.

CUTANEOUS THERMORECEPTORS

A. IGGO

Department of Veterinary Physiology, University of Edinburgh, UK

The discovery by Yngve Zotterman (1935, 1936) that afferent fibres in the lingual nerve respond to changes in temperature of the surface of the tongue, was an important landmark in the study of somatosensory mechanisms. It is a pleasure on this occasion to recall the excitement with which Yngve would recount the circumstances of the discovery, since he was actually investigating the peripheral mechanisms of taste at the time, and probably found the action potentials evoked by his sapid and probably cool solutions, but actually generated in thermoreceptors, to be a great inconvenience.

Good use was made of the discovery, however, particularly when Herbert Hensel brought his strict, methodical and quantitative habits of work to Stockholm. Together with Yngve he published a series of classical papers on the properties of lingual thermoreceptors (Hensel & Zotterman, 1951 a,b,c) that firmly established the properties of cold receptors, especially their ability to encode in their discharge frequency the static temperature of the skin. It is therefore a particular disappointment that Herbert Hensel is not, as originally planned, presenting a paper on thermoreceptors at this Symposium.

Given that the thermoreceptors, both cold and warm, are now well-documented in the literature (see Hensel, 1981 for a comprehensive review) it is necessary to state only their general characteristics, which are :

1. Two kinds of cutaneous thermoreceptor exist - the cold receptors that are excited when the skin temperature falls and - the warm receptors that are, conversely, excited by a rise in the skin temperature.

2. Each category of thermoreceptor has over a given narrow range of temperatures, the capacity to maintain a continuous discharge of impulses, the rate of which is dependent on the temperature. The ranges for the cold and warm thermoreceptors overlap in the middle, but peak sensitivites are separated by about 15°C - at 25/28° for cold receptors and 40/41° for warm receptors i.e. they are maximally excited by non-noxious temperatures.

3. The thermoreceptors have a dynamic sensitivity, that is, they are excited during a rise (warm receptors) or a fall (cold receptors) in skin temperature. The rate of firing depends on the rate of change of temperature, and is always higher than the rate at constant temperatures.

4. The thermoreceptors have a high degree of specificity for temperature and cannot readily be excited by non-thermal natural stimuli.

5. The thermoreceptors detect the temperature and rate of change of temperature, and do not detect the direction across them of a thermal gradient.

6. The afferent fibres are either thin myelinated (1 - 3 µm dia) or non-myelinated (<1 µm dia). The size of the fibres depends on both location of the receptors and the species of animal. Trigeminal cold receptors tend to be myelinated as also are primate cold receptors in general. Warm receptors in general, even in primates, have non-myelinated afferent fibres.

The foregoing catalogue indicates that the original discoveries regarding lingual cold receptors have both been confirmed and, indeed, extended to include similar kinds of receptor on the body surface. They serve also to confirm the hypotheses of 'specificity' of cutaneous receptors, a matter close to Yngve Zotterman's heart.

The recently developed technique of intraneural percutaneous recording from single afferent fibres in conscious man has, quite naturally, raised the question whether specific thermoreceptors are present in human skin. These techniques, as reported to this Symposium by Dr. Vallbo, are capable of recording from the large myelinated as well as the non-myelinated fibres in peripheral nerve. It has, therefore, been surprising that, despite the efforts of several laboratories, there should be such a limited amount of published information on human cutaneous thermoreceptors. They certainly exist (Hensel & Boman, 1960) and indeed in some conditions the cold receptors generate the typical bursting discharge of non-human primate cold receptors (R. Johansson, personal communication). One suggestion coming from the studies on monkeys (Iggo, 1969) was

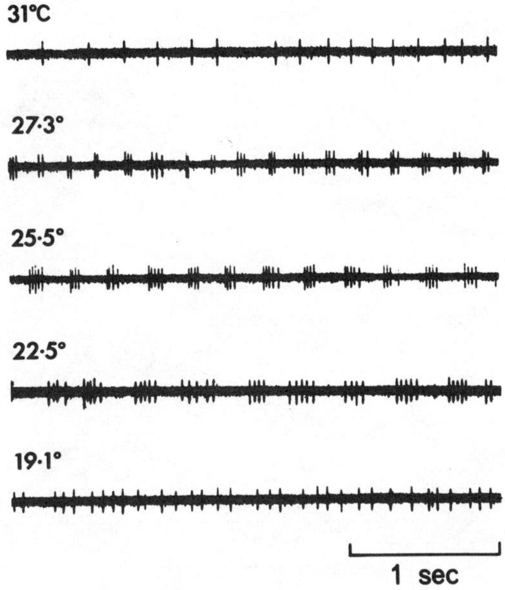

Fig. 1. Discharge of impulses in a cold receptor in the hairy skin of a vervet monkey during continuous cooling (at a skin temperature rate of 0.3°/sec) of the skin. The successive records, which were taken from a continuous record, show the gradual emergence of strongly marked groups of impulses, as the skin was cooled from 42° to 15°C (from Iggo & Iggo, 1971).

that the burst pattern of discharge might encode more securely the temperature of the skin since over a range of about 10°C the skin temperature was inversely proportional to the number of impulses in a burst. Until further work is done this hypothesis cannot be tested in man. Nor can the sensory rôle of the thermoreceptors, although the results of intraneural electrical stimulation of human mechanoreceptor afferent fibres (reported to the Symposium by Dr Vallbo) certainly raise the prospect that they will have a specific sensory function. One is left at this stage with the expectation based on psychophysical studies, that the cold and warm spots in human skin (Blix, 1882) are indeed innervated by specific thermoreceptors.

LABILITY OF BURST DISCHARGE IN COLD RECEPTORS

The cold receptors in monkey skin characteristically discharge, over a range of temperatures from about 32° to 20°, with bursts of impulses. This is illustrated in Fig. 1 which shows the discharge of impulses in a monkey cold fibre recorded while the temperature

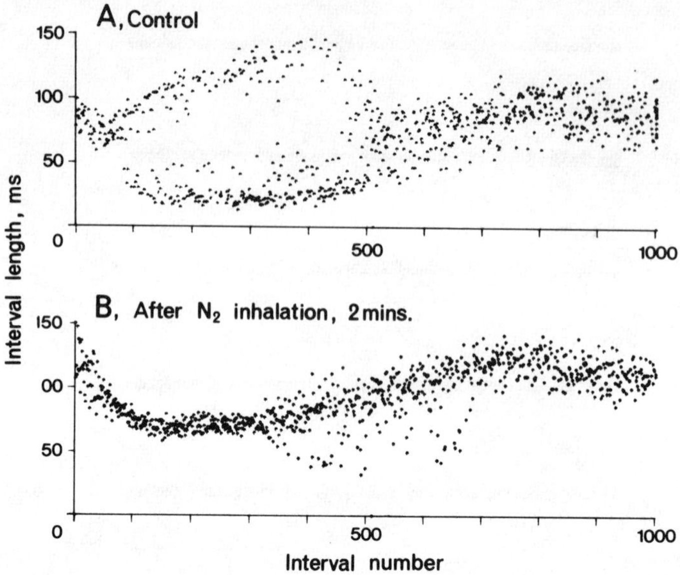

Fig. 2. Computer-derived graphs of the discharge of a monkey cold fibre during ramp cooling of the skin at approximately 0.15°C/sec. A. Normal conditions, with monkey breathing air. B. Two minutes after inhalation with nitrogen was begun. In the normal animal an alternation of long and short intervals began as in Fig. 1, soon after cooling began. This bursting discharge is almost totally absent in the anoxic state (figure by courtesy of A. Iggo and A.S. Paintal).

was falling steadily at 0.3°C/sec. An alternative way of displaying similar information is seen in Fig. 2A, where a computer plot of instantaneous intervals is displayed during a similar thermal ramp. In the early stages of the ramp, when the temperature was falling from 38° to 25° there was an initial decrease in the inter-impulse interval, with regular inter-impulse intervals, as in Fig. 2A, then the appearance of alternate long and short intervals. That is, bursts occur. As the temperature continues to fall the bursts are separated by longer intervals, but the number of impulses in each burst increases. Then after about 500 intervals, when the temperature had fallen to about 25°C, the bursts coalesced and a more or less continuous, but irregular, discharge was present.

The ability of the cold receptors to generate the bursts is heavily dependent on their metabolic state (Hensel, 1953; Iggo & Paintal, 1977). When the oxygen supply to the receptors is decreased there is, within a short time, a change in the properties of the receptors. Three changes are documented 1) the threshold for excitation is raised, so that a larger change of temperature is required before the receptor begins to discharge, 2) the dynamic sensitivity is less, so there is a fall in the total number of impulses discharged during a quantitatively controlled thermal ramp, and 3) the bursts of impulses characteristic of normal cold receptors is absent (Fig. 2B). These changes are not a consequence of an inability of the afferent fibre to conduct impulses, since when electrical stimuli are delivered to the nerve it can still conduct impulses at rates higher than occur in the normal bursts.

A possible involvement of increased sympathetic outflow in the diminished excitability of cold receptors was investigated in two ways in chloralose-anaesthetized rhesus monkeys (Iggo & Paintal, 1977). Since it is known that anoxia causes a massive increase in sympathetic activity, via the adrenals as well as via the peripheral

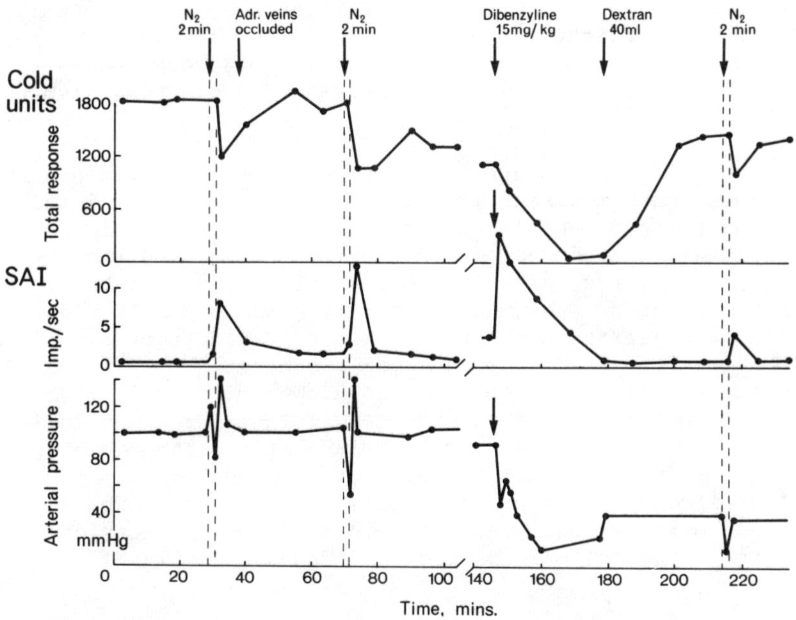

Fig. 3. The effect, on the response of a monkey cutaneous cold units to standardised cold stimuli, of inhalation of nitrogen before and after a) the permanent occlusion of the adrenal veins b) the intravenous injection of the alpha-adrenergic blocking agent, dibenzyline and c) the infusion of dextran. Each filled circle (●) is the total discharge to a cooling ramp. SA I - the discharge in a slowly-adapting type I mechanoreceptor in the same nerve strand. (figure by courtesy of A. Anand, A. Iggo and A.S. Paintal).

sympathetic nerves, two procedures were adopted. First,the adrenal veins were exposed and snares passed around them. After control tests of the effect of anoxia on the discharge of a cold receptor, the snares were tightened to occlude venous drainage from both adrenal glands. As Fig. 3 illustrates this procedure did not prevent a subsequent trial with anoxia from depressing the excitability of the cold receptor. Next an intravenous injection of dibenzyline (15 mg/Kg) an alpha-blocking agent was given, in order to block locally released catecholamine. This caused a precipitate fall in arterial blood pressure, from a mean of 90 mm Hg to less than 20 mm Hg. There was a parallel decrease in response of the cold receptor to the standard cooling test, almost to the point of complete inexcitability. Subsequent i.v. infusion of dextran led to a partial recovery of the arterial pressure, to about 40 mm Hg. This led to a dramatic restoration of the ability of the cold receptor to respond to cooling. The effect of anoxia was also restored. Since dibenzyline causes a persistent block of alpha receptors, it can be concluded that the catecholamines are not involved directly, if at all, in the reduced capacity of cold receptors during ischaemia. Propanalol (1 mg/Kg),a beta-adrenergic, was also without effect.

In separate experiments, using nembutal anaesthetized monkeys, it was observed that direct pressure on the skin with a metal thermode could reduce or abolish the static discharge in cold receptors. This effect was independent of the thermode temperature and was associated with a reduction in blood supply, evidenced by blanching of the skin under the thermode or a flat glass rod. In these deeply nembutal-anaesthetized monkeys the skin of the fore or hindlimbs often became cold, 25°C compared with the normal 33°C of conscious animals or animals anaesthetized with alpha-chloralose. The cold skin, in a laboratory at about the same temperature, was indicative of greatly reduced vascular supply to the skin, although blood flow was not actually measured. The ease with which the normal brisk discharge of cold receptors, especially at 25°C, could be abolished in these animals by light pressure, is a further indication of the importance to the cold receptor of an adequate supply of oxygenated blood.

The conclusion to be drawn from these experiments is that the cold receptors (and presumably also the warm receptors) are heavily dependent on an adequate supply of oxygenated blood and that their full capability is only expressed when this is available. Given this conclusion, it is now relevant to consider the results obtained by Long (1977) on the sensitivity of cold fibres to heat. He found, in pentobarbital-anaesthetized monkeys that the presence of a paradoxical discharge (another of Yngve Zotterman's interests) in cold fibres in glabrous skin of the hand was conditional on body temperature. It was present, or at least much more pronounced, when the rectal temperature was 39.0 to 39.5°C than when it was 37.0 to 37.5°C. Among mechanisms that could account for this striking effect a possible direct catecholamine action was ruled out by the observation that after the alpha-adrenergic blocking agent phenoxybenzamine was administered, the paradoxical response was more conspicuous than before. Acute section of the peripheral nerve supplying the skin containing the cold receptors, also increased the paradoxical response. An elevation of the skin temperature followed both nerve section and phenoxybenzamine administration. The influence of the anaesthetic, pentobarbital, must also be considered. Long states that at the dose levels used, there were only negligible effects on skin temperature, in contrast to the conditions used by Iggo & Paintal (vide supra) so it is unlikely that blood flow was seriously impaired in his basal experimental conditions. Nevertheless, some degree of cutaneous vasoconstriction existed, since both nerve section and alpha-adrenergic blockade led to an increase in skin temperature of about 2°C, an effect attributable to removal of peripheral vasoconstriction. All these results taken together led to the conclusion that the influence of rectal (i.e. deep body) temperature on the paradoxical discharge can be attributed to the increased skin blood flow that would occur in the animals with high core temperatures, although Long himself discounted this possibility.

The results discussed in the foregoing section establish that the transducer mechanisms in cold receptors depend much more heavily on oxidative metabolism than does conduction in the afferent fibre. They provide indirect support for the hypothesis (Pierau et al, 1974) that active ionic pumps underly the discharge of primate cold receptors. Thus the availability of ATP, produced in the receptor by oxidative metabolism, could be a factor determining the level of activity of the electrogenic Na^+/K^+ pump. A lesser degree of activity in the hypoxic or anoxic state could lead to a decrease in the rate of pump turnover, i.e. lower electrogenesis with a consequent depolarisation of the receptor membrane, that in turn would lead to a diminished size of further depolarisation on application of the cooling stimulus, and therefore a net reduction in excitation of the receptor. Such a sequence of events could reduce the amplitude of the postulated oscillations in membrane potential (Iggo & Young, 1975; Pierau et al, 1975; Braun et al, 1980) that are the current model in generation of bursts of impulses. These effects are in contrast to the action of calcium ions, investigated by Hensel and colleagues. Their results (Braun et al, 1980) are consistent with the interpretation that i.v. calcium ions affect the amplitude but not the frequency of the postulated oscillatory waves, in static thermal conditions. Thus after i.v. calcium infusion, the pre-existing bursts of impulses in cat nasal cold fibres, are replaced by single impulses, at the same interval as the previous bursts. A converse effect was reported when EDTA was infused i.v, while recording from a cold fibre that lacked bursts. These now appeared, but the interburst interval corresponded to the original intervals of the steady discharge. The effect of calcium ions can be attributed to indirect effects on the electrogenic Na^+/K^+ pump. Thus an increase in extracellular calcium could lead to a rise in intracellular calcium that would enhance the activity of the pump, leading to a hyperpolarisation of the receptor membrane. This, if the oscillations underlying the bursts, were otherwise unaffected, could lead to a failure of the depolarisation phase to reach threshold for more than one action potential at the spike initiation site. The converse would, on this argument, be expected to hold for the effect of EDTA.

ORIGIN OF BURST DISCHARGE

The hypotheses of Iggo & Young (1975), Pierau _et al_, (1975) and Braun _et al_, (1980) all envisage an oscillating membrane potential in the receptor, that in turn generates the pulsatile bursting discharge of action potentials from the spike initiation site and thence along the afferent fibre. On this basis, and on the assumption that action potentials do not invade the transducer regions in cold receptors, an attempt was made by Anand _et al_, (1982) to separate events at the transducer and the spike initiation sites. The procedure adopted was to fire impulses antidromically from stimulating electrodes interposed between the recording electrodes and the receptor (Iggo, 1958) while recording from a single cold fibre dissected from the median nerve of an anaesthetized rhesus monkey. Conditions were chosen so that the cold unit was firing regularly in bursts as in Fig. 4A. When a burst of antidromic impulses were sent along the fibre, at 100 Hz for 250 msec (i.e. 25 impulses) the normal burst discharge was suppressed for about one second. At least two bursts of impulses would normally have occurred in this time. This result could have arisen because of post-excitatory depression at the spike initiation site, a process consistent with the recorded discharge of only a single impulse, instead of a pair as would normally occur. Further analysis, however, suggests another explanation. When only a small number of antidromic stimuli are inserted (1 to 8) and timed in relation to the bursts so that impulses can invade the receptor and not simply collide with orthodromic impulses, then the result depends on the numbers of both antidromic and orthodromic impulses. If there is a regular train of single impulses (i.e. no bursts) then a single antidromic volley delays the initiation of the next expected orthodromic impulse, and re-sets the train of subsequent orthodromic impulses (Table I). The interval between the arrival of the antidromic impulse at the receptor and the initiation of the next orthodromic impulse is, however, the same as in the normal train. As the number of impulses in the antidromic volley was increased, there was a progressive

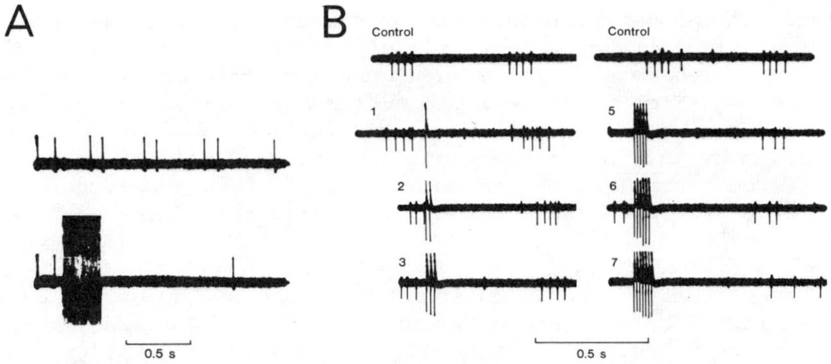

Fig. 4.A. The effect of antidromic invasion of a cold receptor on the 'burst' discharge. The receptor, upper trace, was discharging doublets, and (lower trace) on antidromic invasion from a train of impulses at 100 Hz for 250 msec, there was a prolongation of the interburst interval and the discharge of only a single impulse in the burst. B. Records of the discharge of a cold fibre, normally firing bursts of 4 impulses when trains of antidromic stimuli (1 to 7) were sent down to the receptor. There is both a prolongation of the interburst interval and a reduction and slowing of impulses within the delayed burst, that is dependent on the number of antidromic impulses (figure by courtesy of A. Anand, A. Iggo and A.S. Paintal).

TABLE I

EFFECT OF TRAINS OF ANTIDROMIC STIMULI ON A COLD RECEPTOR WITH A REGULAR INTER SPIKE INTERVAL

Antidromic train	Normal ISI msec	Post-stimulus ISI first	Post-stimulus ISI subsequent (av. of 4)
1	128 ± 11	127	130 ± 13.8
2	123 ± 9.8	157	132 ± 8
3	122 ± 12	176	126 ± 14
5/6	127 ± 7	208	136 ± 14
7/8	127 ± 13	236	142 ± 10
mean ± S.E.	126 ± 0.06 msec.		

(a = antidromic impulse)

increase in the delay of the next orthodromic impulse, and even a small but significant prolongation of the ensuing intervals. The result was even more striking when bursts of orthodromic impulses were present (Fig. 4B). A single antidromic volley, sent in shortly after an orthodromic burst was still capable of delaying the next burst of four orthodromic impulses. With greater numbers of antidromic impulses in the burst there is both a re-setting and an interference with the internal time-relations of the next orthodromic burst.

These results clearly establish that the rhythm of the bursts can be altered by antidromic invasion of the receptor and imply that the postulated oscillation of the transducer is either accessible to the antidromic impulses, or that the grouped discharge arises, in part, from the properties of the spike initiation site. As the number of impulses in a burst increase so, in parallel, does the interburst interval (Iggo, 1969). This then is further evidence that the spike initiation site influences the timing of discharges in the afferent fibre. A possible explanation for these effects is that the discharge of impulses from the spike initiation site actually, by drawing current from the transducer region, influences the postulated rhythmical oscillation in membrane permeability and potential at the transducer site. One test for this hypothesis would be to record the postulated oscillating membrane potential, in the presence and absence of TTX, to test whether the discharge of impulses did indeed have the predicted effect. Such an experiment has yet to be attempted, and indeed, no such postulated localised transducer has yet been recorded.

REFERENCES

Anand, Ashima., Iggo, A. & Paintal, A.S. (1982). Genesis of the 'burst' discharge in primate cutaneous cold receptors. J.Physiol., 327, 66-67P.

Blix, M. (1882). Experimentela bidrag till lösning af frågan om hudnervernas specifika energi. I. Uppsala Läkare Förening Förehandlinger, 18, 87-102.

Braun, H.A., Bade, H. & Hensel, H. (1980). Static and dynamic discharge of bursting cold fibers related to hypothetical receptor mechanisms. Pflüg. Arch. ges. Physiol., 386, 1-9.

Hensel, H. (1953). Das verhalten der Thermoreceptoren bei Ischämie. Pflug. Arch. ges. Physiol., 257, 371-383.

Hensel, H. (1981). Thermoreception and temperature regulation. Monographs of the Physiological Society No. 38. Academic Press, London.

Hensel, H. & Boman, K.K.A. (1960). Afferent impulses in cutaneous sensory nerves in human subjects. J.Neurophysiol. 23, 564-578.

Hensel, H. & Zotterman, Y. (1951a). The response of the cold receptors to constant cooling. Acta physiol. scand., 22, 96-113.

Hensel, H. & Zotterman, Y. (1951b). Quantitative Beziehungen zwischen der Engladung einzelner Kältefasern und der Temperatur. Acta physiol. scand., 23, 291-319.

Hensel, H. & Zotterman, Y. (1951c). Action potentials of cold fibres and intracutaneous temperature gradient. J. Neurophysiol., 14, 377-385.

Iggo, A. (1958). The electrophysiological identification of single nerve fibres, with particular reference to the slowest conducting vagal afferent fibres in the cat. J.Physiol., 142, 110-126.

Iggo, A. (1969). Cutaneous thermoreceptors in primates and sub-primates. J.Physiol., 200, 403-430.

Iggo, A. & Iggo, B.J. (1971). Impulse coding in primate cutaneous thermoreceptors in dynamic thermal conditions. J.Physiol. (Paris), 63, 287-290.

Iggo, A. & Paintal, A.S.(1977). The metabolic dependence of primate cutaneous cold receptors. J.Physiol., 272, 40-41P.

Iggo, A. & Young, D.W. (1975). Cutaneous thermoreceptors and thermal nociceptors. In The Somatosensory System. (ed. H.H. Kornhuber). Thieme, Stuttgart. pp. 5-22.

Long, Randall,R. (1977). Sensitivity of cutaneous cold fibers to noxious heat : paradoxical cold discharge. J.Neurophysiol., 40, 489-502.

Pierau, F-K., Torrey, P. & Carpenter, D.O. (1975). Mammalian cold receptor afferents : role of an electrogenic sodium pump in sensory transduction. Brain Res., 73, 156-160.

Pierau, F-K., Torrey, P. & Carpenter, D. (1975). Effect of Ouabain and potassium-free solution on mammalian thermosensitive afferents in vitro. Pflüg. Arch. ges. Physiol., 359, 349-356.

Pierau, Fr.-K., Ullrich, J. & Wurster, R.D. (1977). Effect of Ca^{++} and EDTA on the bursting pattern of lingual cold receptors in cats. Proc.Int. Union physiol.Sci. 13, 597.

Zotterman, Y. (1935). Action potentials in the glossopharyngeal nerve and in the chorda tympani. Skand. Arch.Physiol. 72, 73-77.

Zotterman, Y. (1936). Specific action potentials in the lingual nerve of the cat. Skand. Arch. Physiol. 75, 105-119.

POSSIBLE REDUNDANCE OF SPINAL PATHWAYS FOR BEHAVIOURAL THERMOSENSITIVITY

ULF NORRSELL

Department of Physiology, University of Göteborg, S-400 33 Göteborg, Sweden

Studies of animal's ability to utilize particular sensory stimuli before and after restricted lesions of the central nervous system employ a traditional method of sensory neurophysiology. The method is not accepted uncritically, however, and a certain disenchantment stems from findings of rapid postoperative recovery, which have indicated a somewhat unexpected, functional redundance of many central nervous structures. Nevertheless, we have used that method for several years at our laboratory for the purpose of locating the spinal, ascending, thermosensory pathways of the cat. The background was formed by numerous reports of thermosensory anaesthesia which had been caused by restricted lesions in the ventral half of the spinal cord of human patients, and indicated a possible lack of redundance in the distal parts of the thermosensory system. The symptoms and the lack of redundance have been attributed to interruption of a unique, thermosensory, spino-thalamic pathway (cf. Norrsell, 1979). Since thermal sensations constitute cutaneous sensory qualities, studies of the thermosensory system for above mentioned reasons might provide a profitable approach for extension of our knowledge about the somatosensory system.

Cats were trained to utilize thermal cues for finding their way to a food reinforcement in an automatic, modified T-maze (Norrsell, 1974). The floors of its two testing alleys were independent thermodes which could be set to different predetermined temperatures above (warm cues) and below (cool cues) the ambient temperature by means of circulating water. The thermodes were made of brass like all of the other floor surfaces inside the maze which were kept together with the entire ambient temperature inside the soundproofed testing cage at approximately 25°C. Barriers and adjustable surface covers were used to restrict the reception of the cues to the paws of one or both body halves. Both forepaws and hindpaws could be used for evaluating the cues, but the forepaws were more important in this respect for most cats. Only one thermode could be touched at a time, and the cats were allowed to move freely back and forth between the thermodes until, but not after making the decision which determined whether food would be served or not. Continuous records were made, not only of the thermode temperatures and the result of each trial, but also each contact with the thermodes throughout the experiments. Such a way it was possible to recognize chance correct decisions, as well as to determine the effort and strategy behind the decisions (cf. Norrsell, 1978).

The cat's behavioural thresholds for reactions to temperature decreases as well as increases was found to be about equal and around 1° for this technique (Finger and Norrsell, 1974). The discriminatory behaviour was found to develop differently, however, for cats who were trained for cool cues (cool cats) compared to that of other cats who were trained for warm cues (warm cats). Subsequent work provided further evidence of differences between the cat's ability to utilize cool and warm cues, and is illustrated in Table 1. The training of the cool and warm cats, whose behaviour went into the table, was made as mirror images for the two groups. Both groups were started with a difference of 20° between the positive and negative cues, which straddled the ambient temperature, but in opposite directions for the two groups. After a cat had satisfied a predetermined criterion (cf. Table 1) the negative cue was reduced in predetermined steps until becoming equal to the ambient temperature (Δ 10°, Δ 4°, Δ 2°, Δ 1°, Δ 0°), and the cat must satisfy the criterion for each step. Consequently, all of the cats were performing a mixed cool/warm discrimination at the beginning, which then was transformed gradually into a pure either cool or warm discrimination, which was reached at the final step. Failures to satisfy the criterion caused introduction of extra training sessions (cf. Table 1), and the speed of the cats' progress through the training sequence consequently could be expressed as multiples of the minimal, criterion number of sessions. The table shows that the two groups advanced through the first 4 steps with mixed cues (1st column) with comparable speeds, although the warm cats possibly were somewhat slower. When they were moved into the final step, however, a marked difference appeared (Table 1, 2nd column). All of the cool cats performed the pure cool discrimination of Δ 10° straight away, while the warm cats required more than 6 times as long to master a pure warm discrimination of equal magnitude. Thus, whereas the cat's behavioural thresholds for cool and warm discriminations were similar, there were marked differences beween its aptitudes to utilize the two types of cues. Warm cues

Table 1. Comparison between two groups of cats submitted to a two-choice task of temperature discrimintion. The average durations ($\overline{x}$) and standard deviations (SD) of learning are shown, and are expressed in multiples of a predetermined, minimal (criterion) number of sessions. The initial tasks of both groups contained positive and negative cues which consisted of 10° deviations from the ambient temperature (25°C) in opposite directions. The positive cue for one group was always 10° below (cool) and for the other group 10° above (warm) the ambient temperature. The training consisted of stepwise (5 steps) diminuition of the negative cues until equal to the ambient temperature after the cats had produced at least 69 % correct responses for 3 consecutive sessions (32 trials/session) of a certain task with purely thermal cues. Failure to reach 69 % for a particular session caused a temporal addition of an auditory, accessory, positive cue for the following session, and consequently also an increase of learning time.

	4 initial steps. (mixed cool/warm discrimination)	Final step. (pure cool or warm discrimination)
Cats with cool positive cues. N = 4	$\overline{x}$ = 1.1 SD: 0.2	$\overline{x}$ = 1.0 SD: 0.0
Cats with warm positive cues N = 6	$\overline{x}$ = 1.3 SD: 0.5	$\overline{x}$ = 6.5* SD: 4.3

* training was interrupted after 11.7 times the minimal number of sessions for one cat who had failed to reach the criterion.

had comparatively less "signal value" in the present context (cf. also Norrsell, 1983).

For people cool and warm stimuli are different, but at the same time part of a conceptual continuum; warmer stimuli at the same time are less cool and vice versa. The same appeared to be true for cats who were submitted to pseudo-reversal experiments, and is illustrated in Fig. 1. The figure shows the results of 5 consecutive sessions for a cat who was allowed to use the paws of one body half at a time, and on alternate days. The cat had always been presented with a warm positive cue which was 10°

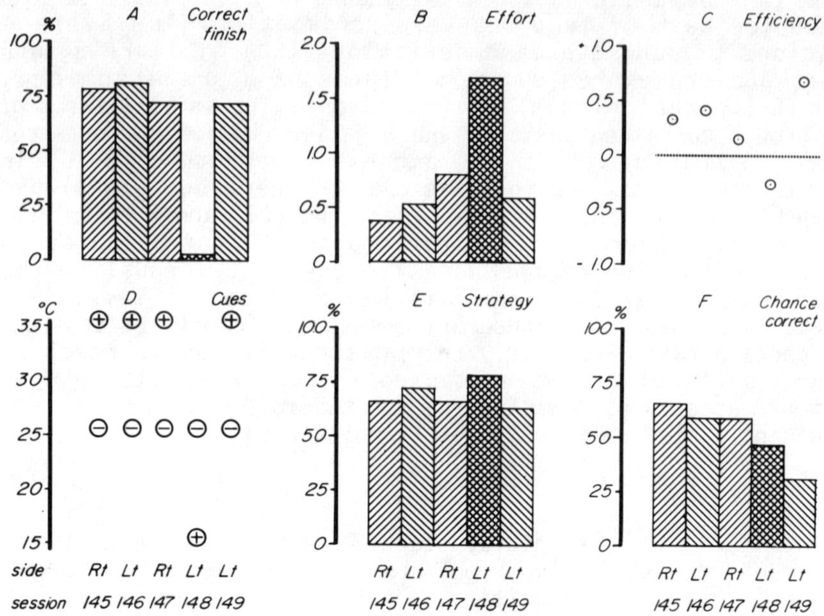

Fig. 1. Series of 5 sessions of two-choice behaviour with thermal cues which were presented unilaterally to the paws of either body half on alternate days. The stimulated side and the session numbers are indicated below the diagrams. At session no. 148 (cross pattern bars) the positive cue was changed from its standard value 10° above the ambient temperature to 10° below, and the cat responded by obeying the negative cue after an increased behavioural effort. A. Per cent correct responses (32 trials/session). B. Average no. of changes between cues/trial. C. Difference between no. of reinforced trials and no. of trials with immediate encounters with positive cue divided by no. of changes between cues. D. Temperatures of positive (+) and negative (-) cues. E. Per cent initial entries in right alley. F. Per cent immediate encounters with positive cue.

above the ambient temperature, and the originally cool negative cue had been moved to become equal to the ambient temperature (stimulus parameters, Fig. 1, D). For one session (Fig. 1, no. 148, cross pattern bars) the positive cue was changed to 10° below the ambient temperature. The term pseudo-reversal was motivated by the fact that the true negative cue was unaltered, whereas the novel, temporary, positive cue corresponded to an earlier negative cue, which had been abandoned months previously. The cat's behaviour, on the other hand, was like that of a true stimulus reversal. The per cent correct responses (Fig. 1, A) during that session dropped from around 75 to a value close to 0, i. e. when presented with two traditionally negative cues the cat did not just become confused, but significantly preferred the one which was less cool, despite the lack of reinforcements. The finding virtually prompted the idea that the cat had learnt to recognize the importance of, not the alley with the warm cue, but the alley with the perceptually warmer cue.

The attempts to interfere with the thermosensory transmission of the spinal cord, to begin with, were made by studying bilateral thermosensitivity before and/or after various chronic bilateral lesions of the upper cervical segments. The lesions of 14 such cats are shown in Fig. 2. The deep lesions of the middle of the lateral funiculi for technical reasons were

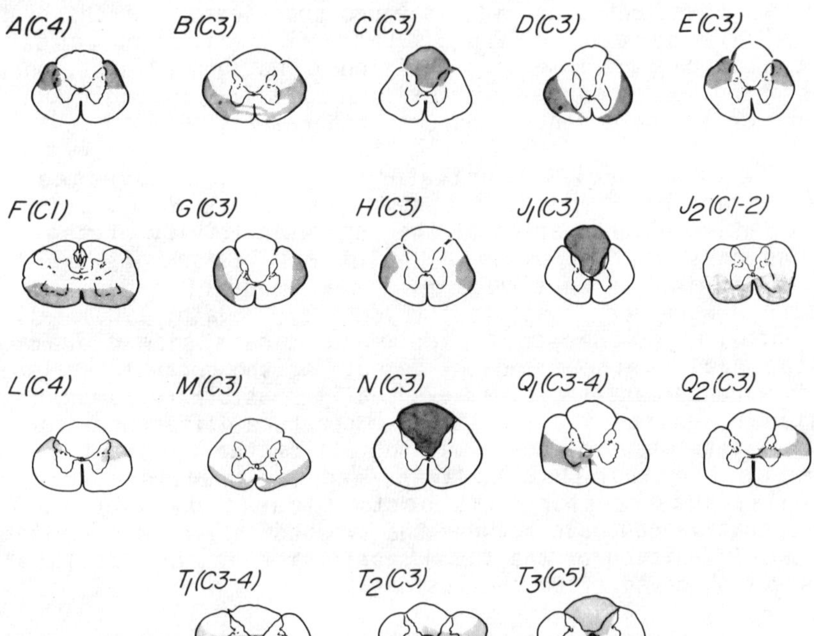

Fig. 2. Reconstructions of spinal cord lesions based on serial sections. Initial letters denote different cats and digits after that letter indicate repeated operations of the same animal. The codes inside the parentheses indicate the segmental levels of the lesions. The figure was published previously (Norrsell, 1979).

made without intermediate testing in two stages at adjacent segments (Fig. 2, cats Q and T). Two cats received combined, double lesions and were tested for thermosensory defects between the operations (Fig. 2, cats J and T). The findings may be summarized rapidly, because none of the lesions of the figure caused thermoanaesthesia. Within the limits of the procedures it was not even possible to establish a conclusive thermosensory deficiency for any of these cats. The figure shows that the different lesions together covered the entire transverse surface area of the spinal cord, and the findings therefore argued against the hypothesis of a unique, spinal, ascending, thermosensory pathway for the cat (cf. Norrsell, 1979).

Some additional cats of the same study were restricted to using the paws of one body half at a time in sample sessions, and such a way it was possible to obtain an estimate of their relative, unilateral thermosensitivity. Unilateral lesions were made in these cats, and it was found that lesions which covered the middle part of a single lateral funiculus caused a contralateral thermosensory deficiency. No such effects, on the other hand, were caused by lesions covering a single dorsal column or a single ventral quadrant (Norrsell, 1979).

The results of the unilateral lesions encouraged the development of a modified testing procedure which permitted quantitative determination of the thermosensitivity of the paws of one body half at a time. Unilateral lesions of the entire lateral funiculus were made in a new group of cats, and the findings were very consistent (Norrsell, 1980a,b, 1983). All of the animals, whether cool cats or warm cats showed dramatic postoperative, thermosensory defects of the contralateral paws which were permanent. It was eventually possible to demonstrate residual sensitivity for large temperature differences for the contralateral paws of most, but not all of the cats. It varied between the animals, however, and was transient for some animals, and therefore was best summarized as poor. The postoperative contrast between the two body halves was striking, and no diminuition of the thermosensitivity of the ipsilateral paws was observed after the lesions.

During the postoperative testing the cats were tried with the paws which were defective and normal with regard to thermosensitivity on alternate days. There was no covariance, however, between the results from the two body halves, as might have been expected on the basis of so called stimulus generalization (Norrsell, 1980a). Possibly the lesions had removed one important, thermosensory input, but not all of the thermosensory inputs from the defective body half to the supraspinal, neural centres in such a way, that the thermosensory cues from the two body halves had become behaviourally incompatible. So called thermosensory dysaesthesia would be the anthropomorphic analogy of the suggested condition, and has been observed in human patients with partial lesions of the spinal cord (Kinnier Wilson, 1927; Davison and Schick, 1935). The hypothesis could go part of the way to explain the differential effects of unilateral and bilateral lesions for the cats, i. e. supposing that cats with pertinent, bilateral lesions would be less disturbed in the T-maze because of the absence of reference to the preoperative, thermosensory conditions. There were also, however, further observations to consider.

The modified technique, which permitted quantitative determination of the thermosensitivity of individual body halves was used, not only for cats with complete transections of one lateral funiculus, but also for cats with partial, unilateral, lateral funiculus lesions. Partial lesions covering the dorsal, or the ventral thirds of the funiculus, however, in contrast to the complete lesions did not cuse any thermosensory disturbances at all (Norrsell, 1980a, unpublished observations). The area in the middle third of the lateral funiculus consequently became most interesting, and the findings motivated reappraisal of some previous observations from one cat which are illustrated in Fig. 3. After preoperative training which included unilateral testing in sample sessions, the left dorsal column was transected, and the cat's postoperative results are shown in the diagrams to the left of the figure. Per cent correct responses are shown in the top row, and the leftmost, white bar of that row shows that the cat still was able to recognize a warm cue of 1°, when she was permitted to use all four paws. Subsequently she was tested with unilateral access of either body half (Fig. 3, striped bars: Left bar, left paws. Right bar, right paws) for an easier cue of 7°, and no difference between the sides was observed. A second lesion covering the middle third of the left lateral funiculus then was made, and the cat still was able to recognize a 1° warm cue when using all paws (Fig. 3, middle diagram, left white bar). A difference between the body halves now had appeared, however, and she was selectively unble to benefit from a 7° warm cue for solving the two-choice problem, when it was presented to the right paws (Fig. 3, middle diagram, righthand striped bar). Finally, a third lesion was made which more or less mimicked the second lesion, but on the opposite side. The cat now became slightly erratic, insofar that she no longer produced an even behavioural performance for large numbers of successive sessions. She eventually showed ability to recognize a warm cue of 2°, however, when she was allowed to use all paws (Fig. 3, rightmost white bar). The thermosensory difference between the body halves had disappeared (Fig. 3, righthand diagram, striped

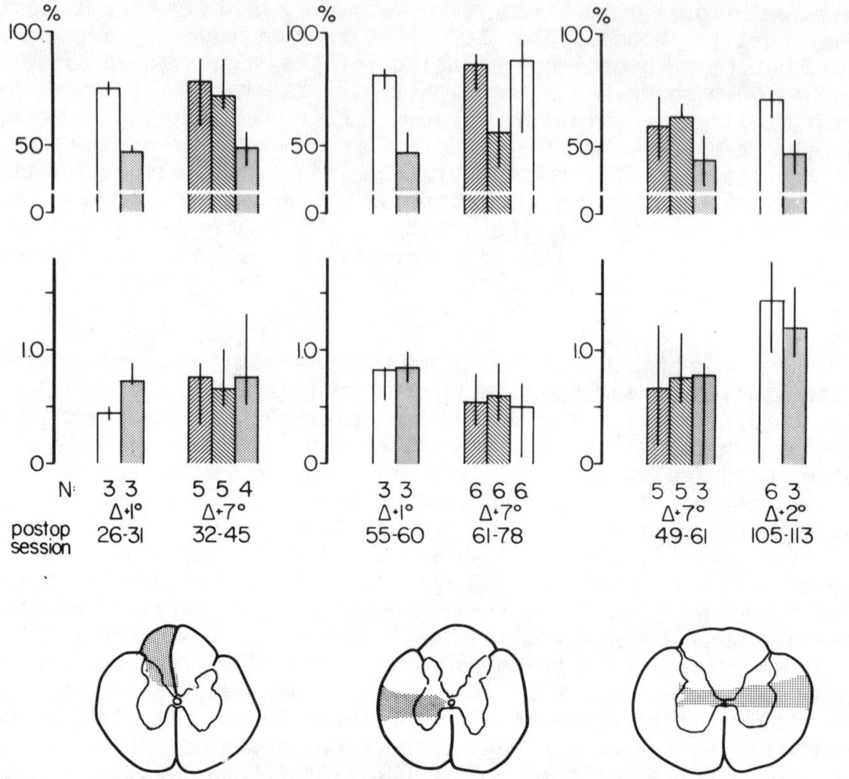

Fig. 3. Thermodiscriminatory behaviour of a warm cat after 3 successive lesions which are shown at the bottom of the figure below a corresponding set of diagrams. The lefthand lesion was put between the 3d and 4th cervical segments after 167 sessions. The middle lesion was put at the 3d cervical segment after 212 sessions. The righthand lesion was put at the 4th cervical segment after 290 sessions. The top row of bars shows the per cent correct responses/session, and the bottom row shows the mean no. of changes beween cues/trial/session (cf. also Fig. 1). Thick bars show median values and thin bars indicate ranges (sample sizes below diagrams). The cat was permitted to use all paws in some sessions (white bars), and the left paws (left striped bars), or the right paws (right striped bars) in other sessions, and was denied access to the cues in control sessions (gray bars). Different magnitudes of thermal cues were used and are shown below the diagrams. See text for further explanation. The data derive from a previously published experiment (Norrsell, 1979; cat U).

bars), but in addition her performance for unilateral, 7°, warm cues appeared to be inferior to her own performance, when she was presented with a 2° warm cue bilaterally. In fact, whereas she was able to satisfy the criterion of 3 successive sessions with at least 69 % correct responses for a 2° warm cue presented bilaterally, she always was found to be unble to satisfy that criterion for unilaterally presented 7° warm cues.

The observation could indicate the presence of bilateral, thermosensory, spatial summation, and if so may suggest a second explanation for the somewhat discordant findings after unilateral and bilateral lesions. Thermosensory spatial summation across the body midline originally was demonstrated in man, and has been investigated carefully (Hardy and Oppel, 1937, 1938; Banks, 1973; Marks and Stevens, 1973; Rózsa and Kenshalo, 1977). Furthermore, bilateral spatial summation, which was shown to occur at the spinal segmental level, was found in electrophysiological recordings from an ascending thermosensory pathway in the rat (Hellon and Mitchell, 1975). It may be assumed that information about the spatial, qualitative and quantitative characteristics of a peripheral, thermosensory event is transmitted to the forebrain via ascending pathways in the spinal cord. Thermosensory spatial summation occurring already at the segmental level, however, should make at least two functionally different pathways necessary for the sake of fidelity. A lesion which destroyed only one of the two hypothetical pathways, and left that pathway intact which transmitted spatially integrated, presumably quantitative information, might cause a rather special thermosensory deficiency. The deficiency might be more easy to demonstrate with a unilateral testing technique involving small receptive fields, than with the corresponding bilateral technique, if it doubled the receptive field sizes.

Nearing the end, it must be noted that peripheral sensory deficits which are caused by spinal cord lesions can be attributed, not only to damage of axons belonging to ascending pathways, but also to damage of systems which control the transmission of such pathways. Damage of control systems alone (cf. Norrsell, 1979, 1983) cannot explain the postoperative, thermosensory deficiencies which have been described, although it may possibly to unknown extent complicate the interpretation of the findings. Nevertheless, the difficulty to demonstrate thermosensory deficiencies after bilateral lesions have indicated a redundance of spinal, ascending, thermosensory pathways in the cat, and the diversity of yet available, relevant electrophysiological observations may point in the same direction (Perl et al., 1962; Simon, 1972; Kumazawa et al., 1975; Price and Browe, 1975). The reliable, contralateral deficiencies which were caused by unilateral, lateral funiculus lesions in contrast have indicated a functional lack of redundance which provides a pivot for current investigations proceeding on basis of the above suggested hypotheses.

Work supported by the Swedish Medical Research Council (project no. 2857).

REFERENCES

Banks, W.P. (1973). Reaction time as a measure of summation of warmth. Percept. Psychophys., 13, 321-327.

Davison, C. and Schick, W. (1935). Spontaneous pain and other subjective sensory disturbances. A clinicopathologic study. Arch. Neurol. Psychiat., 34, 1204-1237.

Finger, S. and Norrsell, U. (1974). Temperature sensitivity of the paw of the cat: a behavioural study. J. Physiol. (Lond.), 239, 631-646.

Hardy, J.D. and Oppel, T.W. (1937). Studies of temperature sensation. III. The sensitivity of the body to heat and the spatial summation of the end organ responses. J. clin. Invest., 16, 533-540.

Hardy, J.D. and Oppel, T.W. (1938). Studies in temperature sensation. IV. The stimulation of cold sensation by radiation. J. clin. Invest., 17, 771-778.

Hellon, R.F. and Mitchell, D. (1975). Convergence in a thermal afferent pathway in the rat. J. Physiol. (Lond.), 248, 359-376.

Kumazawa, T., Perl, E.R., Burgess, P.R. and Whitehorn, D. (1975). Ascending projections from marginal zone (Lamina 1) neurons of the dorsal horn. J. comp. Neurol., 162, 1-11.

Marks, L.E. and Stevens, J.C. (1973). Spatial summation of warmth: Influence of duration and configuration of the stimulus. Am. J. Psychol., 86, 251-267.

Norrsell, U. (1974). An automatic T-maze for temperature discrimination in the cat. Physiol. Behav., 12, 297-300.

Norrsell, U. (1978). Testing procedures and the interpretation of behavioral data. In Recovery from brain damage. (ed. S. Finger). Plenum, New York.

Norrsell, U. (1979). Thermosensory defects after cervical spinal cord lesions in the cat. Exp. Brain Res., 35, 479-494.

Norrsell, U. (1980a). Ipsi- and contralateral cold sensitivity after unilateral lesions of the lateral spinal funiculus of the cat. Acta physiol. scand., 109, 16A.

Norrsell, U. (1980b). Thermosensitivity after unilateral transection of the lateral spinal funiculus of the cat. Neurosci. Letters, Suppl. 5, S479.

Norrsell, U. (1983). Unilateral behvioural thermosensitivity after transection of the lateral funiculus in the cervical spinal cord of the cat. Exp. Brain Res., in press.

Perl, E.R., Whitlock, D.G. and Gentry, J.R. (1962). Cutaneous projection to second-order neurons of the dorsal colulmn system. J. Neurophysiol., 25, 337-358.

Price, D.D. and Browe, A.C. (1975). Spinal cord coding of graded nonnoxious and noxious temperature increases. Exp. Neurol., 48, 201-221.

Rózsa, A.J. and Kenshalo, D.R. (1977). Bilateral spatial summation of cooling of symmetrical sites. Percept. Psychophys., 21, 455-462.

Simon, E. (1972). Temperature signals from skin and spinal cord converging on spinothalamic neurons. Pflügers Arch., 337, 323-332.

Wilson, S.A. Kinnier (1927). Dysaesthesiae and their neural correlates. Brain, 50, 428-462.

CENTRIFUGAL CONTROL OF SOMATOSENSORY INFLOW INTO THE SPINAL CORD

MANFRED ZIMMERMANN

II. Physiologisches Institut, Universität Heidelberg, Im Neuenheimer Feld 326, D-6900 Heidelberg, FRG

INTRODUCTION

In his thesis work Yngve Zotterman studied sensations in humans elicited during limb ischemia used as a method to differentially block peripheral nerves (Zotterman, 1933). To explain some abnormal sensations he considered the possibility that messages in sensory nerves might be inhibited in the central nervous system, a mechanism that could be concluded at that time from work by Sir Henry Head and Otfrid Foerster. Research in animals has shown that inhibition in the spinal dorsal horn descending from the brain is an important mechanism for the modulation of sensory information (reviews by Fields and Basbaum, 1978; Willis, 1982). Early studies on descending inhibition were initiated by Swedish neurophysiologists (Lindblom and Ottosson, 1953; Hagbarth and Kerr, 1954), two of whom contributed to this symposium.

More recently, pain research has led to renewed interest in the inhibitory processes of this kind. A major observation was that wake animals showed diminished responses to noxious stimuli during electrical brain stimulation. At least part of this "stimulation-produced analgesia" was thought to be effected by descending inhibitory influences to the spinal cord. Thus, in anesthetized animals the firing of dorsal horn neurons in response to noxious skin stimuli can be inhibited by concurrent electrical stimulation at various sites in the brainstem. We have performed a series of parametric studies to elucidate the functional charanteristics of this descending inhibition. Our experimental noxious stimulus was skin heating, generating afferent inflow into the spinal cord via heat-sensitive nociceptors. Yngve Zotterman was the first to record from a nociceptor when he exposed the cat's tongue to high temperatures (Zotterman, 1936). His pioneer record is shown in Fig. 1.

Inhibition of dorsal horn neuronal responses to noxious skin heating

We recorded from single dorsal horn neurons in cats during stable anesthesia with nitrous oxide. More than 50% of lumbar dorsal horn neurons with input from the hindleg responded to C-fiber volleys in the posterior tibial and/or superficial peroneal nerves. Most of these units also responded to controlled radiant heat applied to their cutaneous receptive fields. Response thresholds were between 40 and 45oC, a range just below the threshold for heat pain sensations in human subjects. Most probably the excitation of cutaneous nociceptors with afferent C-fibers represents

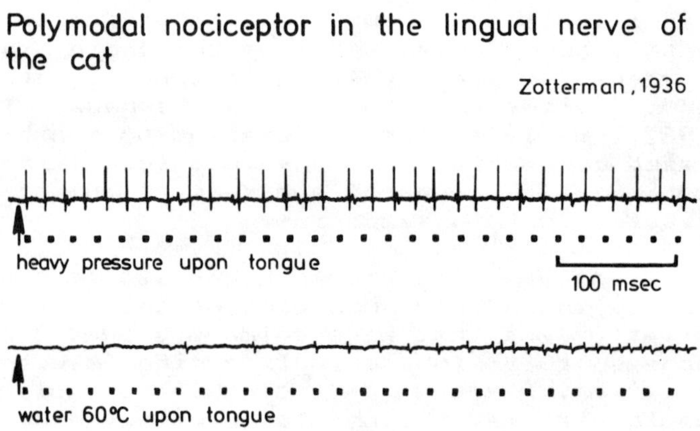

Fig. 1. Single units recorded from the lingual nerve of the cat. The large action potential is from a sensitive mechanoreceptor. The small action potential is from a unit responding to heavy pressure or to a jet of hot water (60 degrees C) on the tongue. This unit was, therefore, called a "pain fibre" (modified from Zotterman, 1936).

the major input for these heat-evoked dorsal horn neuronal responses (Handwerker et al., 1975), although some nociceptors with A-delta-fibers also might be involved (Beck et al., 1974). Characteristically, the responses of dorsal horn neurons to noxious skin heating reveal an afterdischarge outlasting the heat stimulus and the responses of the cutaneous nociceptors (Fig. 2A). This afterdischarge might be the neurophysiological basis for the prolonged aftersensations experienced by human subjects upon noxious skin heating.

Typically, the discharge frequencies of the dorsal horn neurons to graded noxious skin heating were linearly related to the temperature of the skin. This behaviour reflects the encoding characteristics of cutaneous C-fiber nociceptors (Beck et al., 1974). It is conceivable that such graded responses of dorsal horn neurons projecting to the brain via an ascending tract (Willis, 1982) mediate graded responses in perception of and behaviour to such stimuli.

Repetitive electrical stimulation through a focal electrode in the midbrain produced inhibition of the spinal neurons. When the stimulation was performed in the periaqueductqal gray (PAG) of the midbrain, the slope was decreased of the encoding line relating the intensity of the heat stimulus to the discharge rate of the neuron (Fig. 2B). This can be interpreted as a gain control of the dorsal horn neuronal system for the transmission of nociceptive information.

In contrast, stimulation in the lateral reticular formation (LRF) of the midbrain resulted in a parallel shift of the encoding line towards higher intensities of skin heating (Fig. 2B). This influence can be interpreted as a change in setpoint of intensity coding in the dorsal horn neuronal population. These different inhibitory effects suggest separate descending systems (Carstens et al., 1980). It has been proposed that the differential effects on the encoding characteristics may reflect different spatial relationships between excitatory and inhibitory synapses on the same neurons (Carstens et al., 1980), however, alternative explanations can be conceived.

Inhibition of the spinal cord neurons can also be accomplished from other regions of the brainstem, e.g., the raphe nuclei, the locus coeruleus and various regions of the medullary reticular formation

(Mokha et al., 1983; Satoh et al., 1983). On the other hand, stimulation at more rostral sites also produces inhibition in the dorsal horn, e.g., at sites in the hypothalamus, septum, orbital cortex and sensorimotor cortex (Zimmermann, 1976; Carstens et al., 1982; Willis, 1982). It seems that some serial connections exist from rostral to caudal structures to mediate the descending inhibitory influences. However, more than

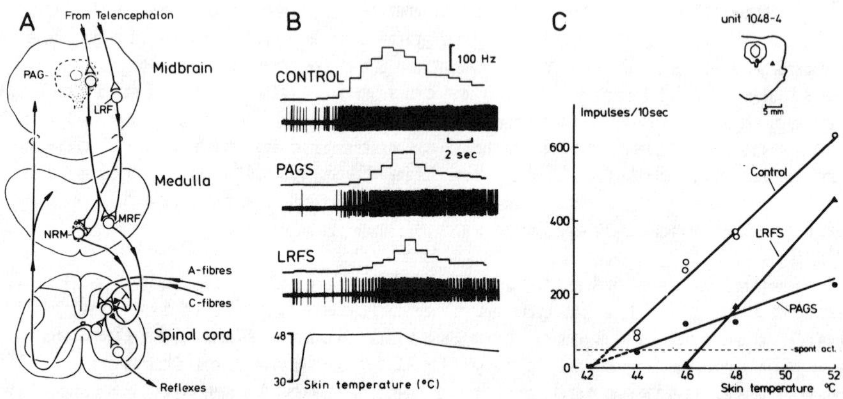

Fig. 2. Two systems of descending inhibition of spinal dorsal horn neurons in the cat. A: Schematic diagram of the possible descending inhibitory connections from midbrain to spinal dorsal horn. Different inhibitory systems can be activated by stimulation in the periaqueductal gray (PAG) and lateral reticular formation (LRF), which are relayed in the nucleus raphe magnus (NRM) and medullary reticular formation (MRF). B: Responses (oscilloscope records and histograms) of a dorsal horn neuron to 48 degrees C heating of the cutaneous receptive field, without midbrain stimulation (control), during repetitive stimulation (450 uA, about 30 Hz) of PAG (PAGS) or LRF (LRFS). Time course of skin temperature is shown below. B: Responses (number of impulses per 10 s) of another unit are plotted versus temperature of skin heating, without midbrain stimulation (control) and during stimulation of PAG (PAGS) or LRF (LRFS). C: Histological section through the midbrain shows localizations of stimulation sites in PAG and LRF (from Carstens et al., 1980).

one system of descending control have been identified by anatomical, physiological and pharmacological means (Willis, 1982).

Spinal inhibition by brain stimulation is not specific for nociceptive information

Inhibition of dorsal horn neurons evoked by stimulation in the brainstem has been reported to suppress specifically the discharges evoked by noxious stimulation, but not to affect the responses to non-noxious cutaneous stimuli (Oliveras et al. 1974). Our results using controlled excitation of sensitive cutaneous mechanoreceptors of the cat hairy skin do not support this contention (Carstens et al., 1981b). We have used a brush moved through the fur in the cutaneous receptive field of a dorsal horn neuron. The brush movement was electronically controlled, the velocity of movement could be varied in a reproducible manner.

Many dorsal horn neurons responded in a graded fashion to such brushing stimuli, they revealed a monotonic stimulus-response function between the velocity of the brush movement and the discharge frequency (Fig. 3A). Presumably, these discharges are due to the activation of low threshold hair follicle receptors with afferent A-beta-fibers (Burgess and Perl, 1973). The majority of the dorsal horn neurons responding to hair follicle afferents also can be excited by noxious skin heating (Fig. 3B), they have been termed multireceptive neurons. Some neurons respond only to low threshold mechanoreceptors. Electrical stimulation in PAG or LRF inhibited these brush-evoked responses in most neurons, only a few remained unaffected. The stimulus-response functions (relating discharge rates to brush velocity) revealed both changes in slope and parallel shifts. However, the attribution of these two effects to stimulation in PAG or LRF was less clear than in the case of activation by noxious skin heating. In most neurons, however, the amount of inhibition of brush-evoked responses was less than that of the heat-evoked discharges. Thus, the non-noxious and noxious inputs to the dorsal horn neurons are affected differentially, but not selectively, by the descending inhibitory systems.

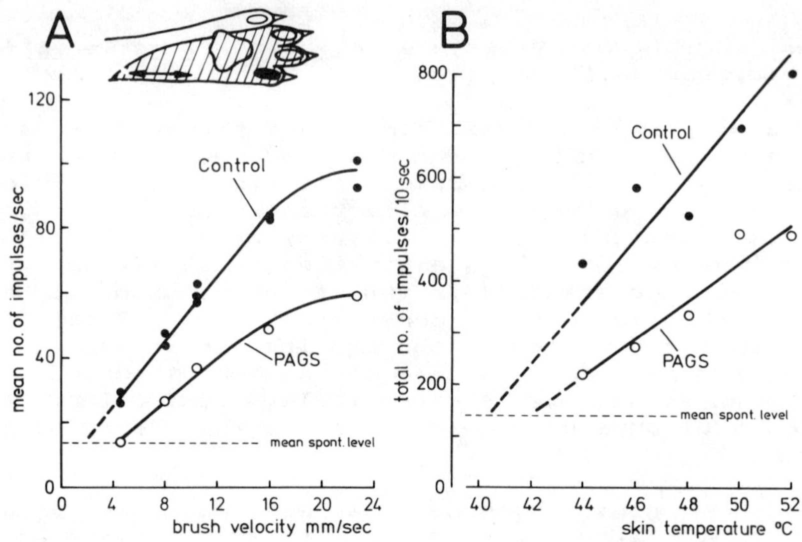

Fig. 3. Coding of hair brush velocity and heat intensity by dorsal horn neurons and effects of stimulation in PAG. A: Responses of a unit to brush stimuli applied at different velocities to the hairy skin, as indicated by arrow in the sketch of the foot. The averaged discharge frequency to the hairbrush is plotted against the velocity of the electronically controlled brush movement, the amplitude was 5 mm. The hatched area is the mechanosensitive receptive field of the unit. The velocity coding was measured before (control) and during (PAGS) repetitive stimulation in the periaqueductal gray (600 uA, about 30 Hz). B: Responses of the same unit as in A to noxious radiant heat stimuli applied to the pad of a toe (black in the sketch of the foot). Heat-evoked responses are plotted against the temperature of the heat stimulus before (control) and during (PAGS) repetitive electrical stimulation in the periaqueductal gray (from Carstens et al. 1981).

Activation of descending inhibition by morphine

Many of the brain structures from where analgesia and descending inhibition can be induced by focal electrical stimulation are rich in opiate receptors. It has been suggested therefore that the analgesic effect of opiates might be due to their activation of inhibitory systems arising from these structures, including the descending influences to the spinal cord. Indeed, injection of minor amounts of morphine at various brainstem sites elicits analgesia (Yaksh and Rudy, 1978) as well as descending inhibition of dorsal horn neurons (Fig. 4, 5).

In a quantitative parametric study morphine microinjected into the PAG has now been shown to probably activate the same system as is activated by focal electrical stimulation (Gebhart et al., 1983). It was crucial in this series of experiments to measure the encoding in dorsal horn neurons of noxious skin stimu

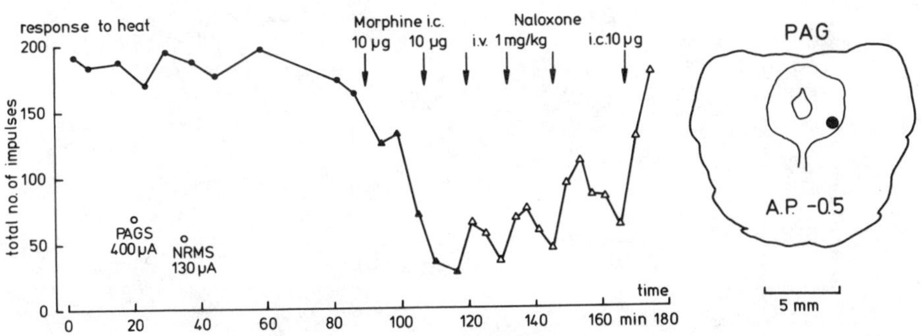

Fig. 4. Inhibition of dorsal horn neuronal responses to noxious skin heating by morphine microinjection into the midbrain. Each symbol plots the integrated discharge of a neuron to noxious skin heating (50 degrees C, 10 s in duration). Inhibition was elicited by repetitive local stimulation in the periaqueductal gray (PAGS) or nucleus raphe magnus (NRMS), and by morphine injection at the same site in the PAG (in histological section). Injection of naloxone i.v. or in the PAG (i.c.), as indicated by arrows, antagonized the effect of morphine. (From Gebhart et al., 1983).

li given at various intensities. As with electrical stimulation in the PAG (cf Fig.2, 3B), morphine injection resulted in a decrease in the slope of the intensity coding line, i.e., morphine also reduced the gain of spinal nociceptive transmission (Fig.5). Morphine injection in the mesencephalic lateral reticular formation (LRF) did, however, not activate descending inhibition. This differential effect of morphine in PAG and LRF is additional evidence to support the contention that different inhibitory systems are triggered from either region.

The activation of descending inhibitory control by local morphine in the PAG is blocked by local or systemic naloxone (Fig.4) whereas the inhibition produced by electrical stimulation in the PAG is not, in otherwise identical experimental conditions (Carstens et al., 1979). A likely hypothesis to explain

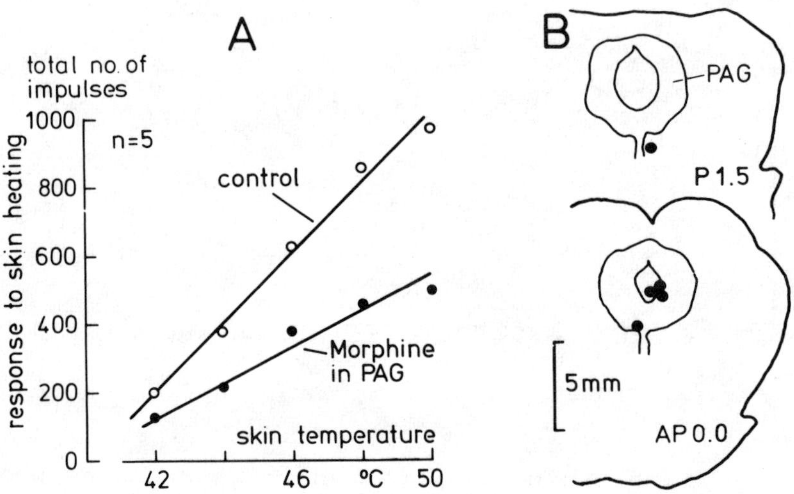

Fig. 5. Activation of descending inhibition by morphine. A: Mean number of impulses/20s of 5 dorsal horn neurons are plotted versus temperatures of skin heating, before (control) and after microinjection of morphine (10-15 microgram) into the PAG. B: Sites of the 5 morphine injections indicated on histological sections of the midbrain. From Gebhart et al. (1983).

these findings is that neurons with opiate receptors exist in the PAG, which are excited by local morphine, this effect being blocked by naloxone. On the other hand, the direct excitation of PAG neurons by electrical stimulation does not depend on opiate receptors, and therefore no effect of naloxone is observed.

It has been suggested that descending inhibition initiated by the action of morphine in the PAG is a basis of opiate analgesia. However, this is probably not the only mechanism of opiate analgesia. Opiate receptors have also been found in other areas of the brain, and in the dorsal horn of the spinal cord. The spinal opiate receptors mediate inhibitory effects on dorsal horn neurons transmitting nociceptive information. It is likely that these same opiate receptors are used also by enkephalin released from inhibitory interneurons located in the dorsal horn (Zieglgaensberger, 1983).

The recently acquired knowledge on spinal opiate mechanisms and their functional implications (Yaksh, 1981) has led to the discovery of potent analgesia produced by peridural administration of opiates in patients with severe chronic pain. This method avoids many of the serious side effects that patients must bear during treatment of pain by systemic opiate administration (Zenz, 1981). However, the risks of the chronic peridural catheter preclude this method of spinal analgesia from being used as a routine procedure.

Functions of descending inhibition

It is not known to what extent the various inhibitory systems cooperate and interact and what their normal physiological functions are. Presumably, they are not only involved in mechanisms of pain and analgesia. They may be part of a system by which the brain actively scans the information contained in the discharge pattern in a population of spinal neurons processing and transmitting sensory information from the skin. How can the inhibitory systems be activated other than by focal electrical stimulation or morphine microinjection? Electrical brain stimulation is highly artificial and unphysiological, it produces considerable vegetative and emotional side effects, being described by patients as disagreeable (Iacono and

Nashold, 1982). Thus, future research should devise more physiological and natural methods to activate the descending control of sensory information.

ACKNOWLEDGEMENTS

I wish to thank Mrs Hannelore Ehlers for editing and typing the manuscript, and Mrs Almuth Manisali for the graphic work. The author's work is supported by the Deutsche Forschungsgemeinschaft (grants Zi 110).

LITERATURE

Beck, P.W., Handwerker, H.O. and Zimmermann, M. (1974). Nervous outflow from the cat's foot during noxious radiant heat stimulation.
Brain Res., 67, 373-386.

Burgess, P.R. and Perl, E.R. (1973). Cutaneous mechanoreceptors and nociceptors.
In Handbook of Sensory Physiology, Vol. 2 - Somatosensory System. (ed. A. Iggo). Springer, New York, pp. 29-78.

Carstens, E., Klumpp, D. and Zimmermann, M. (1979). The opiate antagonist, naloxone, does not affect descending inhibition from midbrain of nociceptive spinal neuronal discharges in the cat.
Neurosci. Lett. 11, 323-327.

Carstens, E., Klumpp, D. and Zimmermann, M. (1980). Differential inhibitory effects of medial and lateral midbrain stimulation on spinal neuronal discharges to noxious skin heating in the cat.
J. Neurophysiol. 43, 332-342.

Carstens, E., Bihl, H., Irvine, D.R.F. and Zimmermann, M. (1981a). Descending inhibition from medial and lateral midbrain of spinal dorsal horn neuronal responses to noxious and nonnoxious cutaneous stimuli in the cat.
J. Neurophysiol. 45, 1029-1042.

Carstens, E., MacKinnon, J.D. and Guinan, M.J. (1982). Serotonin involvement in descending inhibition of spinal nociceptive transmission produced by medial preoptic and septal stimulation.
J. Neurophysiol. 48, 981-991.

Fields, H.L. and Basbaum, A.I. (1978). Brainstem control of spinal pain-transmission neurons.
Ann. Rev. Physiol. 40, 217-248.

Gebhart, G.F., Sandkuehler, J., Thalhammer, J.G. and Zimmermann, M. (1983). Inhibition in the spinal cord of nociceptive information by electrical stimulation and morphine microinjection at identical sites in the midbrain of the cat.
J. Neurophysiol. (submitted).

Hagbarth, K.E. and Kerr, D.I.B. (1954). Central influences on spinal afferent conduction.
J. Neurophysiol. 17, 295-307.

Handwerker, H.O., Iggo, A. and Zimmermann, M. (1975). Segmental and supraspinal actions on dorsal horn neurons responding to noxious and non-noxious stimuli.
Pain 1, 147-165.

Iacono, R.P. and Nashold, B.S. (1982). Mental and behavioral effects of brain stem and hypothalamic stimulation in man.
Human Neurobiol. 1, 273-279.

Lindblom, U.F. and Ottosson, J.O. (1953). Effects of spinal sections on the spinal cord potentials elicited by stimulation of low threshold cutaneous fibres.
Acta Physiol. Scand. 29, Suppl. 106, 191-208.

Mokha, S.S., McMillan, J.A. and Iggo, A. (1983). Descending influences on spinal nociceptive neurons from locus coeruleus: actions, pathway, neurotransmitters, and mechanisms.
In Advances in Pain Research and Therapy, Vol. 5 . (eds. J.J. Bonica, U. Lindblom and A. Iggo). Raven Press, New York, pp. 387-392.

Oliveras, J.L., Besson, J.M., Guilbaud, G. and Liebeskind, J.C. (1974). Behavioral and electrophysiological evidence of pain inhibition from midbrain stimulation in the cat.
Exp. Brain Res. 20, 32-44.

Satoh, M., Akaike, A., Nakazawa, T., Masuda, C. and Takagi, H. (1983). Different roles of the nucleus reticularis paragigantocellularis and nucleus raphe magnus of the rat in the production of analgesia by microinjection of opioids.
In Advances in Pain Research and Therapy, Vol. 5. (eds. J.J. Bonica, U. Lindblom and A. Iggo). Raven Press, New York, pp. 381-386.

Willis, W.D. (1982). Control of nociceptive transmission in the spinal cord.
In Progress in Sensory Physiology, Vol. 3. (Eds. H. Autrum, D. Ottoson, E.R.Perl and R.F. Schmidt). Springer, Berlin, Heidelberg, New York, pp. 1-159.

Yaksh, T.L. (1981). Spinal opiate analgesia: characteristics and principles of action.
Pain 11, 293-346.

Yaksh, T.L. and Rudy, R.A. (1978). Narcotic analgetics: CNS sites and mechanisms of action as revealed by intracerebral injection techniques.
Pain 4, 299-359.

Zenz, M. (1981) Peridurale Opiat-Analgesie. Gustav Fischer, Stuttgart, New York.

Zieglgaensberger, W. (1983). Actions of opioid agonists on mammalian spinal neurons.
Int. Rev. Neurobiol. (in press).

Zimmermann, M. (1976). Neurophysiology of nociception.
In International review of physiology, Neurophysiology II, Vol. 10. (ed. R. Porter). University Park Press, Baltimore, pp. 179-221.

Zotterman, Y. (1933). Studies in the peripheral nervous mechanism of pain.
Acta Med. Scand. 80, 1-64.

Zotterman, Y. (1936). Specific action potentials in the lingual nerve of the cat.
Scand. Arch. Physiol. 75, 105-119.

DESCENDING CONTROL OF NOCICEPTIVE TRANSMISSION BY PRIMATE SPINOTHALAMIC NEURONS

W.D. WILLIS

Marine Biomedical Institute and Departments of Physiology and Biophysics, and Anatomy, University of Texas Medical Branch, Galveston, Texas 77550-2772, USA

INTRODUCTION

The spinothalamic tract in primates, including man, is thought to play a crucial role in nociception. Interruption of the anterolateral quadrant of the spinal cord, through which the spinothalamic tract ascends, results in analgesia on the contralateral side below the lesion (White & Sweet, 1955). Pain sensation is present when just one anterolateral quadrant of the cord is intact, as shown by the patient reported by Noordenbos and Wall (1976). Thus the anterolateral quadrant is both necessary and sufficient for pain sensation in humans. A comparable pathway also exists in monkeys (Vierck & Luck, 1979).

Recordings from spinothalamic tract (STT) cells in the monkey show that many of these neurons can be excited by the application of noxious stimuli to the skin (Willis et al. 1974), muscle (Foreman et al. 1979) or viscera (Milne et al. 1981). While other pathways in the anterolateral quadrant, such as the spinoreticular tract, may also contribute to nociception (Haber et al. 1982), it seems likely that the spinothalamic projection is of particular importance for pain sensation.

Given that the responses of STT cells to noxious stimuli may play a significant role in pain sensation, it is of interest that these responses can be powerfully inhibited by activation of pathways descending from the brain (Willis, 1982).

Inhibition of Spinothalamic Cells by Stimulation in PAG or NRM

STT cells in the monkey have been shown to be inhibited following stimulation in the nucleus raphe magnus (NRM) or the adjacent medullary reticular formation (Willis et al. 1977; Gerhart et al. 1981). The inhibition produced by stimulation in the NRM is of particular interest because of the observation that comparable stimulation in awake, behaving animals results in analgesia (Oliveras et al. 1975).

Stimulation in the periaqueductal gray also results in analgesia in awake, behaving animals (Reynolds, 1969; Oliveras et al. 1974) and in the inhibition of primate STT cells (Yezierski et al. 1982b).

Possible Role of Raphe-Spinal Neurons in Mediating PAG Inhibition

It has been proposed that the inhibition of dorsal horn nociceptive neurons following stimulation in the PAG (cf., Carstens et al. 1979; Oliveras et al. 1974) is due to the excitation of spinally projecting neurons in the NRM and adjacent reticular formation by pathways descending from the PAG to the medulla (Basbaum & Fields, 1978). Part of the evidence for this proposal is that serotonin appears to be involved in the inhibitory action of descending pathways from the PAG (Guilbaud et al. 1973; Carstens et al. 1981) and the NRM (Bourgoin et al. 1980; Oliveras et al. 1978; Rivot et al. 1980). Furthermore, the principle serotonergic projections to the spinal cord originate in the caudal nuclei of the raphe, including the NRM (Bowker et al. 1981). Since there are abundant connections from the PAG to the medial medulla in several species (Abols & Basbaum, 1981; Chung et al. 1983), raphe-spinal neurons seem to be the most likely neuronal population to serve as the serotonergic link in the descending "analgesia" systems. Analgesia and inhibition from stimulation in the PAG or the NRM depend upon the integrity of axons descending in the dorsal part of the lateral funiculus (DLF; Basbaum et al. 1977); thus the NRM and adjacent magnocellular reticular formation are the most likely source of the descending inhibitory pathway, since neurons in these areas project through the DLF (Basbaum et al. 1978; Martin et al. 1979).

An observation that suggests that the descending pathways may be somewhat more complex than was initially thought is shown by the following experiment (Yezierski et al. 1982b). The serotonin antagonist, methysergide, was administered systemically after the inhibition of primate STT cells was demonstrated from both the PAG and NRM. The inhibitory effect of stimulating in the PAG was nearly completely blocked, whereas the inhibitory action of stimulating in the NRM was only partially reduced. Similar results were obtained with other serotonin antagonists, including metergoline.

A number of explanations for these results can be formulated. One possibility is that there is an excitatory serotonin synapse between the PAG and the lower brainstem, in addition to

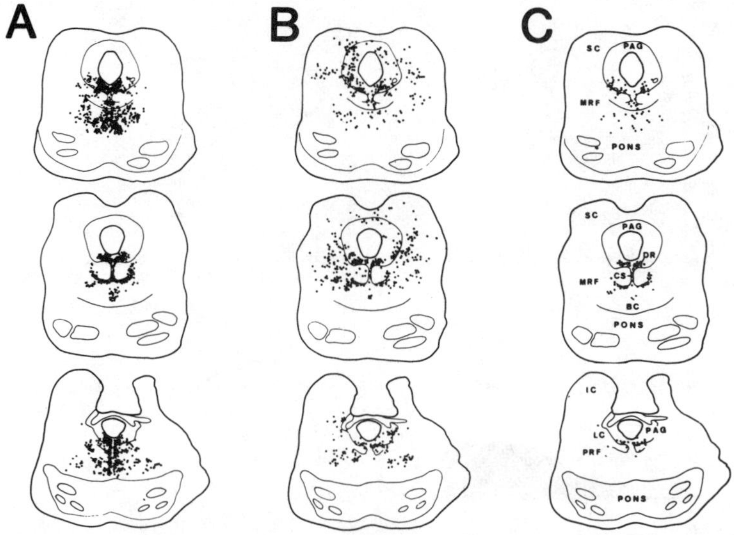

Fig. 1. Serotonin-containing neurons in the midbrain project to the medial medulla. In A are plotted the locations of midbrain neurons containing serotonin-like immunoreactivity. In B are shown neurons labeled retrogradely by horseradish peroxidase injected into the medial medulla (raphe nuclei and adjacent reticular formation). In C are shown the locations of neurons that were double-labeled. Each plot shows all cells on a single 50 µm section. Abbreviations: BC, brachium conjunctivum; CS, central superior nucleus; DR, dorsal raphe nucleus; IC, SC, inferior and superior colliculi; LC, locus coeruleus; PAG, periaqueductal gray; MRF, midbrain reticular formation; PRF, pontine reticular formation. (From Yezierski et al. 1982a.)

an inhibitory one between the NRM and the spinal cord. This possibility is supported by the anatomical study illustrated in Fig. 1 (Yezierski et al. 1982a). In this experiment, neurons in the midbrain were stained immunohistochemically for serotonin (Fig. 1A), and cells were also labeled retrogradely by horseradish peroxidase injected into the medial part of the medulla (Fig. 1B). A number of midbrain neurons were double-labeled (Fig. 1C), indicating that some serotonin-containing neurons in the midbrain raphe nuclei project to the medial part of the medulla. If there is an excitatory serotonergic synapse between the region of the midbrain PAG and the NRM and adjacent medullary reticular formation, and if this synapse is a component of the

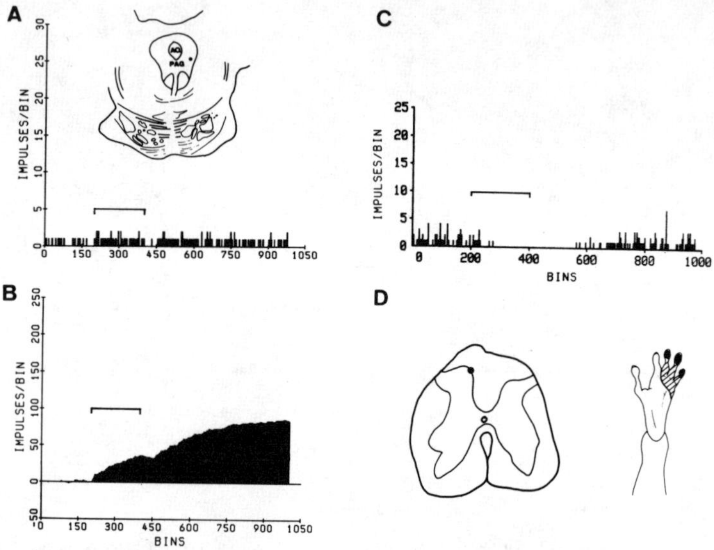

Fig. 2. The peristimulus time histogram in A shows the response of a raphe-spinal neuron to stimulation in the periaqueductal gray (PAG). The stimulus site is shown in the inset. Stimulus trains were applied during the time indicated by the bracket. The same response is shown in B as a cumulative sum histogram. The same stimulus in the PAG inhibited a spinothalamic tract cell, as shown by the histogram in C. The STT cell was located in or near the marginal zone, and its receptive field was on the left foot, as shown by the drawings in D. (From Willcockson et al. 1983.)

descending inhibitory pathway from the PAG to the spinal cord, then a serotonin antagonist might well interfere with the ability of a stimulus in the PAG to produce an inhibition of STT cells or other nociceptive dorsal horn interneurons by blocking the excitation of bulbospinal pathways. The partial antagonism of the inhibition produced by stimulation in the NRM by serotonin blockers could be explained either on the basis of a different serotonin receptor for the inhibitory synapses (cf., Haigler & Aghajanian, 1974; Peroutka et al. 1981) or of release of several transmitters at the spinal cord level (Yezierski et al. 1982b).

We felt that it would be important to determine if there were excitatory serotonergic projections from the PAG onto raphe-spinal neurons in the monkey. Therefore, we recorded from

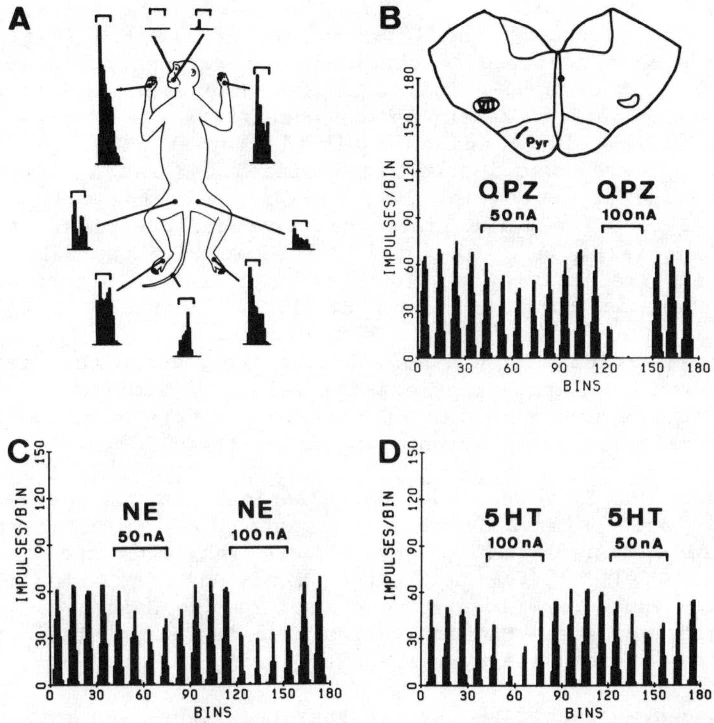

Fig. 3. Drug effects on raphe-spinal cell. The cell was the same as that illustrated in Fig. 2. The receptive field is shown in A; each response is to squeezing the skin at the indicated point using forceps. The location of the cell is shown in the inset in B. The single pass peristimulus time histograms in B-D show the responses of the cell to iontophoretic application of drugs. Glutamate was pulsed from the electrode for 5 s every 10 s to elicit a burst discharge. Quipazine (QPZ), Norepinephrine (NE) and Serotonin (5HT) were released at the times indicated by brackets, using the current shown. (From Willcockson et al. 1983.)

raphe-spinal neurons identified by antidromic activation from the spinal cord and showed that the raphe-spinal cells could be excited by stimulation at points within the PAG that also inhibited STT cells. Then, we released serotonin into the vicinity of the raphe-spinal cells to determine if the serotonin would excite or inhibit them. Contrary to our hypothesis, serotonin proved to be inhibitory. An example of such an experiment is shown in Figs. 2 and 3 (Willcockson et al. 1983).

Recordings were made from a raphe-spinal neuron (Fig. 2A,B; Fig. 3) and from an STT cell (Fig. 2C) in the same animal. An electrode was placed in the lateral part of the PAG (Fig. 2A, inset) in such a position that stimulation through it caused an excitation of the raphe-spinal cell (shown by the peristimulus time histogram in Fig. 2A and by the cumulative sum histogram in Fig. 2B; Ellaway, 1978) and an inhibition of an STT cell (Fig. 2C). The STT cell was located in lamina I and had a receptive field on the left foot (Fig. 2D). The raphe-spinal cell had a large excitatory receptive field that extended over most of the body surface (Fig. 3A). The cell was located in the NRM (Fig. 3B; inset). Its activity was low, and so glutamate was released by iontophoretic current pulses at 10 s intervals to allow testing of the activity of other drugs (Fig. 3B-D). When serotonin was released iontophoretically, it reduced the response of the neuron to glutamate pulses (Fig. 3D). A similar response was seen to the iontophoretic release of the serotonin agonist, quipazine (Fig. 3B), and to norepinephrine (Fig. 3C).

None of the 29 raphe- and reticulospinal neurons tested was excited by serotonin; all but 2 were inhibited. Similarly, 12 of the cells tested with quipazine were inhibited and 2 were unaffected. All 16 cells tested with norepinephrine were inhibited. These experiments cast doubt on the hypothesis that PAG stimulation causes the activation of a descending inhibitory pathway from the medulla by release of serotonin.

Although we cannot be certain that the raphe- and reticulospinal neurons from which we recorded have an inhibitory action on STT cells, the responses of these neurons are at least consistent with this possibility. For example, the average latency for the onset of inhibition of a series of STT cells following PAG stimulation was 24.0 ± 7.2 ms (N = 35) (Gerhart et al. 1984). The mean latency for excitation of raphe- and reticulospinal cells following PAG stimulation was 11.6 ms, and the antidromic conduction time averaged 8.2 ms (Willis et al. 1984). Thus, there is ample time for bulbospinal neurons activated by PAG stimulation to produce an inhibition of STT cells. Furthermore, the stimulus strengths required for activation of raphe- and reticulospinal neurons were of the same magnitude as those required for inhibition of STT cells (50 to several hundred μA).

It is of interest that raphe-spinal neurons typically have very large excitatory receptive fields, like that shown in Fig. 3A (Willcockson et al. 1983; Willis et al. 1984). Often, the responses to stimulation of the face are larger than those shown for this particular cell. The nature of the stimulus needed to activate these neurons was tested in many cases. In general, a noxious stimulus was required, although in a few cases there was

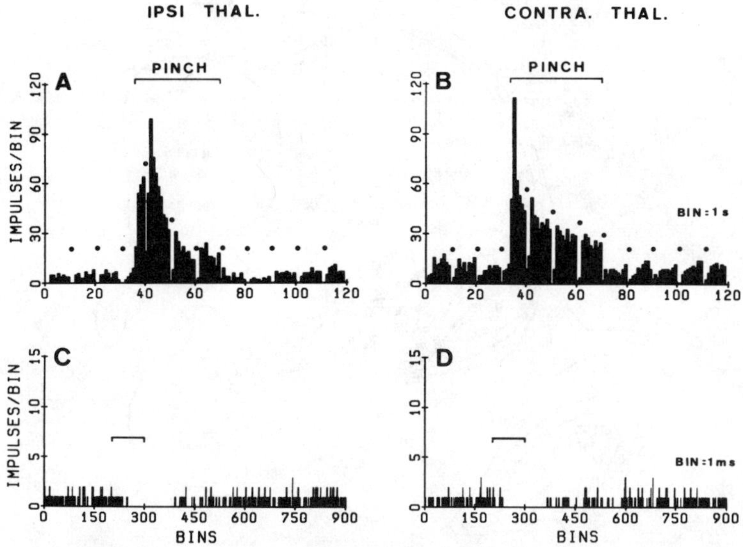

Fig. 4. Inhibition of a spinothalamic tract cell by stimulation in the ventral posterior lateral nucleus on either side of the brain. The single pass peristimulus time histograms in A and B show the effects of intermittent trains of stimuli (2 s trains of 200 μA pulses at 333 Hz every 10 s). The summed peristimulus time histograms in C and D show the effects of 200 ms trains (10 repetitions). (From Gerhart et al. 1981c.)

a response to innocuous mechanical stimulation of the skin as well (Willis et al. 1984). STT cells, like other nociceptive dorsal horn neurons, can often be inhibited by the application of noxious mechanical or thermal stimuli to wide areas of the surface of the body and face (Gerhart et al. 1981b). Besson and his co-workers (Le Bars et al. 1979a,b) have described a system they term "diffuse noxious inhibitory controls" that affects wide dynamic range neurons. This inhibitory system appears to involve a synaptic linkage in the brain stem, perhaps in the NRM, and to require noxious intensities of stimulation for its activation. The suggestion has been made (Le Bars et al. 1979b) that this system helps enhance contrast in the response of the entire population of wide dynamic range neurons, and so the system may be regarded as part of the coding mechanism for nociception, rather than as a negative feedback loop that limits nociceptive input. Our evidence is consistent with this proposal, although we do not see the profound levels of inhibition described for the DNIC system in the rat, nor can we be sure

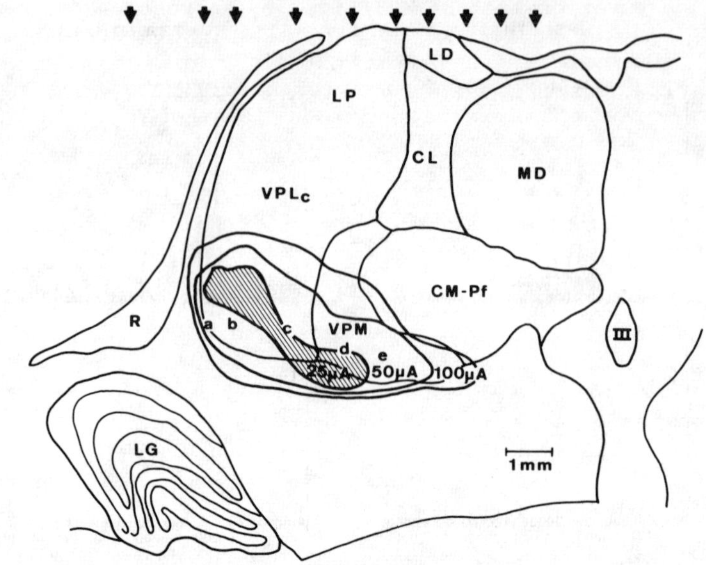

Fig. 5. Map of thalamus showing stimulus strengths required to inhibit a spinothalamic tract neurons. The courses of a series of tracks made by the stimulating electrode are indicated by the arrows. The contours show low threshold areas for the inhibition. The most sensitive area, shown by the hatched zone, had a threshold of less than 25 μA. This area encompassed parts of the ventral posterior lateral and medial nuclei. (From Gerhart et al. 1981b.)

that the raphe-spinal cells from which we record have any relationship to the STT cells.

Inhibition from Thalamus

Although much of our emphasis has been upon the inhibition of STT cells following stimulation in the medial brain stem, we have also noted that inhibition can be produced by stimulation of higher centers of the brain. For example, stimulation in the ventral posterior lateral nucleus of the thalamus on either side of the brain can cause an inhibition of a given STT cell (Gerhart et al. 1981c, 1983). This is shown in Fig. 4.

We were able to map the thalamus to determine the best sites for producing inhibition of STT cells. One such map is shown in Fig. 5. The lowest threshold zone for producing inhibition in this and in several other animals was in the ventrobasal complex.

The observation that stimulation in the ventrobasal complex can result in the inhibition of STT cells is of interest in the light of clinical findings that certain forms of pain can be alleviated by stimulation in the comparable part of the human thalamus or in the internal capsule (Hosobuchi et al. 1973; Mazars, 1975; Tsubokawa et al. 1982).

The pathway responsible for the inhibition of STT cells following stimulation in the ventrobasal complex is not altogether clear. One possibility is a loop through the cerebral cortex. However, an alternative route is by antidromic activation of axons ascending to the ventrobasal complex from the brain stem (or even the spinal cord). In this context, it is of interest that we were able to demonstrate an excitation of raphe-spinal cells following stimulation in the ventral posterior lateral nucleus, confirming the observation of Tsubokawa et al. (1982).

CONCLUSIONS

Like other dorsal horn neurons and like the neurons of the dorsal column nuclei that project to the thalamus (see review by Willis & Coggeshall, 1978), STT cells are under powerful descending controls from the brain stem. Since the descending control systems affect neurons presumed to contribute to various somatic sensations, the descending pathways do not serve only to modulate pain. Rather, these systems are examples of a general organizational feature of sensory pathways which commonly operate under centrifugal control. However, the possibility of an opportunistic use of descending inhibitory pathways for the purpose of interfering with pain make the control systems affecting nociception of particular interest.

ACKNOWLEDGEMENTS

The author thanks his colleagues with whom the experiments described were done, Gail Silver, Helen Willcockson and Griselda Gonzales for their technical assistance, and Phyllis Waldrop for typing. The work was supported by NIH grants NS 09743 and NS 11255 and a grant from the Moody Foundation.

REFERENCES

Abols, I.A. & Basbaum, A.I. (1981). Afferent connections of the rostral medulla of the cat: a neural substrate for midbrain-medullary interactions in the modulation of pain. J. Comp. Neurol., 201, 285-297.

Basbaum, A.I., Clanton, C.H. & Fields, H.L. (1978). Three bulbospinal pathways from the rostral medulla of the cat: an autoradiographic study of pain modulating systems. J. Comp. Neurol., 178, 209-224.

Basbaum, A.I. & Fields, H.L. (1978). Endogenous pain control mechanisms: Review and hypothesis. Ann. Neurol., 4, 451-462.

Basbaum, A.I., Marley, N.J.E., O'Keefe, J. & Clanton, C.H. (1977). Reversal of morphine and stimulus-produced analgesia by subtotal spinal cord lesions. Pain, 3, 43-56.

Bourgoin, S., Oliveras, J.L., Bruxelle, J., Hamon, M. & Besson, J.M. (1980). Electrical stimulation of the nucleus raphe magnus in the rat. Effects of 5-HT metabolism in the spinal cord. Brain Res., 194, 377-389.

Bowker, R.M., Westlund, K.N. & Coulter, J.D. (1981). Origins of serotonergic projections to the spinal cord in rat: an immunocytochemical-retrograde transport study. Brain Res., 226, 187-199.

Carstens, E., Fraunhoffer, M. & Zimmermann, M. (1981). Serotonergic mediation of descending inhibition from midbrain periaqueductal gray, but not reticular formation, of spinal nociceptive transmission in the cat. Pain, 10, 149-167.

Carstens, E., Yokota, T. & Zimmermann, M. (1979). Inhibition of spinal neuronal responses to noxious skin heating by stimulation of mesencephalic periaqueductal gray in the cat. J. Neurophysiol., 42, 558-568.

Chung, J.M., Kevetter, G.A., Yezierski, R.P., Haber, L.H., Martin, R.F. & Willis, W.D. (1983). Midbrain muclei projecting to the medial medulla oblongata in the monkey. J. Comp. Neurol., 214, 93-102.

Ellaway, P.H. (1978). Cumulative sum technique and its application to the analysis of peristimulus time histograms. Electroencephalogr. Clin. Neurophysiol., 45, 302-304.

Fields, H.L. & Anderson, S.D. (1978). Evidence that raphe-spinal neurons mediate opiate and midbrain stimulation-produced analgesias. Pain, 5, 333-349.

Foreman, R.D., Schmidt, R.F. & Willis, W.D. (1979). Effects of mechanical and chemical stimulation of fine muscle afferents upon primate spinothalamic tract cells. J. Physiol., 286, 215-231.

Gerhart, K.D., Yezierski, R.P., Fang, Z.R. & Willis, W.D. (1983). Inhibition of primate spinothalamic tract neurons by stimulation in ventral posterior lateral (VPL_c) thalamic nucleus: possible mechanisms. J. Neurophysiol., 49, 406-423.

Gerhart, K.D., Wilcox, T.K., Chung, J.M. & Willis, W.D. (1981a). Inhibition of nociceptive and nonnociceptive responses of primate spinothalamic cells by stimulation in medial brain stem. J. Neurophysiol., 45, 121-136.

Gerhart, K.D., Yezierski, R.P., Giesler, G.J. & Willis, W.D. (1981b). Inhibitory receptive fields of primate spinothalamic tract cells. J. Neurophysiol., 46, 1309-1325.

Gerhart, K.D., Yezierski, R.P., Wilcox, T.K., Grossman, A.E. & Willis, W.D. (1981c). Inhibition of primate spinothalamic tract neurons by stimulation in ipsilateral or contralateral ventral posterior lateral (VPL_c) thalamic nucleus. Brain Res., 229, 514-519.

Gerhart, K.D., Yezierski, R.P., Wilcox, T.K. & Willis, W.D. (1984). Inhibition of primate spinothalamic tract neurons by stimulation in the periqueductal gray or adjacent midbrain reticular formation. Submitted.
Guilbaud, G., Besson, J.M., Oliveras, J.L. & Liebeskind, J.C. (1973). Suppression by LSD of the inhibitory effect exerted by dorsal raphe stimulation on certain spinal cord interneurons in the cat. Brain Res., 61, 417-422.
Haber, L.H., Moore, B.D. & Willis, W.D. (1982). Electrophysiological response properties of spinoreticular neurons in the monkey. J. Comp. Neurol., 207, 75-84.
Haigler, H.J. & Aghajanian, G.K. (1974). Peripheral serotonin antagonists: Failure to antagonize serotonin in brain areas receiving a prominent serotonergic input. J. Neural Transm., 35, 257-273.
Hosobuchi, Y., Adams, J.E. & Rutkin, B. (1973). Chronic thalamic stimulation for the control of facial anaesthesia dolorosa. Arch. Neurol., 29, 158-161.
Le Bars, D., Dickenson, A.H. & Besson, J.M. (1979a). Diffuse noxious inhibitory controls (DNIC). I. Effects of dorsal horn convergent neurones in the rat. Pain, 6, 283-304.
Le Bars, D., Dickenson, A.H. & Besson, J.M. (1979b). Diffuse noxious inhibitory controls (DNIC). II. Lack of effect on non-convergent neurones, supraspinal involvement and theoretical implications. Pain, 6, 305-327.
Martin, R.F., Jordan, L.M. & Willis, W.D. (1978). Differential projections of cat medullary raphe neurons demonstrated by retrograde labeling following spinal cord lesions. J. Comp. Neurol., 182, 77-88.
Mazars, G.J. (1975). Intermittent stimulation of nucleus ventralis posterolateralis for intractable pain. Surg. Neurol., 4, 93-95.
Milne, R.J., Foreman, R.D., Giesler, G.J. & Willis, W.D. (1981). Convergence of cutaneous and pelvic visceral nociceptive inputs onto primate spinothalamic neurons. Pain, 11, 163-183.
Noordenbos, W. & Wall, P.D. (1976). Diverse sensory functions with an almost totally divided spinal cord. A case of spinal cord transection with preservation of part of one anterolateral quadrant. Pain, 2, 185-195.
Oliveras, J.L., Besson, J.M., Guilbaud, G. & Liebeskind, J.C. (1974). Behavioral and electrophysiological evidence of pain inhibition from midbrain stimulation in the cat. Exp. Brain Res., 20, 32-44.
Oliveras, J.L., Hosobuchi, Y., Guilbaud, G. & Besson, J.M. (1978). Analgesic electrical stimulation of the feline nucleus raphe magnus: development of tolerance and its reversal by 5-HTP. Brain Res., 146, 404-409.
Oliveras, J.L., Redjemi, F., Guilbaud, G. & Besson, J.M. (1975). Analgesia induced by electrical stimulation of the inferior centralis nucleus of the raphe in the cat. Pain, 1, 139-145.

Peroutka, S.J., Lebovitz, R.M. & Snyder, S.H. (1981). Two distinct central serotonin receptors with different physiological functions. Science, 212, 827-829.

Reynolds, D.V. (1969). Surgery in the rat during electrical analgesia induced by focal brain stimulation. Science, 164, 444-445.

Rivot, J.P., Chaouch, A. & Besson, J.M. (1980). Nucleus raphe magnus modulation of response of rat dorsal horn neurons to unmyelinated fiber inputs: partial involvement of serotonergic pathways. J. Neurophysiol., 44, 1039-1057.

Tsubokawa, T., Yamamoto, T., Katayama, Y. & Noriyasu, N. (1982). Clinical results and physiological basis of thalamic relay nucleus stimulation for relief of intractable pain with morphine tolerance. Appl. Neurophysiol., 45, 143-155.

Vierck, C.J. & Luck, M.M. (1979). Loss and recovery of reactivity to noxious stimuli in monkeys with primary spinothalamic cordotomies, followed by secondary and tertiary lesions of other cord sectors. Brain, 102, 233-248.

White, J.C. & Sweet, W.H. (1955). Pain. Its Mechanisms and Neurosurgical Control. Thomas, Springfield.

Willcockson, W.S., Gerhart, K.D., Cargill, C.L. & Willis, W.D. (1983). Effects of biogenic amines on raphe-spinal tract cells. J. Pharmacol. Exp. Therap., in press.

Willis, W.D. (1982). control of nociceptive transmission in the spinal cord. In: Progress in Sesnory Physiology 3. (ed. D. Ottoson). Springer-Verlag, Berlin.

Willis, W.D. & Coggeshall, R.E. (1978). Sensory Mechanisms of the Spinal Cord. Plenum Press, New York.

Willis, W.D., Gerhart, K.D., Willcockson, W.S., Yezierski, R.P., Wilcox, T.K. & Cargill, C.L. (1984). Primate raphe- and reticulospinal neurons: effects of stimulation in periaqueductal gray or VPL_c thalamic nucleus. Submitted.

Willis, W.D., Haber, L.H. & Martin, R.F. (1977). Inhibition of spinothalamic tract cells and interneurons by brain stem stimulation in the monkey. J. Neurophysiol., 40, 968-981.

Willis, W.D., Trevino, D.L., Coulter, J.D. & Maunz, R.A. (1974). Responses of primate spinothalamic tract neurons to natural stimulation of hindlimb. J. Neurophysiol., 37, 358-372.

Yezierski, R.P., Bowker, R.M., Kevetter, G.A., Westlund, K.N., Coulter, J.D. & Willis, W.D. (1982a). Serotonergic projections to the caudal brain stem: a double label study using horseradish peroxidase and serotonin immunocytochemistry. Brain Res., 239, 258-264.

Yezierski, R.P., Wilcox, T.K. & Willis, W.D. (1982b). The effects of serotonin antagonists on the inhibition of primate spinothalamic tract cells produced by stimulation in nucleus raphe magnus or periaqueductal gray. J. Pharm. Exp. Therap., 220, 266-277.

EFFECTS OF SYSTEMIC MORPHINE ON MONKEYS AND MAN: GENERALIZED SUPPRESSION OF BEHAVIOR AND PREFERENTIAL INHIBITION OF PAIN ELICITED BY UNMYELINATED NOCICEPTORS

CHARLES J. VIERCK, Jr., BRIAN Y. COOPER, RICHARD H. COHEN, DAVID C. YEOMANS, and *OVE FRANZÉN

Department of Neuroscience and Center for Neurobiological Sciences, University of Florida College of Medicine, Gainesville, FL, USA
**Department of Psychology, University of Uppsala, Uppsala, Sweden*

After years of extensive usage, morphine is still regarded as the most powerful pharmacological tool for control of pain. Despite this long record of success in the clinics, it is clear that morphine is not an analgesic, in that pain is not obliterated at systemic dosages that leave respiration intact in man (Javert and Hardy, 1951). Clinical patients report that pain can be elicited in the presence of morphine. Also, the effects of morphine on psychophysical ratings of phasically elicited pain are subtle in comparison with subjective estimates of clinical effectiveness (Beecher, 1957). This disparity of clinical and experimental findings has led to a number of hypotheses concerning the primary actions of morphine: (a) Morphine has been claimed to be an anxiolytic, and its action has been likened to the effects of frontal lobotomy, which is said to attenuate the emotional reactions to pain without disturbing the primary sensations of pain (Freeman and Watts, 1948). This explanation suggests that modifications of pain experiences occur by actions of morphine at telencephalic sites rather than via the brain stem - spinal cord circuitry that has been emphasized in recent attempts to understand central mechanisms of opiate hypalgesia (Yaksh, 1981). (b) Morphine has been thought to exert preferential effects on chronic pain, as distinct from phasically elicited pain (Beecher, 1957). Clear definitions of the crucial, distinguishing features of chronic pain or of morphine's effects on pain are not available, but there are a number of possibilities.

Morphine could produce a form of inhibition that reduces temporal summation of low levels of tonic neural discharge, without exerting a discernable effect on high frequency barrages of synchronous activity that are elicited by phasic events (Henry, 1979; Jurna et al., 1973). Also, opiates could curtail the spatial influences of nociceptive inputs, reducing the extent and magnitude of pain and/or eliminating additive complications that can impart an overwhelming quality to otherwise tolerable pain (e.g. from development of arousal and of muscular tension and spasm). Alternatively, morphine might preferentially inhibit input from certain nociceptors that are specialized to discharge in reaction to injury (involving release of algesic chemicals; Randall and Selitto, 1957) or to tonic distortion (activating slowly adapting peripheral elements; Haffner, 1929) or to involvement of receptors innervating certain structures (e.g. deep visceral and muscular nociceptors; Siegmund et al., 1957). Each of these influences could lead to more profound effects of morphine on clinical pain than on pain that is elicited by superficial, phasic stimulation in most experimental settings.

BEHAVIORAL SUPPRESSION BY MORPHINE

It is difficult to determine whether morphine reduces anxiety directly, but our observations in a series of human psychophysical studies of pain sensitivity indicate otherwise. For example, naive subjects reliably experience feelings of anxiety in anticipation of the strongest stimuli delivered in control sessions or following morphine administration. Even experienced subjects report anxiety prior to the strong stimuli at the beginning of sessions preceded by morphine. However, all subjects give evidence of response slowing and lethargy that usually is accompanied by frank dysphoria, and this powerful depression by morphine of behavioral initiative and vigor could easily by interpreted as a lack of anxiety. Regardless of whether this effect is attributed to attenuation of arousal or of emotional or motivational or motoric tone, it indicates that systemic morphine produces a form of behavioral suppression that can be mistaken as indifference or as hypalgesia when the behavior that is reduced or slowed is evoked by painful stimulation.

When monkeys are trained to escape electrical stimulation of the lateral calf at intensities judged by human subjects to range from near pain threshold to strong but tolerable levels, a variety of response measures reflect the generalized suppressive effects of morphine, without providing clear evidence for a direct attenuation of pain intensity (Vierck et al., 1983). For example, the latencies of responses that terminate the electrical stimulation are elevated at dosages of 1 mg/kg and above, but the latencies of responses to a tone signalling the availability of food reinforcement are decreased significantly at the same dosages. Similarly, thresholds for operant reactions that avoid or escape painful levels of stimulation (in paradigms intended to measure pain thresholds) are elevated by dosages of morphine that also increase thresholds for detection of stimuli that are clearly non-painful. Finally, exceptionally high dosages are required to suppress reflexive reactions to pain (generally above 3 mg/kg;

Cooper and Vierck, 1980). For each of these categories of directly elicited reactions to stimuli presumed to be painful to laboratory animal subjects, the minimal effective dosage of systemic morphine is at least an order of magnitude greater than the dosages of morphine administered to humans for control of pain (near 0.1 mg/kg). These results suggest that many of the behavioral effects of morphine that are referred to as "analgesia" are contaminated by or result entirely from a generalized behavioral suppression that can be detected in monkeys at low dosages (e.g., 0.25 mg/kg; Vierck and Cooper, 1983) and is powerfully expressed at the high dosages commonly utilized in studies of CNS mechanisms of opiate hypalgesia.

An important implication of the preceding statement is that the central inhibitory circuits that have been shown to be activated by morphine are interpreted as instrumental in modulation of pain sensitivity on the basis of behavioral measures that do not uniquely assess alterations of pain intensity by morphine (Vierck et al., 1983). The influences of these circuits on motoneurons, on somatic and autonomic reflex transmission and on arousal, attention and motivation must be discriminated from effects on pain intensity before we can identify pathways that modulate pain. In addition, our observations of monkeys reinforce the view that some types of phasic pain are relatively insensitive to systemic morphine, and if this were the case for all phasic pain, humane investigation of opiate hypalgesia in the laboratory would be made difficult, if not untenable. Thus, it is crucial to determine whether the clinical effectiveness of systemic morphine results from selective actions on a subset of inputs or on specialized neural codes that are operative only in chronic pain. If there are certain aspects of pain coding that are preferentially affected by low dosages of morphine, we can appreciate the lack of global analgesia at these levels, and there is a potential to define adequate tests for these specific effects if they occur with phasic stimulation.

THE UTILITY OF ELECTRICAL AND THERMAL STIMULI IN EXPERIMENTS ON MORPHINE HYPALGESIA

Because the methods of stimulation in studies of opiate hypalgesia predominantly involve application of electrical or thermal energy to the skin, human subjects were trained to rate the intensity of pain elicited by heat or electrical stimulation at magnitudes producing sensations spanning much of the pain sensitivity range. The sequence of stimulus intensities and modes was randomized across sessions of 48 trials of stimulation for 3 sec (Fig. 1). Sensations were rated by free magnitude estimation, which involved drawing of lines with lengths in proportion to perceived intensity. In separate tests, the subjects rated the magnitude of pain as it developed after application of an ice cube to the volar wrist. Intramuscular injection of 5 or 10 mg of morphine sulphate did not attenuate the dominant sensations produced by heat (up to 50 degrees C), focal electrical stimulation (up to 1.1 mA/mm^2) or cold (Fig. 1). However, these stimulation procedures activate a variety of nociceptor categories. For example, Zotterman defined separate groups of small myelinated (A delta) and unmyelinated (C) peripheral afferents that are activated by the painful stimuli utilized (Zotterman, 1939). Thus, either systemic morphine at human therapeutic dosages does not attenuate pain elicited in the major classes of nociceptive afferents (e.g. A delta mechanothermal receptors or C polymodal receptors), or effects on a subclass (or subclasses) of nociceptor are masked by uninhibited activity in other subclasses.

To evaluate further the possibility of selective actions on subpopulations of nociceptor afferents, human subjects were trained to attend selectively to the late pain sensations that follow brief trains of electrical stimulation (50 msec of alternating current at 60 Hz) or follow 750 msec of contact of a heated thermode with the shaved, hairy skin of the medial calf. The intensities of second pain sensations were rated by free magnitude estimation. The durations of first and second pain sensations were indicated by monitoring the voltage across a potentiometer that was adjusted by the subjects' variations of finger span to mark the onset, peak and offset of each sensation. The normal latencies and durations of the second pain sensations were consistent with conduction over unmyelinated peripheral afferents. The thresholds of first and second pain differed; and the magnitudes of the two distinct pain sensations were graded with stimulus intensities. The evidence that the second pain sensations from electrical or thermal stimulation represented activity conducted over unmyelinated afferents was: (A) the late onset of the sensations (500 msec or more) that varied with proximal or distal location of the electrodes or thermodes, permitting calculation of maximal conduction velocities of 2 m/sec, (B) the long duration of the sensations (representing dispersion of conduction velocities down to 0.5 m/sec) and (C) a diffuse and variable quality to the second pain sensations, distinguishing them from abrupt and focal first pains. The same subjects that reported no effect of morphine on first pain indicated substantial attenuation or disappearance of the second pain sensations evoked by electrical or thermal stimulation (Fig. 1). This finding suggests a preferential action of systemic morphine on conduction of input from unmyelinated peripheral afferents.

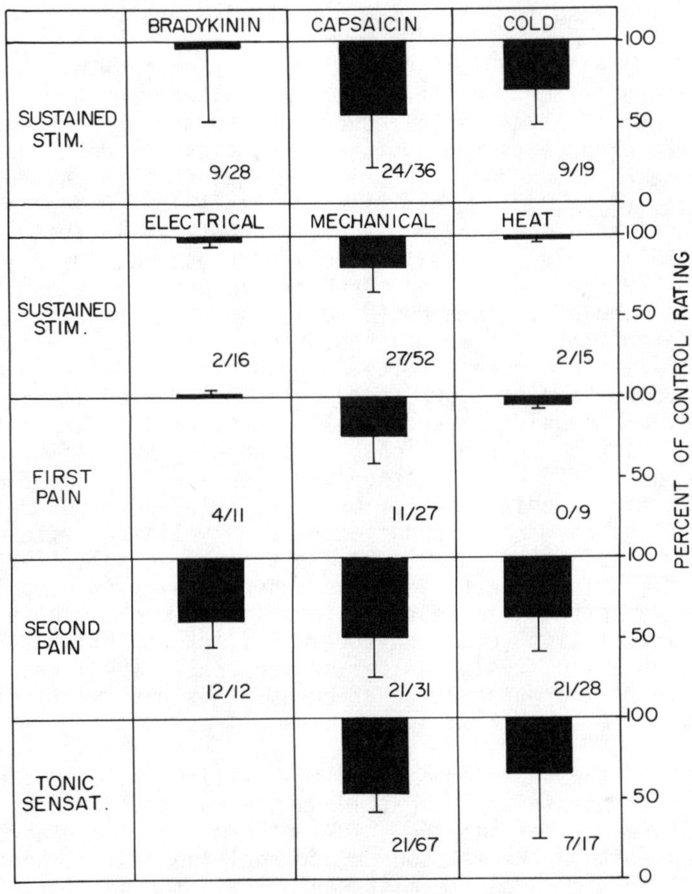

Figure 1. Average percent changes in psychophysical estimates of sensation magnitude are presented for different stimulus conditions and different sensations attended to. Downgoing bars indicate that sensation magnitude was reduced (relative to prior control sessions) by 5 or 10 mg of morphine sulphate, administered 45 min to 2 hr. before any given test. The fractions in the lower right corner of each panel indicate the number of significant reductions of sensation by morphine over the number of comparisons (Tuckey's t-tests; p less than 0.05).

(Fig. 1, cont.) The six varieties of sustained stimulation involved rating of sensation magnitude during ongoing stimulation. The lower panels present ratings of sensations that occurred during brief stimulation (first pain), or began more than 0.5 sec after stimulation and lasted up to 4 sec (second pain) or lingered for 15 to 120 sec (tonic sensations).

Does morphine inhibit central conduction from all unmyelinated afferents, or does it selectively attenuate input from C nociceptors? To address this question, human subjects were trained to produce psychophysical thresholds for detection of warmth, because increases of skin temperature from physiological neutrality (e.g., 33 degrees C) to levels below 40 degrees C should activated non-nociceptive C afferents selectively (LaMotte and Campbell, 1978). Normal detection thresholds for 5 subjects averaged 0.95 degrees C, and morphine did not significanlty affect threshold performance (0.98 degrees C). In addition, magnitude estimates of warmth sensations evoked by thermal stimulation of 1 to 6 degrees C (from an adapting temperature of 33 degrees; rise time of 2.50°C/sec) were not changed significantly by 10 mg. of morphine (2 subjects). Thus, within the continuum of cutaneous thermal sensations, an average dosage of 0.1 mg/kg appears not to affect pain conducted over small myelinated afferents (the first sensation from brief stimulation at 55 degrees C) or sensations of warmth conducted over unmyelinated afferents, but thermal pain attributable to C afferents is selectively inhibited. This suggests that the insensitivity to morphine that has been demonstrated repeatedly in a literature dominated by thermal stimulation (e.g., on the tail flick and hot plate tests; D'Amour and Smith, 1941; Woolfe and MacDonald, 1944) results from activation of thermoreceptive afferents that are relatively insensitive to inhibition by morphine.

Most of the behavioral paradigms utilized to demonstrate morphine hypalgesia apply heat in such a way that the skin warms gradually (e.g., warming the thick epidermis of the glabrous skin by contact with a thermal source, or applying radiant heat). Reactions to warmth can be mistaken as pain thresholds, when the animals are permitted to avoid noxious levels by responding to warmth (Jackson, 1952). When nociceptive sensations do occur, with fast onset or prolonged stimulation above 45 degrees C, escape responses are elicited by the faster conducting, lower threshold, A delta nociceptors. Hence, the trials are completed before the morphine-sensitive discharge in unmyelinated nociceptors reaches the central nervous system, or sustained activation of myelinated nociceptors masks a focal excitation of unmyelinated afferents. This explanation also applies to the relative insensitivity to morphine of reactions to electrical stimulation (Vierck et al., 1983).

EFFECTS OF MORPHINE ON PAIN ELICITED BY ALGESIC CHEMICALS

Given a variety of C nociceptive afferents that have been identified by peripheral axonal recording (Douglas and Ritchie, 1962), there are a number of questions concerning selectivity of morphine's effects that need to be resolved, and the preference for C versus A delta nociceptors should be tested with all stimulus modes that can be controlled within tolerable levels. For example, different algesic chemicals appear to exhibit separable patterns of influence on myelinated and unmyelinated nociceptors. Capsaicin has been shown to selectively stimulate C afferents with cutaneous terminations (Szolcsanyi, 1980; Kenins, 1982), and bradykinin elicits discharge in both A delta and C afferents (Beck and Handwerker, 1974; Fjällbrant and Iggo, 1961; Franz and Mense, 1975). These differences in the spectrum of receptors activated by algesic chemicals provide an opportunity to test for selective effects of morphine. Accordingly, skin on the volar surface of the forearms of human subjects was cleaned with alcohol and lightly abraided with fine sandpaper. In control tests spanning weeks, the subjects received solutions at 1 of 3 concentrations of capsaicin (4.5, 9 or 18 ug/ul) or bradykinin (0.19, 0.38 or 0.75 ug/ul) on 400 mm^2 patches of skin, and they rated the intensities of elicited sensations by magnitude estimation once every 15 seconds for a duration of 5 minutes. Following administration of 10 mg of morphine, i.m., two patches of skin were tested, sequentially. The subjects did not know which chemical or concentration was applied to either patch.

Morphine consistently reduced the intensity of pain elicited by capsaicin, but pain magnitudes that accompanied bradykinin were not reliably reduced (Fig. 1). As was the case with thermal stimulation, sensations attributable to chemically induced activity in C nociceptors were clearly reduced by morphine, but input from myelinated nociceptors appeared to mask any effects of morphine on C afferent input produced by bradykinin. That is, following application of bradykinin, and in the experiment utilizing prolonged thermal stimulation (3 seconds), the subjects rated sensations attributable to activity in both A delta and C afferents, and influences of morphine were not observed. Thus, it appears that the clinical effects of morphine on chronic pain reflect an inhibition of unmyelinated nociceptors, rather than a suppression of maintained discharge by central cells in the pain pathways. The durations of the sensations produced by capsaicin and by bradykinin appeared not to be affected by morphine. The magnitude of the pain elicited by capsaicin was reduced proportionately throughout the observation period. These findings encourage investigation of mechanisms that would predispose different forms of clinical trauma to preferentialy involve unmyelinated nociceptors. For example, C polymodal nociceptors are especially sensitive to naturally occurring algesic substances, and they are prone to sensitization and prolonged discharge (Perl et al., 1976), but these properties can be observed among myelinated nociceptors (Fitzgerald and Lynn, 1977).

EFFECTS OF MORPHINE ON PAIN FROM MECHANICAL DISTORTION

In addition to the release of algesic substances into tissue consequent to injury, clinical pain often involves mechanical disruption and distortion of superficial and deep structures, and therefore it is important to assess the relative effects of morphine on mechanical nociception conveyed by myelinated and unmyelinated afferents. Tonic mechanical stimulation was produced by applying a clamp to the volar skin of the forearm (16.7 gms/mm^2) or by squeezing the distal joint of one thumb with a clamp (fitted with a micrometer, permitting precise control of excursion of the clamp to an average of 2.32 mm from light skin contact). Mechanical stimulation continued for 60 sec, and the subjects rated pain intensity by magnitude estimation at 15 sec intervals. Consistent with the results from the other tests of tonic stimulation that should excite both myelinated and unmyelinated nociceptors, morphine did not reliably decrease the intensity of elicited pain, but a trend was apparent, and one half of the tests produced significant effects (Fig. 1).

In order to evaluate mechanical nociception conveyed by myelinated and unmyelinated afferents, several tests were devised: (A) Either a linear array of 3 sharpened points or a 2 mm by .23 mm edge was placed in contact with the dorsal skin of a distal phalanx (3 mm proximal to the fingernail), as observed microscopically and detected by an electronic circuit. Phasic indentation of the skin (up to 2.45 mm) was produced and monitored by a feedback controlled Ling vibrator, and the indentations occurred at 45 sec intervals, for 40 msec, with an indentation velocity of 28 mm/sec. (B) The jaws of a hemostat lightly gripped the web of skin between two fingers, and at 30 sec intervals, a solenoid closed the hemostat to a predetermined stop (adjusted with a micrometer) and released to the start position within 100 msec. For both types of phasic mechanical stimulation, excursions were determined in preliminary control sessions which produced weak to strong (but tolerable) pain sensations. The time course of elicited pain sensations was charted by instructing the subjects to squeeze a bulb in proportion to pain intensity as it waxed and waned; polygraph records of pressure derived from a pneumatic transducer gave analog readouts of the onsets, peaks and durations of first and second pain sensations. Pain intensities were rated after each trial by magnitude estimation. The results from the different forms of phasic mechanical stimulation were comparable and are combined for the panels in

Fig. 1. The late sensations ordinarily elicted by noxious mechanical compressions were substantially and consistently reduced by morphine. In contrast, an inconsistent reduction of first pain magnitude was obtained. Thus, the results from tests of mechanical nociception were generally consistent with measurements attributed to myelinated, unmyelinated or mixed afferent populations following chemical, thermal or electrical stimulation. Morphine exerted a preferential effect on late sensations. There was a suggestion that morphine reduced first pain from mechanical simulation to a greater extent than first pains from other forms of stimulation, but extensive observations would be required to resolve this issue.

EFFECTS OF MORPHINE ON AFTER SENSATIONS

If unmyelinated input to the central nervous system is selectively attenuated by systemic morphine, and if chronic pain is particularly sensitive to morphine because C nociceptors are especially prone to tonic discharge, then it would be expected that long duration after sensations that can be produced by phasic stimulation would reflect activity in C fibers and would be decreased by low dosages of morphine. Subjects were instructed to attend to and rate the intensity of sensations lingering for 15 seconds to 2 minutes following mechanical or heat stimulation. The after sensations were produced reliably by each mechanical stimulation (e.g., by a 2.45 mm indentation of the dorsal skin of a distal phalanx), but repeated thermal stimuli were required to produce a substantial after sensation; 6 trials of stimulation at 60 degrees C for 750 msec were delivered at 12 sec intervals, and the subjects rated the magnitude of the after sensation at 15 sec intervals for 2 min. Early in the observation period, the after sensations typically had a faintly painful quality, but after 30 sec or so, sensations that were not painful often could be discerned at the region stimulated.

Following either thermal or mechanical stimulation, morphine occasionally attenuated the magnitude of after sensations (Fig. 1). Overall, only 33% of the comparisons of after sensations revealed significant reductions by morphine, but this could be due to the low frequency of clearly painful tonic sensations produced by our stimulus conditions. Full evaluation of the effects on tonic nociception may require stimulus conditions that damage the tissue substantially. The present observations indicate only that systemic morphine does not block all forms of tonic discharge elicited by painful stimulation. The contrast between consistent effects on second pain sensations and variable reductions of tonic sensations suggests that the fundamental effect of morphine is upon inputs from unmyelinated nociceptors and not on the capacity of peripheral or central cells to discharge repetitively. However, it is conceivable that distinct second pain sensations represent after discharge of central cells that is not dependent entirely upon concurrent inputs from unmyelinated afferents. Second pain could reflect release from inhibition, rather than afferent driving by C nociceptors. If morphine enhanced such a central inhibitory process, second pain could be reduced without selective depression of input from C fibers.

EFFECTS OF MORPHINE ON NOCICEPTIVE AUTOINHIBITION

In an attempt to evaluate the effect of morphine on inhibition elicited by afferent input, we determined in pilot experiments that electrical stimulation over a skin site prepared with capsaicin produced a long lasting attenuation of the pain produced by the algesic chemical (Fig. 2). Imposition of 0.75 mA/mm^2 DC pulses (0.5 msec pulses at 200 Hz) for 50 msec produced a brief exacerbation of pain that was followed by a 66% reduction of the distinct pain attributable to capsaicin. The peak reduction of chemically induced pain occurred at 3.2 sec, followed by a gradual return of sensation magnitude for up to 12 sec following the shock stimulus. The subjects received 10 electrical stimuli (one every 30 sec) following the application of capsaicin. The magnitude of the pain from capsaicin was charted continuously by polygraph records of finger span. The slow changes in pain intensity elicited in this experiment were easily followed by the finger span technique.

Consistent with the previous experiment involving algesic chemicals (Fig. 1), morphine attenuated capsaicin-induced pain but did not eliminate it. Thus, a significant pain level remained, and the effects of electrical stimulation could be compared with normal sessions in terms of changes in duration and percent changes in magnitude of capsaicin-induced pain. Surprisingly, morphine significantly reduced the percent magnitude and the duration of inhibition following electrical stimulation (Fig. 2). This experiment does not support the notion that morphine enhances and prolongs an auto-inhibitory effect of pain input onto central cells in the pain pathways (see also LeBars et al., 1981). Furthermore, the long time course of inhibition (12 seconds for control sessions) revealed by superimposing electrical stimulation on a chemically treated patch of skin does not reinforce the possibility that the second pain sensation following electrical stimulation reflects a release from inhibition. Second pain is more likely attributable to late arriving input from unmyelinated afferents (Bishop and Landau, 1958; Torebjörk and Hallin, 1973; Van Hees and Gybels, 1972), and morphine's depression of second pain can be presumed to arise in part from occupation of opiate receptors on afferents in the dorsal horn. These presynaptic receptors are located in the region of termination of nociceptive afferents (LaMotte et al., 1976; Yaksh, 1978).

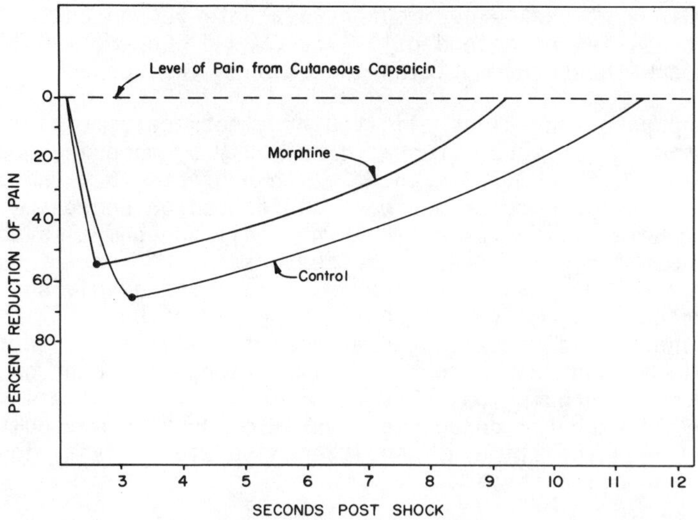

Figure 2. Average percent reduction of capsaicin-induced pain by electrical stimulation over the 20 by 20 mm patch of skin on the lateral calf that was pretreated with capsaicin. In control sessions for 8 subjects, baseline levels of pain were reduced by 66% within 3.5 sec of shock onset, and pain magnitude returned to baseline levels within 12 sec. Prior administration of morphine significantly attenuated the magnitude and duration of the inhibition elicited by electrical stimulation.

SUMMARY AND CONCLUSIONS

From our perspective, the primary goal of the experiments reported here was to determine whether an appropriate experimental model of morphine hypalgesia could be developed using humane methods of phasic stimulation. The results confirm clinical observations that therapeutic dosages of systemic morphine do not produce analgesia, particularly for many forms of phasically elicited pain. This presents a serious dilemma concerning animal models of pain and analgesia, because high dosages of morphine produce a profound suppression of behavior, and the experimental production of chronic pain is questionable, ethically. However, when the full time course of phasic pain is attended to, it is apparent that low dosages of morphine powerfully decrease late events that are referred to as second pain. These findings resurrect the potential to meaningfully investigate morphine hypalgesia in animal models. Importance is attached to evaluation of responding to phasic input over unmyelinated peripheral afferents. Activity in unmyelinated nociceptors can be evoked by stimulus intensities that are tolerated readily by the subjects, and a requirement that experimenters evaluate responses to different intensities of second pain (Vierck and Cooper, 1983) can even be met without traumatizing the skin or the subjects.

Second pain sensations elicited by electrical, mechanical or thermal stimulation were attenuated reliably by morphine dosages ranging from 0.05 to 0.15 mg/kg in human subjects, but the intensities of first pain reactions were unaffected or decreased only slightly. Selectivity was observed not only for unmyelinated afferents but for unmyelinated nociceptors, since sensations of warmth were not decreased by morphine. In tests involving psychophysical judgements of sensations produced by activity in both myelinated and unmyelinated afferents, elicited by sustained electrical, mechanical or thermal stimulation, little or no attenuation by morphine was observed. Thus, suppressive effects on pain by therapeutic dosages of systemic morphine may only be observable when the stimulus conditions isolate activity in unmyelinated nociceptors.

The difference in results with myelinated vs. unmyelinated input serves as a control against suggestive effects in subjects that are aware that morphine has been administered and expect pain to be attenuated. Similarly, the lesser inhibitory effect of electrical stimulation on capsaicin-induced pain in the presence of morphine indicates that the subjects objectively evaluated pain intensity, regardless of a priori expectations. In this regard, overtraining the subjects with repeated control sessions before administration of morphine provided them with an important degree of confidence in their ratings. A significant result that is not revealed by graphs or statistics of psychophysical ratings is that the only effects identified by subjects as unambiguous were on late pain sensations. That is, although quantified ratings of sensations related to myelinated input were occasionally reduced signficantly by morphine in individual subjects, only effects on late sensations were identified as obvious in interviews following sessions with morphine.

The results of the studies reported here indicate strongly that systemic morphine preferentially inhibits input from unmyelinated nociceptors, but the possibility of additional effects on myelinated afferents or upon the discharge capabilities of central cells are not ruled out. The dosage and route of administration of morphine determine the effects that are observed, and at high systemic dosages a variety of dramatic effects are produced, including somatic motoric inhibition, respiratory depression, and anesthesia (Goodman and Gilman, 1975). A number of physiological studies have demonstrated that high dosages of morphine (or activation of central sites that are affected by morphine) can inhibit inputs from small and large myelinated afferents to the spinal cord (Gray and Dostrovsky, 1983; Jurna and Heinz, 1979; LeBars et al., 1976). The point of the present article is to emphasize procedures that will permit us to evaluate the effects of morphine at dosages utilized clinically. This approach is likely also to produce insight into the role of endogenous opiate compounds in the control of pain when released naturally and at physiological levels. Attenuation of input from unmyelinated nociceptors and suppression of behavioral initiative are observed with low dosages of systemic morphine, and these influences would ordinarily be adaptive in situations involving long-term recovery from extensive injuries, where it is important that the organism rest comfortably and avoid strenuous activity. However, to the extent that opiate compounds are released in anticipation of dangerous situations or immediately after acute injury, limiting the impact of afferent inflow from C nociceptors would reduce also the severity of acute pain. Thus, the distinction between acute and chronic pain may be useful in describing the relative contributions of different control systems, but fundamental differences (that would preclude investigation of chronic pain control by acute pain models) should not be assumed until proven conclusively.

Supported by NINCDS grant NS 07261 and NIMH grant MH 25737.

REFERENCES

Beck, P.W. and Handwerker, H.O. (1974). Bradykinin and serotonin effects on various types of cutaneous nerve fibers. Pfluger's Arch., 347, 209-222.

Beecher, H.K. (1957). The measurement of pain: Prototype for the quantitative study of subjective responses. Pharmacol. Rev., 9, 59-209.

Bishop, G.H. and Landau, W.M. (1958). Evidence for a double peripheral pathway for pain. Science, 132, 712-713.

Cooper, B.Y. and Vierck, C.J., Jr. (1980). A comparison of operant and reflexive measures of morphine analgesia. Neuroscience Abstr., 6, 430.

D'Amour, F.E. and Smith, D. (1941). A method for determining loss of pain sensation. J. Pharmacol. Exp. Ther., 72, 74-79.

Douglas, W.W. and Ritchie, J.M. (1962). Mammalian nonmyelinated nerve fibers. Physiol. Rev., 42, 297-334.

Fitzgerald, M. and Lynn, B. (1977). The sensitization of high threshold mechanoreceptors with myelinated axons by repeated heating. J. Physiol., 365, 549-563.

Fjällbrant, N. and Iggo, A. (1961). The effect of histamine, 5 hydroxytryptamine and acetylcholine on cutaneous afferent fibers. J. Physiol., 156, 578-590.

Franz, M. and Mense, S. (1975). Muscle receptors with group IV afferent fibers responding to application of bradykinin. Brain Res., 96, 369-383.
Freeman, W. and Watts, J.W. (1948). Pain mechanisms and the frontal lobes: a study of pre-frontal lobotomy for intractable pain. Ann. Intern. Med., 28, 747-754.
Goodman, L.S. and Gilman, A. (1975). The Pharmacological Basis of Therapeutics. MacMillan, New York.
Gray, B.G. and Dostrovsky, J.O. (1983). Descending inhibitory influences from periaqueductal gray, nucleus raphe magnus, and adjacent reticular formation. I. Effects on lumbar spinal cord nociceptive and nonnociceptive neurons. J. Neurophysiol., 49, 932-947.
Haffner, F. (1929). Experimentelle Prufung Schmerzstillender mittel. Deutsch Med. Nschr, 55, 731-733.
Henry, J.L. (1979). Naloxone excites nociceptive units in the lumbar dorsal horn of the spinal cat. Neurosci., 4, 1485-1491.
Jackson, H. (1952). The evaluation of analgesic potency of drugs using thermal stimulation in the rat. Brit. J. Pharmacol., 7, 196-203.
Javert, C.T. and Hardy, J.D. (1951). Influence of analgesics on pain intensity during labor (with a note on natural childbirth). Anesthesiol., 12, 189-215.
Jurna, I., Grossman, W. and Theres, C. (1973). Inhibition by morphine of repetitive activation of cat spinal motoneurons. Neuropharmacol., 12, 983-993.
Jurna, I. and Heinz, G. (1979) Differential effects of morphine and opioid analgesics on A and C fibre-evoked activity in ascending axons of the cat spinal cord. Brain Res., 171, 573-576.
Kenins, P. (1982). Responses of single nerve fibers to capsaicin applied to the skin. Neuroscience Lett., 29, 83-88.
LaMotte, C., Pert, C.G. and Snyder, S.H. (1976). Opiate receptor binding in primate spinal cord: distribution and changes after dorsal root section. Brain Res., 112, 407-412.
LaMotte, R.H. and Campbell, J.N. (1978). Comparison of responses of warm and nociceptive C fiber afferents in monkey with human judgements of thermal pain. J. Neurophysiol., 41, 509-528.
LeBars, D., Guilbaud, G., Jurna, J. and Besson, J.M. (1976). Differential effects of morphine on responses of dorsal horn lamina V type cells elicited by A and C fibre stimulation in the spinal cat. Brain Res., 115, 518-524.

LeBars, D., Chitour, D., Kraus, E., Clot, A.M., Dickinson, A.H. and Besson, J.M. (1981). The effect of systemic morphine upon diffuse noxious inhibitory controls (DNIC) in the rat: Evidence for a lifting of certain descending inhibitory controls of dorsal horn convergent neurones. Brain Res., 215, 257-274.

Perl, E.R., Kumazawa, T., Lynn, B. and Kenins, P. (1976). Sensitization of high threshold receptors with unmyelinated (C) afferent fibers. Prog. Brain Res., 43, 263-277.

Randall, L.O. and Selitto, J.J. (1957). A method for measurement of analgesic activity of inflamed tissue. Arch. Int. Pharmacodyn. Ther., 111, 409-419.

Siegmund, E.A., Cadmus, R.A. and Lu, G. (1957). A method for evaluating non-narcotic and narcotic analgesics. Proc. Soc. Exp. Biol. Med., 95, 729-731.

Szolcsanyi, J. (1980). Effect of pain-producing chemical agents on the activity of slowly conducting afferent fibres. Acta Phys. Acad. Sci. Hung., 56, 86.

Torebjörk, H.E. and Hallin, R.G. (1973). Perceptual changes accompanying controlled, preferential blocking of A and C fibre responses in intact human skin nerves. Exp. Brain Res., 16, 321-332.

Van Hees, J. and Gybels, J.M. (1972). Pain related to single afferent C fibers in human skin. Brain Res., 48, 397-400.

Vierck, C.J., Jr. and Cooper, B.Y. (1983). Guidelines for assessing pain reactions and pain modulation in laboratory animal subjects. In Advances in Pain Research and Therapy, Vol. 6. (Eds. L. Kruger and J. Liebeskind). Raven Press, New York.

Vierck, C.J., Jr., Cooper, B.Y., Franzen, O., Ritz, L.A. and Greenspan, J.D. (1983). Behavioral analysis of CNS pathways and transmitter systems involved in conduction and inhibition of pain sensation and reactions in primates. In Progress in Psychobiology and Physiological Psychology, Vol. 10. (Eds. J. Sprague and A. Epstein). Academic Press, New York.

Woolfe, G. and MacDonald, A.D. (1944). The evaluation of the analgesic action of pethidine hydrochloride (Demerol). J. Pharmacol. Exp. Ther., 80, 300-307.

Yaksh, T.L. (1978). Analgetic actions of the intrathecal opiates in cat and primate. Brain Res., 153, 205-210.

Yaksh, T.L. (1981). Spinal opiate analgesia: Characteristics and principles of action. Pain, 11, 293-346.

Zotterman, Y. (1939). Touch, pain and tickling: An electrophysiological investigation on cutaneous sensory nerves. J. Physiol., 95, 1-28.

Session V

SOMATOSENSORY DYSFUNCTIONS

THE HYPERPATHIC SYNDROME: A CHALLENGE TO SPECIFICITY THEORY

PATRICK D. WALL

Cerebral Functions Group, Department of Anatomy, University College London, Gower Street, London WC1E 6BT, UK

Yngve Zotterman influenced our way of thinking in many ways. One was the tactic he shared with his master, Lord Adrian, which was to start in the periphery and to attribute as many sensory phenomena as possible to particular types of afferent fibres. This eminently sensible initial tactic has been carried to an extreme by a number of authors, into the claim that the sole origin of pain is activity in a particular type of fibre, the nociceptor. Since they have assigned the monopoly of extracting this quality of sensory experience to a specific fibre type in the periphery, the only role they can allow to central synaptic processes is to control the amplitude of sensation without affecting its quality or location. The central nervous system is seen to consist of a series of labelled line pain specific pathways whose gain may change but not their information content. The adherents of the specificity theory are wrong for many reasons, one of which is the subject of this paper, the hyperpathic syndrome. Like any syndrome, this consists of a grouping of signs and symptoms, each present in varying degrees in particular cases. It is necessary to propose a theory to explain the syndrome as an entity and not to explain one fraction such as primary hyperalgesia and then to imply that the whole is explained. We have to explain why the following characteristics are grouped together.

Characteristics of the Hyperpathic Syndrome:

Symptoms

1. Ongoing pain
2. Raised threshold
3. Dysaesthesia
4. Allodynia
5. Hyperalgesia
6. Summation
7. Delay and overshoot

8. Radiation of sensation from the point of stimulus
9. Spread of the affected area

Other factors

10. Affected by sympathetic efferents
11. Affected by local anaesthesia
12. Imitated by central lesions
13. Unaffected by nerve graft
14. Affected by large fibre stimulation
15. Speed of onset

Causes

The most florid and classical origins of the syndrome are from nerve injury as in causalgia or from peripheral nerve disease as in post-herpetic neuralgia. However many aspects appear with purely cutaneous lesions even experimentally produced (Lewis, 1942; Hardy, Wolff and Goodell, 1952). Deep lesions of soft tissue and of viscera may result in a similar condition (Travell and Simons, 1983; Procacci and Zoppi, 1984). Of crucial importance for any explanation is the fact that central lesions can exactly imitate the sensory symptoms (Pagni 1984). These central hyperpathic syndromes cannot be simply explained by involving the central terminals of peripheral afferents in the lesion as might be suggested to explain the pain of multiple sclerosis in the spinal cord. Some of the causative lesions are clearly in the rostral brain stem far distant from afferent terminals. The frequent occurrence of the intractable syndrome makes its explanation a matter of some clinical urgency although such an explanation may also assist the elucidation of pain mechanisms in the nervous system by demonstrating the validity or not of rival theories.

The symptoms

The most concise description of symptoms useful to both clinician and scientist remains that of Noordenbos (1969). The ongoing pain is variable and often has a deep diffuse quality uninfluenced by peripheral stimulation. The raised threshold is usual and might be expected from any partial nerve lesion but is not present in trigeminal neuralgia. In some cases it is not present with thermal stimuli as Lindblom writes elsewhere in this book. Dysaesthesia refers to abnormal unpleasant sensations which are not imitated by the acute pains evoked by mixtures of noxious stimuli. Allodynia is a useful new term for pain evoked by innocuous stimuli. Hyperalgesia refers to the production of excessive pain by a normally painful stimulus. An important characteristic is that repeated or prolonged stimuli slowly build up by summation to produce initially minor and later unbearable sensations. Elsewhere in this book, Lindblom describes the prolonged reaction time of patients with this syndrome. Once the stimulus has ceased, the sensation is likely to continue with a marked overshoot. The first evoked sensation may be relatively well localised but rapidly radiates or shoots to involve unstimulated areas. As time goes by, the area spreads from which these abnormal sensations can be evoked. For example, hyperpathia limited to part of a finger after a digital nerve lesion may spread to involve the whole hand and lower arm although the centre of gravity remains on the initially affected finger. It is quite apparent that this syndrome involves more than the production of pain by normally painless stimuli since the perception of the temporal and spatial aspects of the stimulus are as disturbed as is the quality of the stimulus.

Other factors.

In addition to the symptoms reported by patients we must add the results of manipulations which add to the definition of the problem. Allodynia is decreased by sympathectomy (Loh et al. 1980). This must be due to the removal of sympathetic efferent action in the periphery since alleviation is produced by the local peripheral application of antisympathetic drugs (Hannington Kiff, 1974). The condition can be imitated by central surgical lesions such as mesencephalic tractotomy (White & Sweet, 1969; Pagni, 1984). Seven cases of chronic causalgia were treated by nerve resection and by grafting to replace the damaged nerve (Noordenbos & Wall, 1981). All 7 cases reconstructed their hyperpathic syndrome when the nerve had regenerated through the graft. This means that complete surgical destruction of the formerly partially damaged nerve fails to abolish the generator of the sensory disorder. It has been known since the time of the First war that local anaesthesia peripheral to a region of nerve damage abolishes the pain. This means that the nerve impulses which trigger the abnormal sensations do not only originate from the area of nerve damage but peripheral to it. When we introduced transcutaneous nerve stimulation as a test of the gate control theory (Wall & Sweet, 1967) it was with considerable trepidation that we tried this on cases of causalgia since it might reasonably be predicted that we would make the pain worse. As it turns out, low level stimulation proximal to the lesion decreases the hyperpathia (Meyer & Fields, 1972). Even more interesting from the viewpoint of explanation, large fibre stimulation abolishes the hyperpathia and returns the threshold for evoking pain to normal (Lindblom & Meyerson, 1975). Finally we must take into account the natural history of the pain. It can certainly arise very rapidly after injury. In causalgia from nerve injury, the majority of cases show clear signs within hours, (White & Sweet, 1969). The reports by patients of pain immediately after injury are extremely variable

(Melzack, Wall & Ty, 1982). We do not know if the condition can appear instantaneously with the injury. There is no doubt that there is a subsequent evolution which proceeds for some months with the gradual spread of hyperpathic area far beyond the territory of the nerves involved in the original injury. It is evident that in explaining this pain state it is necessary to explain the action of nerves other than those damaged in the periphery. These other nerves must include sympathetic efferents, central cells and large diameter peripheral afferents which are stimulated by innocuous stimuli. It is also clear that although the condition can be rapidly initiated it is increased by slow processes.

Now that we have described the key features of the syndrome, let us proceed to enquire if changes in the periphery could provide an adequate explanation. We are seeking some way in which the disease could produce an abnormal peripheral barrage. Since pain is the dominant feature, a peripheral specificity theory would have to find some way in which nociceptors are driven to abnormal activity.

Peripheral sources of abnormal afferent barrages:

- Abnormal receptors
- Changed axonal membrane including sprouts
- Reaction of neighbouring intact axons
- Dorsal root ganglion cells

There is no doubt that after intense stimulation, peripheral endings, which have been fired subsequently change their sensitivity (see LaMotte in this book). However, there remain serious problems in explaining even the simplest primary hyperalgesia in the immediate area of injury. There is no agreement on the relative roles of A delta versus C fibres (Campbell & Meyer, 1983; LaMotte et al. 1982). Thermal stimuli do not necessarily produce mechanical sensitisation of peripheral ends in contradiction to the obvious sensory experience of sun burn. The duration of pain long outlasts the duration of the afferent barrage (LaMotte et al. 1982). However, Woolf and McMahon in work to be reported have recorded a subpopulation of C fibres close to the site of a thermal injury which develop very prolonged after discharges after mechanical stimuli to their receptive fields. Fitzgerald (1979) has shown that C fibres up to 10 mm from the site of injury may be sensitised and that this effect depends on nerve impulses generated in injury stimulated fibres. It is therefore apparent that some fraction of damaged nerve ends may become sensitised to respond to previously innocuous stimuli and that this effect may spread to nearby fibres which were not initially responding to the stimulus. This effect is presumably produced by the axon reflex of Lewis (1942).

There are three other known sources of an abnormal barrage from the periphery following injury. Damaged nerve membrane proximal to the terminals changes its properties. Wall and Gutnick (1974) showed that sprouts growing from cut axons developed three abnormal properties, 1. ongoing discharge, 2. mechanical sensitivity, 3. sensitivity to sympathetic amines. Clearly any injury or disease is unlikely, except in the most superficial injuries, to be limited only to nerve ends and therefore one must include the possibility of abnormal axon reactions as well as terminal receptor changes. A second type of peripheral axon change known to occur is that certain types of intact peripheral sensory fibres sprout into the region denervated by nerve destruction (Devor et al. 1979). Finally, we now know that peripheral axon damage has consequences which spread centrally to the dorsal root ganglia and produce a change of their properties not unlike the changes which occur in sprouts emitted from cut axons (Wall & Devor, 1984).

Four changes in the peripheral have been listed here which may well contribute to the hyperpathic syndrome. Before proceeding to ask if they could possibly provide an adequate explanation in combination, we must ask if there are additional central changes which we could or should also be used for explanation of the syndrome.

Central physiological changes triggered by injury in the periphery:

Secondary allodynia
Loss of central inhibitions
Expanded receptive fields, acute
" " " , chronic
Changed flexor reflexes
Changed autonomic reflexes

Sir Tomas Lewis (1942) in concert with the tactics of his contemporaries Zotterman and Adrian proposed an entirely peripheral mechanism for hyperalgesia following injury. Hardy, Wolff and Goodell (1952) while accepting the peripheral component, showed that peripheral stimuli also triggered central changes to produce secondary hyperalgesia which slowly spread far beyond the boundaries involved in the injury. In their remarkable chapter "The nature of cutaneous hyperalgesia", they summarise the evidence for including central changes. Their most cogent evidence comes from the stimulation of nerves central to a local anaesthetic block where the periphery is excluded and yet secondary hyperalgesia develops. They show a slow half hour spread of hyperalgesia to 100mm from the electrical stimulus point central to a blocked peripheral nerve. Clinical investigators now generally support the contention that central changes must be involved to explain the spatial distribution and time course of secondary hyperalgesia which fails to match the peripheral afferent barrage. This primary peripheral and secondary central change is particularly evident where hyperalgesia is referred to the skin after visceral damage (Malliani, Pagni & Lombardi, 1984; Procacci & Zoppi, 1984).

Physiological investigations now show two examples of changes in spinal cord organisations which follow intense peripheral excitation with a minimal latency of some 5-10 minutes. These changes are triggered by nerve impulses from the periphery but are sustained by central mechanisms. Woolf (1983) has examined the stimuli which produce activity in single flexor motor neurons in decerebrate rats. Initially the motor neurone responds only to a limited area of one leg. However, if this area is damaged the area of skin which will evoke a response slowly expands to include a wide area including the opposite leg. If now the entire area of damage is locally anaesthetised to abolish afferents from that area, the expanded sensitive area including the opposite leg remains. This shows that central changes remain even though the peripheral afferent barrage which induced the central change has been abolished. McMahon & Wall (1983) examined in decerebrate rats a specific type of marginal cell in the dorsal horn which projects to the brain. A burn was placed on the skin well outside the receptive field of such cells. After a period of 10 minutes, the receptive field of these cells slowly expanded to incorporate the damaged area of skin. C fibres were implicated as being necessary to trigger this plasticity since it did not occur in animals where the C fibres had been poisoned by the local application of capsaicin. Central mechanisms were involved in this delayed slow expansion of receptive fields since they could be evoked by stimulation of the central end of a cut peripheral nerve or dorsal rootlet where no change was induced in the peripheral ends of sensory fibres.

We have just described central changes which might account for some of the central increases of sensitivity which spread with a latent period of minutes after the arrival of an intense afferent barrage. There are in addition a further series of central changes with latent periods of days which are to be attributed to peripheral damage but where the peripheral signal is the transport of chemical substances rather than the delivery of nerve impulses. Following the section of a peripheral nerve, there is a decrease of the evoked secondary depolarisation of the afferent terminals (Wall & Devor, 1981) a decrease of inhibitions associated with cells receiving impulses from the damaged afferents (Woolf & Wall, 1982) and an expansion of the receptive field of deafferented cells so that they begin to respond to nearby intact afferents (Devor & Wall, 1981a & b; see also Kaas in this book). This receptive field expansion is also produced by chemical lesions limited to small afferents (Wall, Fitzgerald & Woolf, 1982). It is not produced by blocking nerve impulses (Wall et al. 1982) and Devor will be reporting that the receptive field changes are prevented by the application of transport blockers central to the peripheral nerve lesion.

It is apparent that physiological studies show two different changes of central reactivity following peripheral damage. Both involve signals of the presence of damage delivered by small afferent fibres. One is triggered by impulses and produces, after a delay of minutes, the expansion of the receptive fields of at least two types of cells. The second is triggered by the transport of abnormal concentrations of chemicals after a delay of days and results in a further reorganisation of the way in which afferent signals are handled by spinal cord. A remarkable example of this reorganisation which may have a double significance with respect to hyperpathia has been reported by Blumberg & Jänig (1982). They have recorded from sympathetic efferent fibres in normal nerves and in the same type of fibre which terminates in a nerve which has been chronically sectioned and ligated to form a neuroma. These fibres are undoubtedly affected by the nerve section as far back as their cell bodies (Jänig & McLachlan, 1984) as are the sensory afferents and the motorneurones. However, these fibres have changed their reflex receptive fields and respond in a grossly abnormal fashion to distant inputs by way of the brainstem. It is apparent that sensory interneurones and motor neurones and sympathetic ganglion cells all respond abnormally following peripheral damage.

Why does the hyperpathic syndrome challenge specificity theory? A summary.

We have listed fifteen aspects of this syndrome which need an integrated explanation. Implicit in a number of chapters in this book is the proposal that pain is only perceived if nociceptive afferents and their specific central cells are activated. We have 3 classes of symptoms to incorporate in an adequate theory of the hyperpathic state.

1) Quality, Symptoms 1-5. The type of sensation does not fit the stimulus.
2) Time, Symptoms 6 & 7. The duration of sensation does not match the stimulus.
3) Location, Symptoms 8 & 9. The location of pain does not match the location of the stimulus or of the injury.

Quality. **To preserve peripheral specificity theory it is necessary that nociceptors become sensitized. We know that this occurs in the immediate region of injury but there are two unexplained aspects. First the sensation perceived is an unusual dysaesthesia and not a** combination of sensations evoked by acute noxious stimuli. Second pain is evoked by innocuous stimuli at a considerable distance. For example, hyperpathia occurs in the hand minutes after the onset of cardiac ischaemia.

Time. No recordings from normal or damaged nociceptors have shown the expected signs of summation or long latency and prolonged after discharge which would be required to explain the sensory experience of the patient.

Location. The radiation of the location of sensed pain after a local stimulus has no correlate in peripheral nerve fibres, all of which are characterised by small receptive fields. Even more impossible to explain in the periphery is the spread of the hyperpathic area with time to far distant areas. No functioning widely distributed branched single axons have been shown to exist outside the dermatome of a fibre.

Other factors. Of the 6 other factors listed, nos. 10 and 11 may well indicate the abnormality of the peripheral barrage. Factors 12-15 have no peripheral correlate and clearly direct attention to the CNS. For example large nerve fibre stimulation can not affect the smaller nociceptive afferents directly and yet the ongoing pain disappears and the pain threshold becomes normal (Lindlom & Meyerson, 1975). This result not only suggests an increase of central inhibition but also fails to support a peripheral sensitization of nociceptors.

If it is no longer possible to explain the entire syndrome by sensitization of peripheral nociceptors, is it possible to propose that part of the disease is a disorder of specific central pain cells which normally only respond to nociceptive afferents? This approach leads rapidly to a reductio ad absurdam of specificity. Ongoing pain would be produced by false signals in hypothetical "pain" cells. Evoked pain from innocuous stimuli would be produced by the appearance of novel effective synapses on pain cells which somehow now evoke a novel quality, time course and location of the sensation. How many tortuous untested sequential hypotheses may one invent to defend a classical position? Three observations and a single hypothesis can replace this multiplicity of ad hoc hypotheses as follows. Observation 1, peripheral damage triggers a cascade of change from peripheral receptors at least as far as the spinal cord. Observation 2, these changes increase excitability partly directly and partly by the removal of central inhibitions. Observation 3, the increased excitability produces abnormal temporal and spatial patterns of response in spinal cord cells. The hypothesis, pain is triggered by these abnormal specifiable spatial and temporal patterns of response rather than by the activity of a single specific type of cell.

References

Blumberg, H. and Janig, W. (1981). Skin nerves with experimentally produced neuromata. In Phantom and Stump Pain. (eds. J. Siegfried and M. Zimmerman). Springer Verlag, Berlin.

Campbell, J.N. and Meyer, R.A. (1983). Sensitisation of unmyelinated nociceptive afferents in monkey varies with skin type. J. Neurophysiol., 49, 98-110.

Devor, M., Schonfeld, D., Seltzer, Z and Wall, P.D. (1979). Two modes of cutaneous reinnervation following peripheral nerve injury. J. Comp. Neur., 185, 211-220.

Devor, M. and Wall, P.D. (1981a). The effect of peripheral nerve injury on receptive fields of cells in cats spinal cord. J. Comp. Neurol., 199, 277-291.

Devor, M. and Wall, P.D. (1981b). Plasticity in the spinal cord sensory map following peripheral nerve injury. J. Neurosci., 1, 679-684.

Fitzgerald, M. (1979). The spread of sensitisation of polymodal nociceptors in the rabbit from nearby injury and by antidromic nerve stimulation. J. Physiol., 297, 207-216.

Hannington Kiff, J.G. (1974). Pain Relief. Heinemann, London.

Hardy, J.D., Wolff, H.G. and Goodell, H. (1952). Pain sensations and reactions. Williams & Wilkins, Baltimore.

Jänig, W. and McLachlan, E. (1984). On the fate of sympathetic and sensory neurons projecting into a neuroma of the superficial peroneal nerve in the cat. J. Comp. Neurol., In press.

LaMotte, R.H., Thalhammer, J.G., Torebjörk, H.E. and Robinson, C.J. (1982). Peripheral neural mechanisms of cutaneous hyperalgiesia following mild injury by heat. J. Neurosci., 2, 765-781.

Lewis, T. (1942). Pain. Macmillan, London.

Loh, L., Nathan, P.W., Schott, G.D. and Wilson, P.G. (1980). Effects of regional guanethidine infusion in certain painful states. J. Neurol. Neurosurg. & Psychiat., 43, 446-451.

Lindblom, U. and Meyerson, B.A. (1975). Influence on touch, vibration and cutaneous pain of dorsal column stimulation in man. Pain, 1, 257-270.

Malliani, A., Pagni, M. and Lombardi, F. (1984). Visceral versus somatic mechanisms. In Textbook of Pain. (eds. P.D. Wall and R. Melzack). Churchill Livingstone, London.

McMahon, S.B. and Wall, P.D. (1983). A system of rat spinal cord lamina 1 cells projecting through the contralateral dorsolateral funiculus. J. Comp. Neurol., 214, 217-223.

Melzack, R., Wall, P.D. and Ty, T.C. (1982). Acute pain in an emergency clinic: latency of onset and descriptor patterns related to different injuries. Pain, 14, 33-44.

Meyer, G.A. and Fields, H.L. (1972). Causalgia treated by selective large fibre stimulation of peripheral nerve. Brain, 95, 163-168.

Noordenbos, W. (1969). Pain. Elsevier, Amsterdam.

Noordenbos, W. and Wall, P.D. (1981). Implications of the failure of nerve resection and graft to cure chronic pain produced by nerve lesions. J. Neurol. Neurosurg. Psychiat., 44, 1068-1073.

Pagni, C.A. (1984). Central pain due to spinal cord and brain stem damage. In, Textbook of Pain. (eds. P.D. Wall and R. Melzack). Churchill Livingstone, London.

Procacci, M. and Zoppi, M. (1984). Heart Pain. In, Textbook of Pain. (eds. P.D. Wall and R. Melzack). Churchill Livingstone, London.

Travell, J.G. and Simons, D.G. (1983). Myofascial pain and dysfunction. Williams & Wilkins, Baltimore.

Wall, P.D. & Devor, M. (1981). The effect of peripheral nerve injury on dorsal root potentials and on transmission of afferent signals into the spinal cord. Brain Res., 209, 95-111.

Wall, P.D. and Devor, M. (1984). Sensory afferent impulses originate from dorsal root ganglia as well as from the periphery in normal and nerve injured rats. Pain, in press.

Wall, P.D., Fitzgerald, M. and Woolf, C.J. (1982). Effects of capsaicin on receptive fields and on inhibitions in rat spinal cord. Exp. Neurol., 78, 425-436.

Wall, P.D. and Gutnick, M. (1974). Properties of afferent nerve impulses originating from a neuroma. Nature, 248, 740-743.

Wall, P.D., Mills, R.G., Fitzgerald, M. and Gibson, S.G. (1982). Chronic blockade of sciatic nerve transmission by tetrodotoxin does not produce central changes in the dorsal horn of the spinal cord of the rat. Neuroci. Lett., 30, 315-320.

Wall, P.D. and Sweet, W.H. (1967). Temporary abolition of pain in man. Science, 155, 108-109.

White, J.C. and Sweet, W.H. (1969). Pain and the Neurosurgeon. C.C. Thomas, Springfield.

Woolf, C.J. (1983). Changes in flexor reflex excitability following acute thermal injury in rat - evidence for central mechanisms. J. Physiol., in press.

Woolf, C.J. and Wall, P.D. (1982). Chronic peripheral nerve section diminishes the primary afferent A-fibre mediated inhibition of rat dorsal horn neurones. Brain Res., 242, 77-85.

SPINAL CORD PLASTICITY INDUCED BY PERIPHERAL NERVE INJURY AND SOME CONSEQUENCES FOR SENSORY DYSFUNCTION

MARSHALL DEVOR

Life Sciences Institute, Hebrew University of Jerusalem, Jerusalem 91904, Israel

When a nerve is cut, the CNS loses access to sensory information from a corresponding portion of the periphery. In addition to this loss of input, however, nerve injury also initiates several positive, and less generally appreciated processes. These include the creation of spurious afferent inputs, and the induction of CNS changes which may profoundly distort the neural processing of both spurious and remaining true signals.

Processes contributing to abnormal afferent discharge include: (i) collateral sprouting of certain types of nociceptors from intact neighboring nerves (Devor et al., 1979; Jackson and Diamond, 1981); (ii) the generation of spontaneous discharge in nerve end neuromas, regenerating sprouts, and patches of demyelination (Wall and Gutnick, 1974; Calvin et al., 1982; Devor, 1983a); (iii) mechanosensitivity at these sites; (iv) altered chemosensitivity, and especially the development of sensitivity to sympathetic outflow and to adrenergic agonists (Wall and Gutnick, 1974; Devor, 1983b); (v) crosstalk between injured fibers, including between afferents and efferents (Seltzer and Devor, 1979; Lisney and Devor, 1983); (vi) spontaneous discharge, mechanosensitivity, chemosensitivity and crosstalk in perpherally axotomized dorsal root ganglion (DRG) cells (De Santis and Duckworth, 1981; Wall and Devor, 1983).

In the present paper I shall review some recent observations on the reorganization of spinal somatotopy induced by nerve injury. In addition, I shall discuss the possible significance of this type of reorganization, in conjunction with the abnormal afferent barrage, for sensory disorders that frequently follow nerve injury.

PLASTICITY IN SPINAL SOMATOTOPY

The Spinal Map

The skin surface is represented in the dorsal horn as a 2-dimensional map. In the rostrocaudal direction, rostral dermatomes (e.g. forelimb) are mapped rostrally; caudal dermatomes (e.g. hindlimb) caudally. In the mediolateral direction distal parts of individual dermatomes (e.g. toes and foot) are represented medially; and proximal parts (e.g. thigh) laterally. This pattern is simply established in a series of closely spaced microelectrode penetrations across the width of the dorsal horn. In normal cats and rats, the medial 1/2 - 2/3 of the dorsal grey matter at the level of the lumbar enlargement is devoted to the toes and foot. That is, virtually all cells in this region have small excitatory receptive fields (RF's) below the ankle. Moving laterally, one encounters larger RF's extending onto the thigh, perineum and lower back. A corresponding map is seen with morphological methods (Devor and Claman, 1980; Koerber and Brown, 1982). The transition between the toe-foot and thigh (etc.) zones is sharp and easy to demarcate.

A priori, primary sensory maps are expected to be organized in development and to be functionally rigid in adulthood; all the more so for an evolutionarily conservative structure like the spinal cord. The discovery of substantial plasticity in the layout of the map following partial neurectomy in mature mammals is particularly striking in light of this expectation.

Plasticity Following Neurectomy

Acute transection with tight ligation of the two nerves innervating the foot, the sciatic and saphenous nerves, has the expected effect of eliminating all toe-foot RF's. Medial cells are left without any RF, and activity cannot even be evoked in the background "hash". Lateral cells lose the distal part of their more extended RF's, but continue to respond essentially normally to skin of the thigh and/or perineum. When the mapping procedure is repeated in animals operated days or weeks earlier, however, the picture in the medial dorsal horn is very different (Fig. 1). Instead of the silence of the acute preparations, the whole region from the dorsal column laterally is alive with responsiveness to skin proximal to the knee. Large unit spikes are frequently encountered that respond to hair movement, touch and/or pinch on the proximal leg. Similarly, the background "hash" responds almost everywhere

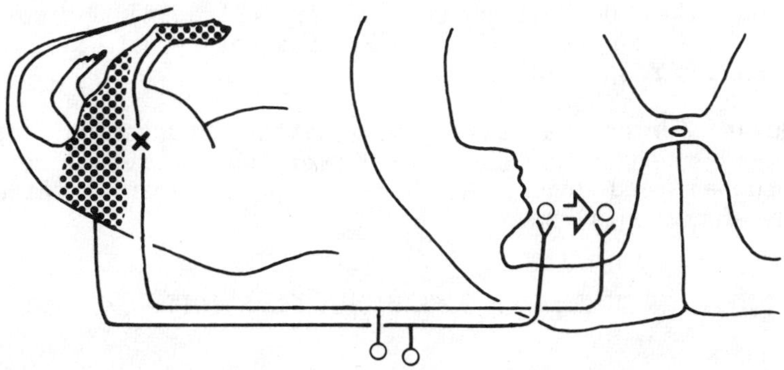

Figure 1: Sketch of the somatotopic projection of the foot and the thigh onto the dorsal horn. The arrow represents the plastic change that results in medial dorsal horn cells responding to cutaneous stimulation of the thigh after transection (X) of the sciatic and saphenous nerves.

(Devor and Wall, 1978, 1981a,b; Lisney, 1983; also see Pomeranz et al., 1983). The structure of the novel RF's resembles that of normal lateral cells. They range widely in size, often feature convergence from high and low threshold cutaneous receptors (including thermal nociceptors) and some, although perhaps fewer than normal, have internal structure.

In cats, the expansion of the former toe-foot RF's onto the thigh takes about four weeks, although early signs are evident in two. In rats, novel proximal RF's appear quite suddenly between the third and fourth postoperative day, with the process mature by the fifth. If the foot nerves are permitted to regenerate, toe-foot RF's return and thigh RF's disappear (Devor and Wall, 1981b; Lisney, 1983). The cells that acquire novel thigh RF's have not yet been characterized in terms of dendritic morphology or efferent connections. They cannot be a rare class of cell, however, as in favorable cases the number of cells with novel proximal RF's encountered per electrode penetration approachs the number with toe-foot RF's in intact animals. Furthermore, at least some respond antidromically to electrical stimulation of the cord at high thoracic levels and are therefore projections cells.

Other Examples of Somatotopic Reorganization

Somatotopic remapping is not peculiar to the dorsal horn following peripheral deafferentation. It is also induced by dorsal rhizotomy, and it occurs at other levels of the neuraxis including the dorsal column nuclei, trigeminal complex, thalamus and cortex (Wall, 1977; Dostrovsky _et al_., 1982; Merzenich and Kass, 1982).

Going further afield, the initial repression (spinal shock) and subsequent recovery and overshoot of spinal reflexes after spinal cord injury may well be another manifestation of the same fundamental spinal lability.

MECHANISM OF SOMATOTOPIC PLASTICITY

CNS or PNS

It is quite certain that the emergence of novel proximal RF's results from spinal synaptic reorganization and not aberrant peripheral regeneration. The most clearcut evidence of this is that re-section of the originally cut nerves does not affect the novel RF's, but they disappear immediately following anaesthetic block or transection of cutaneous nerves serving thigh skin (for other evidence see Devor and Wall, 1981a; Devor, 1983c).

We still lack a comprehensive understanding of the mechanism of the spinal reorganization and of the peripheral events that initiate it. The fundamental problem is to account for the transfer of information from thigh afferents which end primarily in the lateral part of the dorsal horn, to second order cells populating its medial part. For some years now, two candidate possibilities have been invoked in cases such as this: (i) "sprouting," that is, the growth of new neurites and the formation of new synapses and (ii) the strengthening of a pre-existing but functionally ineffective synaptic channel (Wall, 1977).

CNS Sprouting

The classic report by Liu and Chambers (1958) that surviving primary afferents can elongate substantially following rhizotomy, (but see Goldberger and Murray, 1974; Rodin _et al_., 1983) and the now numerous demonstrations of long-distance collateral sprouting after CNS injury in neonates, point to sprouting as the logical explanation. Indeed, by the standards of these reports, the range of sprouting required (several hundred microns) is modest. However, when the spinal distribution of thigh nerve afferents was traced using acid phosphatase histochemistry (Devor and Claman, 1980) and transganglionic transport of lectin conjugated HRP (WGA-HRP, Seltzer and Devor, 1983) there was no indication of even a modest amount of lateromedial sprouting of thigh afferents following transection of the sciatic and saphenous nerves. Furthermore, in Golgi-stained sections of rat spinal cord, there was no indication of mediolateral elongation of dendritic processes (Devor, 1983c). Negative evidence of this sort, naturally, does not rule out the possibility of sprouting. From the robustness of the electrophysiological remapping, however, it is reasonable to have expected robust anatomical remapping (sprouting).

Modulation of Synaptic Efficacy

The reference here is to a functional change in the efficiency of transfer of information from primary to second order neurons, one that does not require large scale changes in axonal or dendritic morphology. Specifically, it is proposed that in normal animals there exists a mono- or polysynaptic channel from thigh skin onto medial dorsal horn cells. Engagement of this channel by natural (mechanical or thermal) stimulation of the thigh, however, is normally too weak to bring the postsynaptic cell to firing threshold. Transection of the foot nerves brings about a strengthening of the relay.

As a heuristic device, the weak channel is sometimes referred to as being made up of "silent" synapses. Unfortunately, this term is somewhat misleading, and it is always irritating to neurophysiological purists, implying as it does a synaptic structure with no postsynaptic action. No such unconventional synapse, of course, has actually been demonstrated in the spinal cord. Absolute "silence," however, is not at issue. Rather, the hypothesis requires only that the synapses be "relatively ineffective" in the sense that they produce small, far subthreshold EPSP's on the postsynaptic cell, and this situation, far from being unconventional, is more or less the rule in CNS neuropil.

There are a large number of factors which determine the efficacy of a synaptic contact. A few pre- and a few post-synaptic examples are illustrated in Figure 2 (see figure legend). Correspondingly, an appropriate change in any of these parameters could strengthen the synaptic contact, that is, increase the likelihood of its evoking an impulse in the postsynaptic cell. Note the distinction drawn between local changes in synaptic size or position, and longer range extension of neurites referred to above under the heading "sprouting".

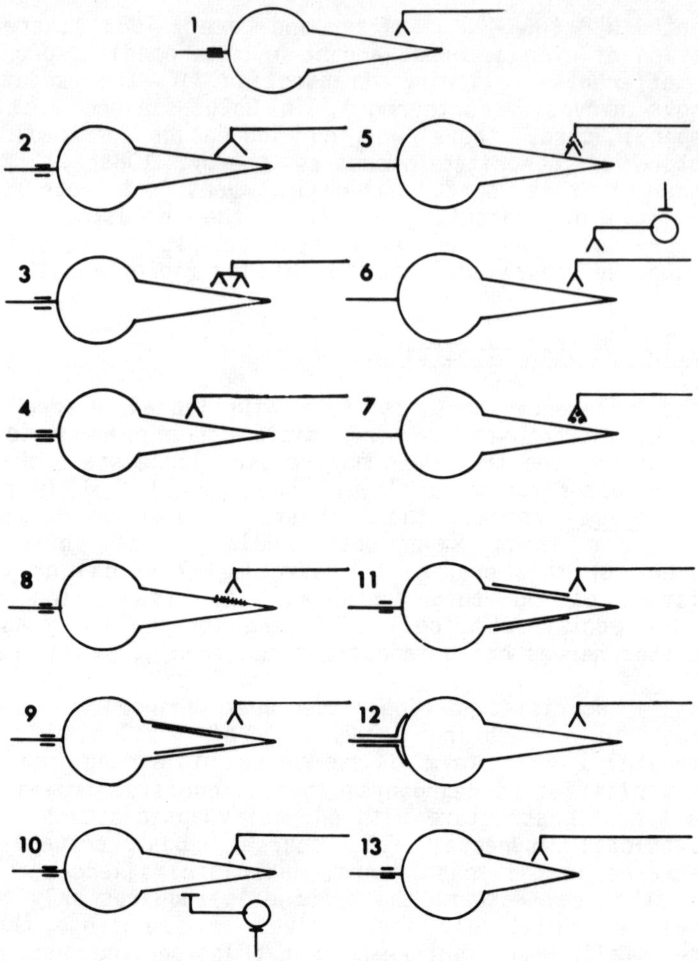

Figure 2: Some presynaptic (2-7) and postsynaptic (8-13) processes capable of changing the effectiveness of a synaptic contact. (1) standard axodendritic synapse; (2) expansion of the presynaptic terminal; (3) local sprouting of additional terminals; (4) translocation of the terminal to a site closer to the spike initiation zone; (5) narrowing of the synaptic cleft; (6) modulation of presynaptic inhibition; (7) increase in amount of transmitter released; (8) increased sensitivity of the postsynaptic element; (9) decrease in dendritic membrane conductance (passive shunting); (10) modulation of postsynaptic inhibition; (11) increased dendritic electrogenicity (e.g. dendritic spikes); (12) increased excitability at the spike initiation zone; (13) dendritic retraction bringing the synapse to a site electrically more favorable.

The strongest evidence for a pre-existing relatively ineffective synaptic channel is the fact that many medial dorsal horn cells with normal toe-foot RF's in intact cats and rats can be driven by electical stimulation of nerves (e.g. the posterior cutaneous nerve of the thigh) that terminate in skin far from the RF. Indeed, latency considerations suggest that many of these contacts are monosynaptic (Devor and Wall, 1981a). The presumption is that with synchronous activation of multiple weak synapses, small EPSP's summate and reach threshold (Merrill and Wall, 1972; Devor, et al., 1977; Devor and Wall, 1981a). In intracellular recordings subthreshold EPSP's can be evoked in a fringe outside of the borders of the natural RF (Mendell et al., 1978, and personal communications), and subliminal EPSP's are no doubt generated over a much larger area. It is not known which of the menagerie of possible mechanisms accounts for the strengthening of these relatively ineffective synapses following peripheral neurectomy. The recent discovery that the dorsal root potential (DRP) and primary afferent depolarization (PAD) collapse at just the time that novel RF's appear (Wall and Devor, 1981; Wall, 1982), suggests that a presynaptic disinhibitory process may be involved (Fig. 2 (6)).

Anatomical Substrates

One of the persistent problems with the modulation hypothesis is the matter of structure. Even though functionally weak, relatively ineffective synapses must still have some morphological substrate. Of course, part of the novel RF's could be activated polysynaptically. There is no morphological problem here given the broad mediolateral spread of local axonal arbors of many second order sensory neurons (Brown, 1981; Egger et al., 1981). Two recent observations may now have resolved the problem for the monosynaptic case. First, many medial dorsal horn neurons have substantial mediolateral dendritic spread. In normal rats, for example, roughly half of neurons in the exclusive toe-foot zone have at least one lateral dendrite long enough to enter the region of dense thigh afferent terminals (Devor, 1983c). Second, tracing studies using WGA-HRP (Seltzer and Devor, 1983) show that many thigh afferents enter the dorsal columns and then sweep laterally through the medial dorsal horn (Fig. 3). Ultrastructural studies will be required to establish whether or not synaptic contacts are formed en passant, and whether such synapses change detectably (e.g. by local sprouting) following nerve section.

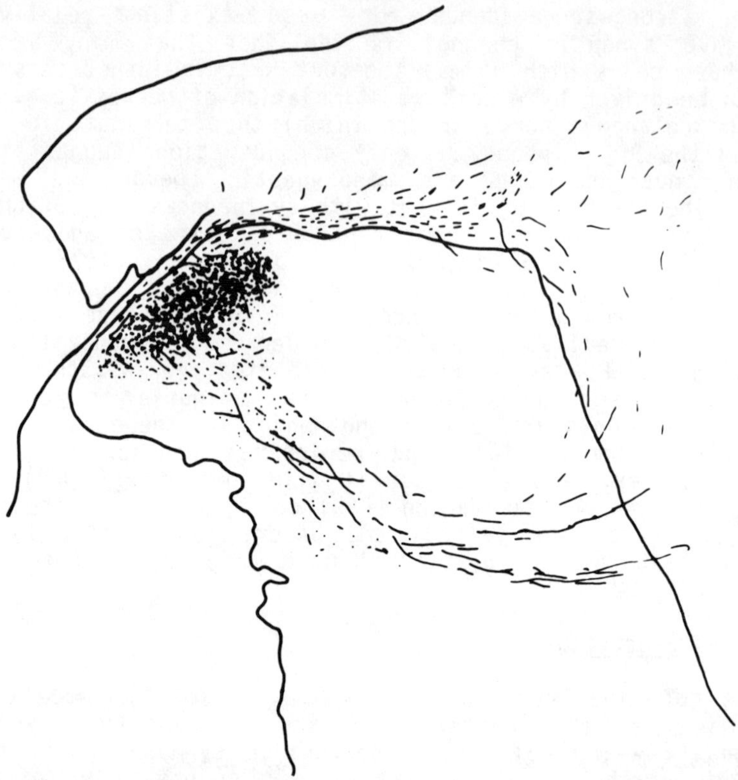

Figure 3. Camera-lucida drawing of the trajectory of thigh afferent fibers entering the lateral dorsal horn (from Seltzer and Devor, 1983).

The Retrograde Signal

The nature of the retrograde signal that induces the central change resulting in RF expansion remains unclear. The simplest possiblity, of course, is that nerve section blocks afferent impulse traffic. Several factors weigh against this possibility, however. First, impulses are blocked instantaneously but synaptic reorganization in the dorsal horn requires days or weeks. Futhermore, at the time novel RF's appear in rats at least, the nerve is not silent. As mentioned above, the nerve end neuroma produces a massive ectopic afferent barrage. Third, nerve crush and cut have the same effect on impulse traffic. Nerve crush, however, does not induce novel RF's (Devor and Wall, 1981b). Finally, chronic nerve conduction block with tetrodotoxin does not bring about RF expansion (Wall et al., 1982).

Recent data hint that the signal may, in fact, involve a trophic substance(s) rather that impulses. Specifically, when colchicine in concentrations adequate to block anterograde and retrograde axoplasmic flow is applied to the sciatic and saphenous nerves proximal to the point of nerve section, the emergence of novel RF's is largely prevented (Devor, 1983c). A trophic signal could also account for the difference between nerve cut and crush, and perhaps also for the difference in the timing of spinal reorganization in small (rats) and large (cats) animals. Obviously the source and nature of the trophic substance, if one indeed exists, has still to be determined.

SOME CONSEQUENCES FOR SENSORY DYSFUNCTION AND PAIN

Sensory Dysfunction

The shift in dorsal horn somatotopy, and allied physiological changes, shed new light on a number of curious and unexplained sensory abnormalities in nerve injury patients. Some examples: (i) In amputees, (and after other injury to major nerve trunks) cutaneous stimulation on the stump and torso sometimes produces sensations referred to the phantom (e.g. Cronholm, 1951; Howe, 1983). This, of course, is a straightforward prediction given the novel response of medial dorsal horn cells to proximal afferents. (ii) The quality of such referred sensations may also be seriously distorted. Weak stimuli such as hair movement, for example, may cause intense pain in the phantom (Cronholm, 1951; Howe, 1983). This could occur if central cells that originally signalled pain acquire novel inputs from low threshold afferents. (iii) Although stimuli on the stump may be distorted in terms of location and quality, overall many more dorsal horn neurons than normal now serve this region. Such anomalous over-representation should permit finer distinctions to be made between nearby stimuli. This could contribute to the improvement in two-point discrimination known to characterize amputation stumps (Teuber et al., 1949; Haber, 1955). (iv) Following nerve damage DRP and PAD onto the central terminals of cut nerve fibers is suppressed (Wall and Devor, 1981; Wall, 1982). This is expected to reduce spatial (surround) and temporal inhibition (Woolf and Wall, 1982) and therefore to intensify paraesthesias and pain associated with spontaneous and evoked discharge originating in and near neuromas, in axotomized DRG cells, and perhaps also in cut fibers that have regenerated abnormally. (v) A possibly important consequence of expanded RF's is that due to increased RF overlap, a given stimulus activates many more dorsal horn neurons than previously. The exaggerated ascending volley that results could contribute to post-traumatic hyperaesthesia and hyperalgesia.

Conclusion

Synaptic organization in the dorsal horn, and elsewhere in the somatosensory system as well, shows substantial lability in the face of partial deafferentation. The resulting distortion of central processing, in conjunction with the abnormal afferent signals produced by nerve injury, create a potential for sensory disorders which cannot be fully understood by consideration of peripheral or central processes in isolation.

ACKNOWLEDGEMENTS

The work described here was done in collaboration with D. Claman, I. Frank, M.-L. Obermeier, Z. Seltzer and P.D. Wall. Supported by the Thyssen Foundation and the Israel Academy of Sciences. I thank E. Glazer, M. Kaitz, A. Schilling and S. Tsarnas for their help with the manuscript.

REFERENCES

Brown, A.G. (1981). Organization in the Spinal Cord. Springer Verlag, Berlin.

Calvin, Wm. H., Devor, M. and Howe, J.F. (1982). Can neuralgias arise from minor demyelination? Spontaneous firing, mechanosensitivity and afterdischarge from conducting axons. Expt. Neurol., 75, 755-763.

Cronholm, B. (1951). Phantom limbs in amputees. Acta. Psychiat. Neurol. Scand., Suppl. 72, 1-310.

DeSantis, M. and Duckworth, J.W. (1982). Properties of primary afferent neurons from muscle which are spontaneously active after a lesion of their peripheral processes. Expt. Neurol., 75, 261-274.

Devor, M. (1983a). The pathophysiology and anatomy of damaged nerve. In: Textbook of Pain. (eds. P.D. Wall and R. Melzack) Churchill Livingston, London, In Press.

Devor, M. (1983b). Nerve pathophysiology and mechanisms of pain in causalgia. J. Autonom. Nerv. Syst., 7, 371-384.

Devor, M. (1983c). Plasticity of spinal cord somatotopy in adult mammals: Involvement of relatively ineffective synapses. In: Nervous System Regeneration. (eds. A.-M. Giuffrida-Stella, B. Haber, G. Hashim and J.R. Perez-Polo) Alan R. Liss, New York, In Press.

Devor, M. and Claman, D. (1980). Mapping and plasticity of acid phosphatase afferents in the rat dorsal horn. Brain Res., 190, 17-28.

Devor, M. and Bernstein, J.J. (1982). Abnormal impulse generation in neuromas: Electrophysiology and ultrastructure. In: Abnormal Nerves and Muscles as Impulse Generators. (eds. J. Ochoa and Wm. Culp) Oxford University Press, New York. pp.363-380.

Devor, M., Merrill, E. and Wall, P.D. (1977). Dorsal horn cells responding to stimulation of distant dorsal roots. J. Physiol., 270, 519-531.

Devor, M. Schonfeld, D., Seltzer, Z. and Wall, P.D. (1979). Two modes of cutaneous reinnervation following peripheral nerve injury. J. Comp. Neurol., 185, 211-220.

Devor, M. and Wall, P.D. (1978). Reorganization of the spinal cord sensory map after peripheral nerve injury. Nature, 276, 75-76.

Devor, M. and Wall, P.D. (1981a). The effect of peripheral nerve injury on receptive fields of cells in the cat spinal cord. J. Comp. Neurol., 199, 277-291.

Devor, M. and Wall, P.D. (1981b). Plasticity in the spinal cord sensory map following peripheral nerve injury in rats. J. Neurosci., 1, 679-684.

Dostrovsky, J.D., Ball, G.J., Hu, J.W. and Sessle, B.J. (1982). Functional changes associated with partial tooth pulp removal in neurons of the trigeminal spinal tract nucleus, and their clinical implications. In: Anatomical, Physiological and Pharmacological Aspects of Trigeminal Pain. (eds. B. Matthews and R.G. Hill). Excerpta Medica, Amsterdam. pp. 293-310.

Egger, M.D., Freeman, N.C.G., and Proshansky, E. (1981). The significance of laminar arrangement. In: Spinal Cord Sensation. (eds. A.G. Brown and M. Rethelyi) Scottish Academic Press, Edinburgh.

Goldberger, M.E. and Murray, M. (1974). Restitution of function and collateral sprouting in the cat spinal cord: The deafferented animal. J. Comp. Neurol., 158, 37-54.

Haber, W.B. (1955). Effects of loss of limb on sensory functions. J. Psychol, 40, 115-123.

Howe, J.F. (1983). Phantom limb pain - a re-afferentation syndrome. Pain, 15, 101-107.

Jackson, P.C. and Diamond, J. (1981). Regenerating axons reclaim sensory targets from collateral nerve sprouts. Science, 214, 926-928.

Koerber, H.R., and Brown, P.B. (1982). Somatotopic organization of hindlimb cutaneous nerve projections to cat dorsal horn. J. Neurophysiol., 48, 418-489.

Lisney, S.J.W. (1983). Changes in the somatotopic organization of the cat lumbar spinal cord following peripheral nerve transection and regeneration. Brain Res., 259, 31-39.

Lisney, S.J.W. and Devor, M. (1983). Interactions between fibers in damaged peripheral nerve in the rat. In Press.

Liu, C.N. and Chambers, W.W. (1958). Intraspinal sprouting of dorsal root axons. Arch. Neurol. Psychiat., 79, 46-61.

Mendell, L.M., Sassoon, E.M. and Wall, P.D. (1978). Properties of synaptic linkage from long ranging afferents onto dorsal horn neurons in normal and deafferented cats. J. Physiol., 285, 299-310.

Merrill, E.G. and Wall, P.D. (1972) Factors forming the edge of a receptive field. The presence of relatively ineffective afferents. J. Physiol., 226, 825-846.

Merzenich, M.M. and Kaas, J.H. (1982). Reorganization of mammalian somatosensory cortex following peripheral nerve injury. TINS, Dec., 433-436.

Pomerantz, B.H., Markus, H. and Krushelnycky, D. (1983). Spread of saphenous somatotopic projection map in spinal cord and behavioral effects after chronic sciatic denervation in adult rat. In Press.

Rodin, B.E., Sampogna, S.L. and Kruger, L. (1983). An examination of intraspinal sprouting in dorsal root axons with the tracer horseradish peroxidase. J. Comp. Neurol., 215, 187-198.

Seltzer, Z. and Devor, M. (1979). Ephaptic transmission in chronically damaged peripheral nerves. Neurology, 29, 1061-1064.

Seltzer, Z. and Devor, M. (1983). Effect of nerve section on the spinal distribution of neighboring nerves. In Press.

Teuber, H.L., Krieger, H.P. and Bendell, M.B. (1949). Reorganization of sensory function in amputation stumps: Two-point discrimination. Fed. Proc., 8, 156.

Wall, P.D. (1977). The presence of ineffective synapses and the circumstances which unmask them. Phil. Trans. Roy. Soc. B., 278, 361-372.

Wall, P.D. (1982). The effect of peripheral nerve lesions and of neonatal capsaicin in the rat on primary afferent depolarization. J. Physiol., 329, 21-35.

Wall, P.D. and Devor, M. (1981). The effect of peripheral nerve injury on dorsal root potentials and on transmission of afferent signals into the spinal cord. Brain Res., 209, 95-111.

Wall, P.D. and Devor, M. (1983). Sensory impulses originate in dorsal root ganglia in normal and nerve injured rats. In press.

Wall, P.D. and Gutnick, M. (1974). Ongoing activity in peripheral nerves: The physiology and pharmacology of impulses originating from a neuroma. Exp. Neurol., 43, 580-593.

Wall, P.D., Mills, M., Fitzgerald, M. and Gibson, S.J. (1982). Chronic blockade of sciatic nerve transmission by tetrodotoxin does not produce central changes in the dorsal horn of the spinal cord of the rat . Neurosci. Lett. 38, 315-320.

Woolf, C.J. and Wall, P.D. (1982). Chronic peripheral nerve section diminishes the primary afferent A-fiber mediated inhibition of rat dorsal horn neurons. Brain Res., 242, 77-85.

SENSIBILITY ABNORMALITIES IN NEURALGIC PATIENTS STUDIED BY THERMAL AND TACTILE PULSE STIMULATION

HEINRICH FRUHSTORFER and *ULF LINDBLOM

Institute of Physiology, University of Marburg, D-3550 Marburg, FRG
**Department of Neurology, Karolinska Hospital, S-104 01 Stockholm, Sweden*

SUMMARY

Six patients with neuralgia and disturbed thermosensibility were examined with graded thermal and mechanical pulses which were applied in the pain area and in contralateral normal skin for comparison. In five patients stimuli in the non-noxious range evoked pain. Reaction times were modality specific which suggests that the pathophysiology of the abnormal stimulus-evoked pain sensations was central although the original lesion was in the peripheral nerve. With repetitive stimulation habituation of sensation was observed in five patients and abnormal summation in two. It was concluded that pulsed natural stimuli may be valuable in further analyses of the mechanisms of sensory dysfunction and pain, and possibly also in clinical evaluation.

INTRODUCTION

In the affected skin of patients suffering from neuralgia, shifts in threshold or altered gradients of the intensity function for mechanical and/or thermal stimulation can usually be detected in various combinations (Lindblom and Verillo 1979). Often pain thresholds are lowered to such an extent that even the weakest stimuli evoke pain (i.e. allodynia; see Merskey 1982). If elicited by identical tactile pulses, pain in the affected skin and touch on the contralateral side appear with the same short latency indicative of a fast peripheral conduction. Thus, the abnormal pain responses seem to be due do a secondary central dysfunction rather than to an abnormal discharge at the site of the lesion in the peripheral nerve.

The aim of the present investigation was a further analysis of the altered pain response in the same type of patients by means of graded warm and cold pulses. This procedure enabled the measurement of reaction times also to thermal stimuli and helped to define the abnormal sensations with respect to changes in quality, spatial distribution and time course. It is hoped that the present results and further studies with a similar approach will contribute to the understanding of the pathophysiology of hyperpathia and spontaneous pain of neurologic type.

METHODS

Patients. The patient group consisted of six female patients aged 36-55 years. They were selected from a larger group of neuralgia patients who had been examined with quantitative sensory tests (Lindblom and Verillo, 1979) and who had displayed an abnormal thermal sensibility. They suffered from continuous neuralgic pain due to entrapment or trauma (Table I). Two patients had signs of dysautonomia with bluish skin colour, reduced skin temperature and

Table I. Duration, region and type of spontaneous pain in the six patients.

Patient No.	Age	Duration (years)	Region	Type	Affected Nerve	Aetiology
1	47	4	Thigh	Aching Burning	Cut.lat.fem.	Entrapment
2	36	2	Thigh	Aching	Cut.lat.fem.	Entrapment
3	44	4	Knee	Aching Burning	Saphenous	Entrapment
4	54	5	Knee	Aching Stabbing	Saphenous	Trauma
5	55	10	Foot	Aching Burning	Sciatic	Trauma
6	54	4	Foot	Burning Throbbing	Sural	Trauma

oedema. None of the patients had responded to sympathetic blocking procedures.

Pain scaling. The intensity of the spontanous pain was recorded at irregular intervals before, during, and after sensory examination; for this the patients had to fill in a visual analogue scale (no pain - maximal pain).

Thermal thresholds. Cold, warm and thermal pain thresholds were determined by the Marstock method (Fruhstorfer et al.,1976) in the painful area and on the corresponding contralateral site. The size of the thermode was 25 x 50 mm.

Thermal pulses. Thermal stimuli were presented to both painful and corresponding normal skin areas by a water-circulated thermode of 18 mm diameter which allowed rapid temperature changes (Fruhstorfer and Detering, 1974). Usually, 10 identical thermal pulses of 2.5 s duration were presented at a rate of 1/30 s. The amplitude of the pulses was 5, 10, or 15 °C either to the warm or cold side starting from an adapting temperature of 30 or 35 °C. In special cases stimuli of smaller amplitude or shorter duration were used.

Tactile pulses. Tactile pulses (half sine-wave) of 10 ms duration were presented to abnormal or corresponding contralateral normal skin by the plastic probe (diameter 2mm) of a mechanostimulator (Lindblom, 1974). In the reaction time measurements a stimulus intensity of 2-3 times the threshold amplitude for perception was chosen. Series of 10 stimuli were presented at a rate of 1/30 s.

Reaction times and sensations. Reaction times to thermal and tactile pulses were recorded with a precision of 1 ms and from each trial of 10 stimulations the median reaction time was calculated. The sensations were noted with respect to their quality, spatial distribution and duration.

RESULTS

Spontaneous pain. During their stay in the laboratory all patients had continuous spontaneous pain of a burning or aching nature in the distribution of the injured nerve. The spontaneous pain was not changed in a systematic way as a consequence of sensory testing (Fig.1). In one patient (case 3) both thermal and mechanical stimulation produced increasingly painful aftersensations which outlasted sensory testing by more than two hours. In contrast to this, in another patient (case 6), when determining the heat tolerance level, the heat application gave a several hours long

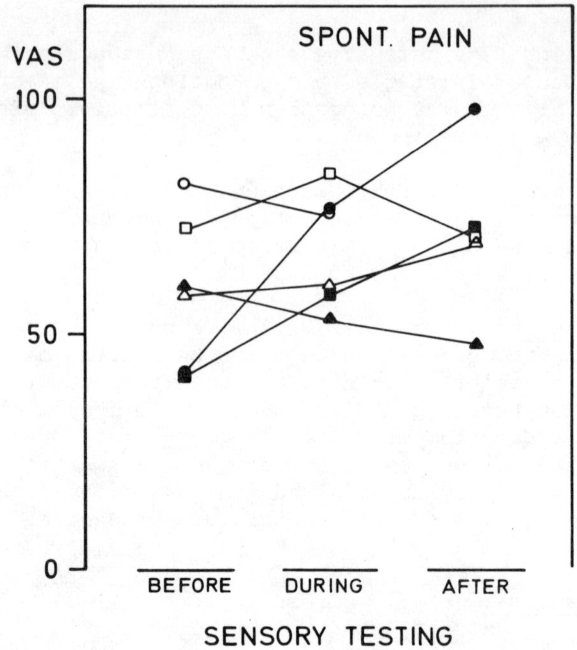

Figure 1. Visual analogue scale (VAS) ratings of spontaneous pain before, during and after sensory testing in the six patients.

lasting relief of the spontaneous pain; this effect could be repeated in the same patient on later ocasions.

Thermal thresholds. In five patients (case 1 to 5) thermal sensibility was altered in the painful skin area as revealed by the initial Marstock test (Fig.2). The indifferent temperature range between cold and warm thresholds was either wider (case 1, 3 and 5), smaller (case 4) or of normal size (case 2 and 6). The most conspicuous finding was a reduction in thermal pain thresholds on the abnormal side for either cold (case 1 to 4) or both cold and warmth (case 1 and 4). In cases 2, 3 and 4 the cold pain was evoked at temperatures which were close to skin temperature and clearly outside the noxious range.

Reaction times. In contrast to the thermal thresholds, the reaction times to thermal pulse stimulation (Fig.3) did not show clear side differences except for one patient (case 4). She had considerably shorter reaction times for warm stimulion the abnormal side. One patient (case 6) had unusually short reaction times for

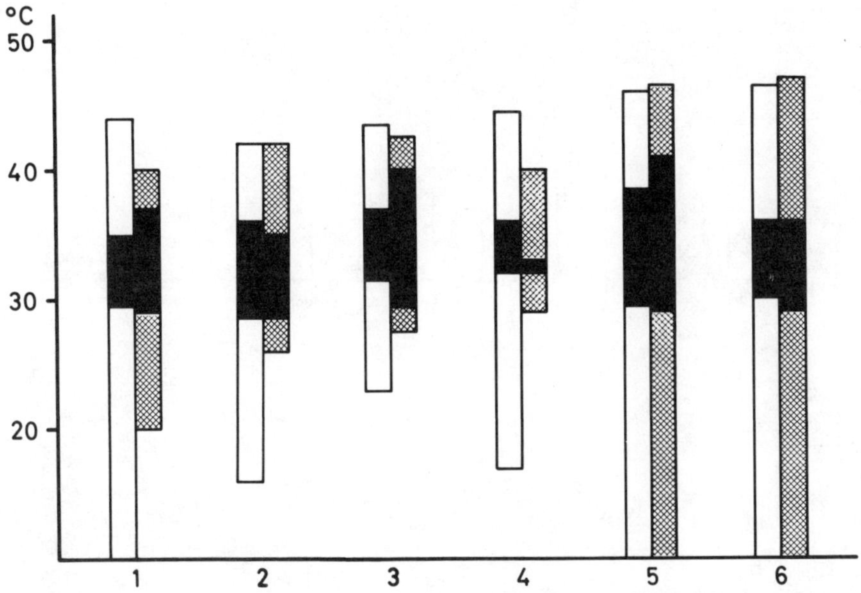

Figure 2. Cold and warm thresholds (filled columns) together with cold pain and heat pain thresholds (open columns) of the six cases. Cold pain thresholds below 10 °C were not determined. Abnormal side hatched.

thermal as well as tactile stimulation. The tactile reaction times were largely symmetrical except for one patient (case 1) who had longer reaction times for touch on the non-painful side; the reason for this was obscure.

Alterations in quality, distribution and time course of evoked sensations. Usually both thermal and tactile pulses evoked qualitatively abnormal sensations. They were described as "different", "stronger" or painful. When the sensation was painful there was often a reminiscent tactile or thermal sensation immediately before the pain or as an aftersensation. All instances of abnormal pain sensations evoked by tactile, cold or warm stimuli are shown in table II. The abnormalities with respect to the spatial and temporal dimensions are also displayed in table II. Radiation of sensation outside of the area of the stimulus probe could occur to a large area such as the whole lower leg. The radiation was usually rapid at the onset of stimulation, and if there was an appreciable aftersensation, the distribution reversed and shrank successively as the

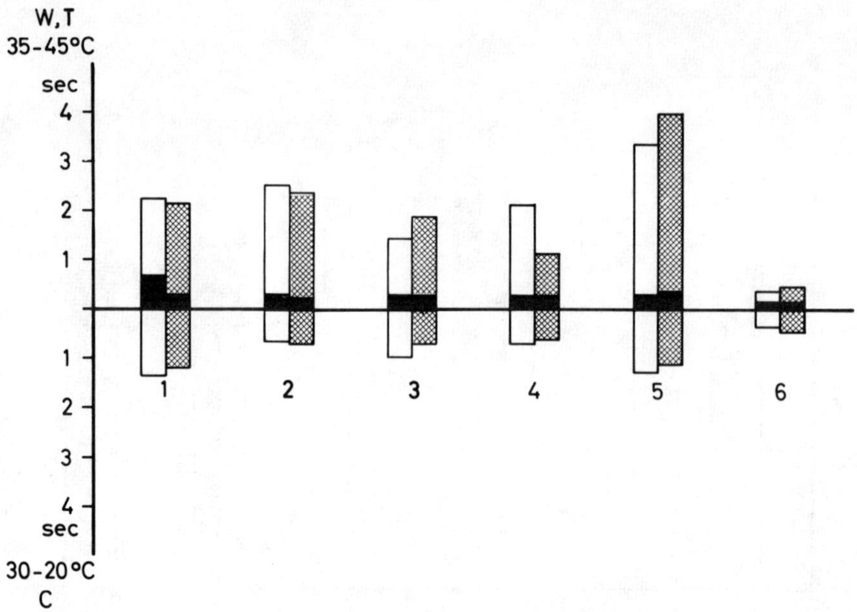

Figure 3. Reaction times to tactile (T, filled columns), warm (W, upper part of open columns) and cold (C, lower part of open columns) stimulation in the six patients. Abnormal side hatched.

sensation faded. The most common temporal abnormality were aftersensations which were recorded in seven instances. The duration varied from a few seconds to some minutes. On repetitive stimulation with constant or increasing pulse duration, it was observed in six instances that the painful sensation habituated and left only the basic modality of sensation. On the other hand, the reverse phenomenon, i.e. summation, occurred in three instances; here successive stimulations resulted in increasing pain and could eventually lead to a prolonged exacerbation of pain. In one patient (case 5) the warm pulses delivered to the non-painful foot evoked a sensation in the pain area (mirror sensation). In another patient (case 6) warm pulses on the abnormal side were felt as cold (paradoxical sensation).

DISCUSSION

One of the most intriguing questions in nerve injuries with chronic pain is to what extent the pain originates from an abnormal impulse

Table II. Occurrence of pain, radiation and abnormal time course of sensations evoked by innocuous tactile (T), cold (C) and warm (W) stimulation.

Patient No.		1	2	3	4	5	6
Pain	T	+		+	+	+	
	C	+	+	+	+		
	W	+			+		
Radiation	T	+		+	+	+	+
	C	+				+	
	W	+					
Abnormal Time Course	T	+		+	+		
	C	+	+		+		
	W	+		+	+		+

discharge in nociceptive afferent fibers in the periphery. Relapses of pain after section of the afferent nerve, even when a temporary block had suggested a peripheral origin, indicate that some central dysfunction was responsible for the production of pain even though the original lesion was peripheral. Signs of exaggerated sensibility in the pain area such as aftersensation and radiation are apparently best explained on a central basis, e.g. facilitation of the subliminal fringe of relay interneurones. Allodynia might be due to disinhibition of the polymodal interneurones which respond to both noxious and innocuous stimuli. On the other hand, all these phenomena of disturbed sensibility could alternatively be produced by amplification of discharge and cross-talk between touch and pain fibers in the peripheral lesion. They are therefore not reliable indices for differentiation between peripheral and central mechanisms.

In a recent investigation of neuralgia patients (Lindblom and Verillo, 1979), the reaction times of the tactile allodynia were too short to explain this type of pain response as due to ephapses between touch and pain fibers in the periphery. In the present investigation it was found that the reaction times of thermal allodynia were different for warm and cold pulses and, as with tactile stimulation, of the same order as for the respective normal sensations. That suggests that the afferent impulses were conducted in the tactile, cold and warm fibers, respectively, at least along the peripheral pathway up to the spinal cord. Thus all instances of allodynia observed in this study were most probably due to a central dysfunction. If the spontaneous pain from which all patients suffered was produced by the same mechanism, a peripheral neurotomy would not be a logical procedure for pain relief in these cases. It is suggested that reaction time determination should tentatively be tried as an additional test in the clinic for prognostic evaluation of neurotomy or neurolytic blocks.

Another outcome of the present study was the potential value of using natural, pulsed stimuli for a closer description and analysis of various sensory abnormalities. Phenomena like summation, aftersensation and radiation are well known and easily recognized with conventional bedside sensibility tests. But the rather frequent observation of habituation of the response was unexpected and would hardly have been made without using stimulation with pulses of variable strength and duration. It is tempting to speculate that summation and habituation reflect antagonistic forms of a disturbed central adaptation. The small material of the present study precludes a closer discussion of the pathophysiology and of the question of a common basis for hyperpathia and resting pain. The irregular display of various abnormal phenomena, which appears in table II, necessitates a large material for systematic observations.

ACKNOWLEDGEMENTS

We are grateful to Berit Lindblom for technical and clinical assistance. The study was supported by Karolinska Institutets Fonder and Vivian L. Smith Foundation for Restorative Neurology (U.L.), and Deutsche Forschungsgemeinschaft (H.F.).

REFERENCES

Fruhstorfer, H. and Detering, I. (1974). A simple thermode for rapid temperature changes. Pflügers Arch. ges. Physiol., 349, 83-85.

Fruhstorfer, H., Lindblom, U. and Schmidt, W. G. (1976). Method for quantitative estimation of thermal thresholds in patients. J. Neurol. Neurosurg. Psychiat., 39, 1071-1075.

Lindblom, U. (1974). Touch perception threshold in human glabrous skin in terms of displacement amplitude on stimulation with single mechanical pulses. Brain Res., 82, 205-210.

Lindblom, U. and Verillo, R. T. (1979). Sensory functions in chronic neuralgia. J. Neurol. Neurosurg. Psychiat., 42, 422-435.

Merskey, H. (1982). Pain terms: A supplementary note. Pain, 14, 205-206.

THE ROLE OF C-NOCICEPTORS IN CUTANEOUS HEAT PAIN AND HYPERALGESIA

ROBERT H. LAMOTTE

Department of Anesthesiology, Yale University School of Medicine, New Haven, Connecticut 06510, USA

INTRODUCTION

Neurophysiological studies have demonstrated the existence of cutaneous nociceptors that respond with graded discharge to noxious heat stimuli of different temperatures. The studies have also shown that certain nociceptors can be sensitized to heat after cutaneous heat injury (Beitel and Dubner, 1976a,b, Bessou and Perl, 1969, Campbell et al., 1979, Fitzgerald and Lynn, 1977, Iggo, 1959, Kumazawa and Perl, 1977, Lynn, 1979, 1980, Meyer and Campbell, 1981, Perl et al., 1976, Torebjörk and Hallin, 1974). These results indicate a role of cutaneous nociceptors in heat pain and the hyperalgesia that develops after cutaneous injury. However, no clear relationship between nociceptor responses and the sensation of pain can be established until combined neurophysiological and psychophysical studies are carried out in the same or related species using the same set of experimental stimuli. With this idea in mind, my colleagues and I have investigated some peripheral neural mechanisms of cutaneous heat pain and hyperalgesia (LaMotte et al., 1982, 1983, Torebjörk et al., 1984). The goals of these studies were to determine the following: (1) which set of afferent fibers and which aspects of their discharge contribute to the threshold, magnitude, and duration of cutaneous heat pain; (2) after heat injury, which cutaneous receptors and which characteristics of their discharge contribute to hyperalgesia - the latter characterized as a lower than normal heat pain threshold and greater than normal magnitude ratings and durations of heat pain.

The magnitude of pain during each heat stimulation changes as a function of time. Thus, in order to obtain measures of the temporal profiles of magnitude judgments of pain, we required our subjects to make continuous magnitude ratings of pain by moving a lever along a line intended to represent the magnitude of pain (Figure 2A). The bottom of the line was no pain, and the top of the line was maximum pain in relation to a standard delivered only once at the beginning of the experiment. All test stimuli were presented on a base temperature of 38°C, maintained between stimulations. Stimuli of 39 to 51°C, each of 5 s duration, were presented in order of ascending temperature in steps of 2°C with an interstimulus interval of 25 s. Then, a mild heat injury was produced by a conditioning stimulus (CS) - usually either 50°C, 100 s duration, or 48°C, 360 s. It was a "mild heat injury" (see LaMotte et al, 1982), because the CS produced only erythema with no visible edema, which disappeared after several days. After the CS, stimuli of 39 to 47°C were given starting 0.5 min and again 5 min after the CS. Stimuli of 39 to 51°C were given starting 10 min after the CS and, in some experiments, again 20 min after.

The pain ratings of the stimuli delivered prior to the CS were regarded as judgments obtained from normal skin, since these test stimuli were found not to produce hyperalgesia or heat sensitization of nociceptors (LaMotte et al., 1982). Responses following the CS indicated the development of hyperalgesia. Separate tests were made on the hairy skin of the volar forearm, calf or foot, and the glabrous skin of the hand (thenar eminence).

The same series of heat stimuli were presented to the cutaneous receptive fields of four types of heat sensitive receptors, as recorded from hairy and glabrous skin of the anesthetized monkey: (1) C-fiber mechanoheat (CMH) nociceptors (C-polymodal nociceptors); 2) A-fiber mechanoheat (AMH) nociceptors; 3) low-threshold "warm" receptors (with C fibers), and 4) low threshold "cold" receptors (with A-delta fibers).

In a subsequent study, the same heat stimuli were delivered to the hairy skin of the calf or foot on the receptive fields of single CMH nociceptive afferent fibers recorded in awake humans while the subjects were simultaneously rating the magnitude of pain.

THE PERIPHERAL NEURAL EVENTS UNDERLYING JUDGMENTS OF PAIN DURING HEATING OF NORMAL SKIN

The results we obtained from normal skin supported the hypothesis that the mean frequency of discharge of CMH nociceptors provided the major peripheral contribution to the threshold, magnitude and duration of heat pain. The evidence for this is summarized in the following:

(1) The distributions of the response thresholds of only CMH nociceptors and not those of any other cutaneous receptors, as recorded in the monkey, overlapped the distributions of heat pain thresholds of humans, both for hairy and for glabrous skin. For the hairy skin (Figure 1), a mean threshold of 45.5^{o}C was obtained for CMHs and 44.3^{o}C for pain sensation. For glabrous skin, mean thresholds of 44.5 and 43.3^{o}C were obtained for CMHs and pain sensation, respectively. The CMHs recorded from human hairy skin had thresholds of 41-43°C, but we think it likely that CMHs with higher thresholds exist in humans since those with thresholds as high as 47°C have been recorded from hairy skin of the arm in humans (Van Hees and Gybels, 1981).

The response properties of CMHs in the anesthetized monkey and the awake human were similar. For CMHs with comparable heat thresholds (41-43°C), the mean frequency of discharge to each heat stimulus (averaged for all CMHs tested) was virtually identical for CMHs in monkeys and humans, both in the overall magnitude of discharge and in the temporal profile of discharge.

Most of the AMH nociceptors recorded from the money had heat thresholds of 51°C or greater. However, our sample did not include the more sensitive A-delta heat nociceptors known to exist in monkey (Dubner et al., 1977) and man (Adriaensen et al., 1983). While these more sensitive AMH nociceptors may well have been active during heat stimulations in our studies, they did not evoke pain response latencies that were too short to be explained by conduction along C fibers, and they did not elicit consistent double sensations of first and second pain.

(2) A compression block of conduction in A fibers in humans did not alter either the pain thresholds or the magnitude ratings of pain nor did it lengthen the response latencies of pain ratings during heat stimulations of normal skin. These results indicated that AMH nociceptors did not play an essential role in mediating heat pain sensations in our experiments.

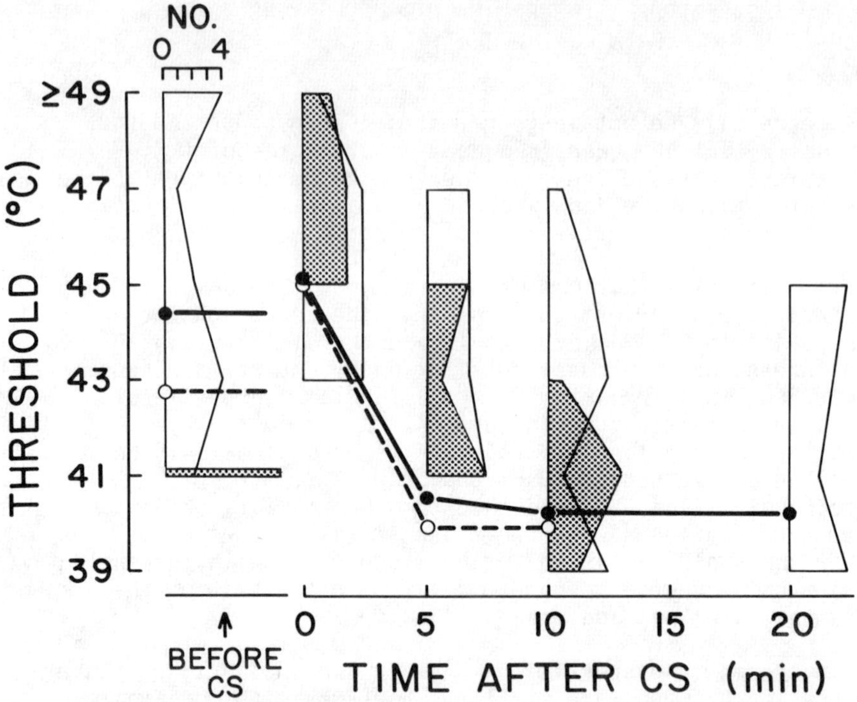

Figure 1. The mean pain threshold and the distributions of response thresholds of CMH nociceptors before and after a mild heat injury of the hairy skin. The heat injury was produced by a conditioning stimulus (CS) which, in most cases, was 50^{0}C and 100 s duration. The data are indicated for the times at which a sequence of test stimuli began (see main text). The time course of the development of heat sensitization of CMH nociceptors: The number (No) of CMH nociceptors having the indicated threshold temperatures is represented by the width of the histogram at each temperature. At each time of testing, two histograms are shown, one representing the distributions of CMH in monkey (solid line), the other distributions of CMHs in humans (shaded area). CMHs in humans were tested only up to 10 min after the CS. The time course of the development of hyperalgesia: The closed circles (solid, thicker lines) represent the mean pain thresholds of 13 human subjects tested on the forearm. The open circles (dashed lines) are the mean pain thresholds of eight subjects tested on the back of the hand during a compression block of conduction in myelinated peripheral nerve fibers. This block, which effectively eliminated a contribution to pain from AMH nociceptors, failed to alter the time course of hyperalgesia. From data presented in LaMotte et al., 1982.

(3) The suprathreshold responses of the individual CMH nociceptor and the magnitude ratings of pain made by the individual subject commonly increased monotonically with stimulus temperature. This is illustrated in Figure 2 by the ratings of one subject and the pattern of discharge in a single CMH nociceptor recorded in the monkey. In general, greater magnitude ratings of pain were associated with longer durations of pain, and pain duration could outlast the end of a higher temperature stimulus by as much as 10 sec or more. In contrast, the responses of each CMH nociceptor in human as well as in monkey stopped within 1-2 sec after the onset of cooling (downward arrow in Figure 2).

The recordings of CMH activity in awake humans allowed us to obtain simultaneous measures of magnitude ratings of pain and the responses of the single CMH nociceptor. It was found that the relation between the subject's scaling function and the intensity-response function of the CMH nociceptor varied somewhat from one experiment to the next regardless of whether the results were obtained from the same or from different subjects. However, when averages were computed for all 14 CMH nociceptors tested, there was a linear relationship between the mean number of impulses evoked in CMHs by each heat stimulus and the median maximum ratings of pain (over the stimulus range of 45 to 51°C). Thus, it appeared that the magnitude of heat pain sensation was more closely related to the magnitude of response in a population of CMH nociceptors than in responses of any individual nociceptor.

A linear relation between the mean number of impulses evoked by each stimulus in CMHs and median ratings of pain was also found in comparisons between discharges in monkey CMHs and human pain ratings for the hairy skin of the arm and also for discharges in monkey CMHs and pain ratings during heat stimulation of the glabrous skin. The linearity did not necessarily hold between the number of impulses evoked in the single CMH and the magnitude ratings of any individual subject, even when both measures were obtained simultaneously in awake human. The intensity-response functions (total evoked impulses vs. stimulus temperature) differed, as did response thresholds, for different CMHs, particularly for the monkey CMHs. Further, the recordings of CMHs in humans indicated that discharges in single CMHs with thresholds of 41°C typically began several degrees below the stimulus temperature that evoked a threshold rating of pain in the subject. Thus, heat pain sensation required input from more than one CMH and, as the stimulus

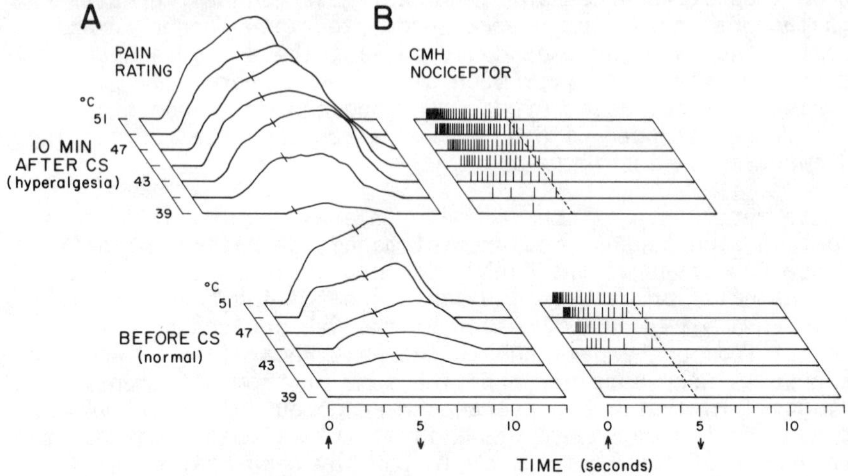

Figure 2. The magnitude ratings of pain by a human subject and the evoked responses of a CMH nociceptor in the monkey during heat stimulations of the hairy skin before and 10 min after a mild heat injury (produced by a conditioning stimulus or CS). Each horizontal line represents the passage of time during the first part of the 30-s trial in which a heat stimulus was delivered. Stimulus temperature ranged in steps of 2°C from 39 to 51°C. The upward arrow marks the onset of the stimulus while the downward arrow and horizontal tic marks indicate the termination of the stimulus. A: The ratings of the magnitude of pain sensation. The vertical axis (not shown) represents the magnitude of pain as rated continuously throughout the test during stimulations of the volar forearm. B: The responses of a CMH nociceptor in the monkey whose receptive field was located on the hairy skin of the middle finger. Each vertical mark represents a single action potential (From LaMotte et al., 1983 with permission from J. Neurophysiol.).

temperature was increased, both the number of active CMHs, as well as the frequencies of their discharge, increased. Our data did not allow us to factor out the relative importance of the proportion of active CMHs vs. the mean frequencies of discharge as mechanisms of coding the magnitude of painful heat.

(4) We calculated the temporal profiles of the mean frequencies of discharge in CMHs that should occur at entry to

the spinal cord. This analysis took into account the temporal dispersions of evoked discharge in CMH afferents due to differences in their conduction velocities and, thus, differences in conduction time along the human leg or arm. For most subjects, pain ratings began after CMH activity first reached the spinal cord and continued at least several seconds (sometimes up to 10 s) after the termination of all CMH activity. A conduction block of activity in A fibers did not increase pain response latencies. The duration of pain was typically longer than the duration of CMH activity.

It is known that evoked discharge in dorsal horn neurons receiving input from CMH afferent fibers may continue well beyond the termination of discharge in CMH afferent fibers (Handwerker et al., 1975, Price et al., 1978). It is also known that a repetitive electrical synchronous activation of discharge in C fiber afferents evokes a progressively greater and longer discharge in dorsal horn neurons ("wind-up") (Mendell and Wall, 1965). It is likely that as the temperature of a stimulus increases, the increasing discharge frequencies of CMHs are integrated over time by dorsal horn neurons resulting in afterdischarges that can continue beyond the termination of CMH activity and, accordingly, may account for the aftersensations of pain following the termination of heat stimuli.

The above findings (3 and 4) support the conclusion that central mechanisms play an important role in the intensive, spatial and temporal integration of heat evoked discharge in nociceptive afferent fibers and that the frequencies of discharge in the CMH population provide the most relevant information in coding the threshold, magnitude and duration of heat pain sensation.

THE PERIPHERAL NEURAL BASIS FOR HYPERALGESIA FOLLOWING A MILD HEAT INJURY

The results obtained following the conditioning stimulus supported the conclusion that only the heat sensitization of CMH nociceptors contributed to the development of hyperalgesia. This conclusion was based primarily on the close similarities in the time course of changes in pain ratings and changes in CMH responses after the CS and on the failure of a compression block of conduction in A fibers to alter the development of hyperalgesia (Figure 1). The results are summarized in the following:

(1) Magnitude ratings of pain underwent changes that were similar in time course to changes in the responses of most CMH nociceptors in monkeys and humans. Immediately after the CS, pain thresholds were elevated for many subjects, but by 5-10 min after the CS, thresholds of all subjects were lowered by 4 to 6°C below their pre-CS (normal) values. Also, within 5-10 min after the CS, magnitude ratings of pain elicited by most suprathreshold stimuli were greater than pre-CS values. Parallel changes were seen in the response thresholds of CMH nociceptors in monkeys and humans and in the number of impulses evoked in CMHs by suprathreshold stimuli.

The finding that the responses of CMH nociceptors can be fatigued prior to their sensitization to heat illustrates the two ways in which the responses of CMHs to heat can be altered by preceding stimulations. In both normal and heat sensitized skin, prolonged or repetitive stimulations at sufficiently short interstimulus intervals can reduce CMH responsiveness (fatigue) whereas sufficiently intense or injurious stimulations may enhance (i.e. sensitize) their responses (Beitel and Dubner, 1976a,b, Bessou and Perl, 1969, Gybels et al., 1979, Kumazawa and Perl, 1977, LaMotte et al., 1982, LaMotte and Campbell, 1978, Torebjörk and Hallin, 1977). However, the enhanced responsiveness of CMHs may not be manifested until there has been a sufficiently long period without stimulation.

Neither AMH nociceptors nor low threshold warm or cold receptors exhibited changes in their responses after the CS that were similar to changes in the responses of CMH nociceptors. Only a small number of AMH nociceptors and cold receptors were sensitized after the CS, and then, only transiently so and of insufficient magnitude to contribute to changes in pain ratings. The responses of warm receptors were greatly suppressed after the CS and only partially recovered 20 min later.

Many subjects reported an unpleasant or mildly painful sensation when the hyperalgesic skin within the area of injury was gently rubbed. Similarly, it was found that a number of CMH nociceptors developed a weak response to gentle cutaneous rubbing with a piece of gauze (J.G. Thalhammer and R.H. LaMotte, unpublished observations). Since most CMHs did not respond to the same stimulus applied prior to the CS, it is possible that this sensitization to mechanical rubbing may contribute to the dysaesthesia elicited by rubbing the skin after a cutaneous heat injury.

(2) A compression block of conduction in A fibers did not alter the time course of the development of hyperalgesia in humans, nor did it alter the magnitude of changes in their pain thresholds or their magnitude ratings of pain. These findings further supported the hypothesis that heat sensitization of nociceptors with C fibers, probably CMH nociceptors, contributed to the development of hyperalgesia after the mild heat injury.

The heat sensitization of AMH nociceptors may be a major determinant of hyperalgesia following more intense heat injuries than those in our experiments. AMH nociceptors can be greatly sensitized by a prolonged heat stimulus of 53°C (e.g. 30 s) (Campbell et al., 1979, Fitzgerald and Lynn, 1977, Meyer and Campbell, 1981). The same stimulus can suppress the responses of CMH nociceptors to heat - at least those of CMHs with receptive fields within the area directly heated. The possibility of heat sensitization of CMHs with receptive fields outside this area has yet to be explored. It has also been shown that an ischemic block of conduction in A fibers reduces, although does not eliminate, the magnitude of hyperalgesia after the injury produced by the 53°C stimulus (Meyer and Campbell, 1981). These results indicate a role for the heat sensitization of AMH nociceptors following stronger heat injuries without ruling out a minor contribution from sensitized CMH nociceptors.

Although our findings indicated that the heat sensitization of CMH nociceptors contributed to hyperalgesia after heat injury, the sensitization did not fully account for the magnitude of hyperalgesia. The increases over normal in the magnitude of CMH responses (mean number of impulses per stimulus), after heat injury of either hairy or glabrous skin, were not in exact proportion to increases in the magnitude ratings of pain. For glabrous skin, although changes in the thresholds of many monkey CMHs paralleled changes in human pain threshold, the responses to suprathreshold stimuli did not increase a sufficient amount to account for the increases in the magnitude ratings of pain. However, neither could the response alterations of any other type of cutaneous receptor as studied in the monkey account for the magnitude of hyperalgesia observed for the glabrous skin in humans. A study of the heat sensitization of glabrous skin CMHs in awake humans has yet to be made.

For hairy skin, the magnitude of sensitization in monkey CMHs or human CMHs in relation to the magnitude of hyperalgesia was less at lower stimulus temperatures and greater at middle or higher temperatures than would be predicted from the linear relation in normal skin between median pain ratings and the mean number of impulses in CMHs. For example in Figure 2A, the subject's rating of 43°C in hyperalgesic skin was similar to his rating of 51°C in normal skin. In contrast, the responses of the CMH nociceptors to 43°C did not undergo as great an increase following heat sensitization (Figure 2B). Nevertheless, it was clear that the enhanced magnitude ratings of pain in hyperalgesic skin were due, at least in part, to the enhanced discharge frequencies of CMH nociceptors.

There are a number of possibilities that might be explored in future studies in order to explain the mismatches we observed between the magnitude of hyperalgesia and the magnitude of CMH sensitization. First, there is always the possibility of a missing class of C fiber nociceptors, although we believe this unlikely based on the results of extensive searches for heat sensitive C fiber afferents carried out in previous studies (Beitel and Dubner, 1976b, Bessou and Perl, 1969). There is also a possibility that our sample of CMHs did not provide a complete representation of the sensitization of CMHs that occured following the CS. For example, we found that the receptive fields of some CMHs and AMHs increased in size by expanding into an area of heat injury. This was possibly due to the existence of branches of the parent axon of the nociceptor that extended outside the receptive field and became responsive to heat only after their terminal endings were sensitized by a chemical substance released from heat injured tissue (Thalhammer and LaMotte, 1982). There is also the possibility that CMHs with receptive fields even more remote to an area of heat injury may become sensitized, although evidence for this is conflicting (Croze et al., 1976, Perl et al., 1976, Thalhammer and LaMotte, 1982). Thus, the responses of adjacent or remote CMHs to heat stimuli of 39-51°C might be minimal or absent in normal skin but, after heat injury and remote sensitization, be increased substantially at the lower stimulus temperatures yet saturated at the higher temperatures. If true, this might account for the reduced slopes of intensity response functions of spinothalamic and VPL neurons driven by CMH input following heat injury of the skin in monkeys. The responses of these neurons were enhanced more to lower than to higher stimulus temperatures after heat injury (Kenshalo et al., 1979, 1980).

Another reason to be concerned about an adequate sample of nociceptors is that the changes in responses of CMH nociceptors to heat after heat injury were not uniform. This was true for both monkey and human CMHs. Some were suppressed and did not subsequently exhibit sensitization, while those that were sensitized differed as to the amount of their sensitization or in the nature of their sensitization (e.g. some exhibiting increased responses to suprathreshold without a change in threshold, others changing in threshold but not in suprathreshold responses, etc.). It was also found that the capacity of CMHs or AMHs to become sensitized to heat or mechanical stimuli following a heat injury of 57°C could differ for two regions within the same receptive field (Thalhammer and LaMotte, 1982). One hypothesis is that the terminal endings of nociceptors differ as to the type, number, and location of receptor sites for algogenic substances as released by injured tissue. In any case, the variations between individual nociceptors in their capacities for sensitization make it advisable that in a future study, a more representative sample of sensitized CMHs be obtained from a larger number of CMHs including those with receptive fields that are located varying distances away from the site of injury.

A third possibility for discrepancies between the magnitude of sensitization in CMHs and the magnitude of hyperalgesia is that there are central transformations of information provided by the nociceptive afferent population. Aside from the capacities of dorsal horn neurons for transient spatial, intensive and temporal integration of nociceptive input, there is also the possibility that these neurons may themselves become sensitized for a prolonged period of time. This may occur, for example, due to the release of chemical transmitter substances such as substance P (Henry, 1976) from CMH afferent endings or from terminals of associated interneurons intensely activated during or following tissue injury. If such a central sensitization resulted in decreased thresholds along with decreased slopes of the intensity response functions of dorsal horn neurons, then after heat injury a slight increase over normal in CMH afferent input at lower temperatures might be magnified much more than a larger increase at higher temperatures. This effect, purely speculative at the moment, might account for the observed greater changes in magnitude ratings of lower than higher stimulus temperatures after heat injury (LaMotte et al., 1983).

Despite certain differences observed in responses of CMH nociceptors to heat and human judgments of pain, our data clearly support two conclusions: 1) that the mean frequency of discharge in the CMH nociceptor population provides a major peripheral source of information about the magnitude and duration of heat pain, and (2) the heat sensitization of CMH nociceptors following a mild heat injury of the skin contributes to the development of hyperalgesia within the injured area.

REFERENCES

Adriaensen, H., Gybels, J., Handwerker, H.O., and Van Hees, J. (1983). Response Properties of Thin Myelinated (A) Fibres in Human Skin Nerves. J. Neurophysiol., 49, 111-122.

Beitel, R.E., and Dubner, R. (1976a). Sensitization and Depression of C-Polymodal Nociceptors by Noxious Heat Applied to the Monkey's Face. In Advances in Pain Research and Therapy. (eds. J.J. Bonica, and D. Albe-Fessard), Raven Press, New York, pp. 149-153.

Beitel, R.E., and Dubner, R. (1976b). The Response of Unmyelinated (C) Polymodal Nociceptors to Thermal Stimuli Applied to the Monkey's Face. J. Neurophysiol., 39, 1160-1175.

Bessou, P., and Perl, E.R. (1969). Response of cutaneous sensory units with unmyelinated fibers to noxious stimuli. J. Neurophysiol., 32, 1025-1043.

Campbell, J.N., Meyer, R.A., and LaMotte, R.H. (1979). Sensitization of Myelinated Nociceptive Afferents That Innervate Monkey Hand. J. Neurophysiol., 42, 1669-1679.

Croze, S., Duclaux, R., and Kenshalo, D.R. (1976). The Thermal Sensitivity of the Polymodal Nociceptors in the Monkey. J. Physiol., 263, 539-562.

Dubner, R., Price, D.D., Beitel, R.E., and Hu, J.W. (1977). Peripheral Neural Correlates of Behavior in Monkey and Human Related to Sensory-Discriminative Aspects of Pain. In Pain in the Trigeminal Region. (eds. D.J. Anderson and B. Matthews). Elsevier, Amsterdam, pp. 57-66.

Fitzgerald, M., and Lynn, B. (1977). The Sensitization of High Threshold Mechanoreceptors with Myelinated Axons by Repeated Heating. J. Physiol., 265, 549-563.

Gybels, J., Handwerker, H.O., and Van Hees, J. (1979). A Comparison Between the Discharges of Human Nociceptive Nerve Fibres and the Subject's Ratings of his Sensations. J. Physiol., 292, 193-206.

Handwerker, H.O., Iggo, A., and Zimmermann, M. (1975). Segmental and Supraspinal Actions on Dorsal Horn Neurons Responding to Noxious and Non-Noxious Skin Stimuli. Pain, 1, 147-165.

Henry, J.L. (1976). Effects of Substance P on Functionally Identified Units in Cat Spinal Cord. Brain Res., 114, 439-451.

Iggo, A. (1959). Cutaneous Heat and Cold Receptors with Slowly Conducting (C) Afferent Fibres. Q. J. Exp. Physiol., 44, 362-370.

Kenshalo, D.R. Jr., Leonard, R.B., Chung, J.M., and Willis, W.D. (1979). Responses of Primate Spinothalamic Neurons to Graded and to Repeated Noxious Heat Stimuli. J. Neurophysiol., 42, 1370-1389.

Kenshalo, D.R., Jr., Giesler, G.J., Leonard, R.B., and Willis, W.D. (1980). Responses of Neurons in Primate Ventral Posterior Lateral Nucleus to Noxious Stimuli. J. Neurophysiol., 43, 1594-1614.

Kumazawa, T., and Perl, E.R. (1977). Primate cutaneous sensory units with unmyelinated (C) afferent fibers. J. Neurophysiol., 40, 1325-1338.

LaMotte, R.H., Thalhammer, J.G., Torebjörk, H.E., and Robinson, C.J. (1982). Peripheral Neural Mechanisms of Cutaneous Hyperalgesia Following Mild Injury by Heat. J. Neurosci., 2, 765-781.

LaMotte, R.H., Torebjörk, H.E., Robinson, C.J., and Thalhammer, J.G. Time-Intensity Profiles of Cutaneous Pain in Normal and Hyperalgesic Skin: A Comparison with C-Nociceptor Activities in Monkey and Human (submitted for publication).

LaMotte, R.H., Thalhammer, J.G., and Robinson, C.J. (1983). Peripheral Neural Correlates of the Magnitude of Cutaneous Pain and Hyperalgesia: A Comparison of Neural Events in Monkey with Sensory Events in Human. J. Neurophysiol., 50, 1-26.

LaMotte, R.H., and J.N. Campbell (1978). Comparison of the Responses of Warm and Nociceptive C-Fiber Afferents in Monkey with Human Judgments of Thermal Pain. J. Neurophysiol., 41, 509-528.

Lynn, B. (1979). The Heat Sensitization of Polymodal Nociceptors in the Rabbit and its Independence of the Local Blood Flow. J. Physiol., 287, 493-507.

Lynn, B. (1980). Heat Pain Sensitivity of Human Skin After Mild Heat Injury and its Lack of Dependence on the Local Blood Flow. Pain, 8, 189-196.

Mendell, L.M. and Wall, P.D. (1965). Responses of Single Dorsal Cord Cells to Peripheral Cutaneous Unmyelinated Fibers. Nature, 206, 97-99.

Meyer, R.A., and Campbell, J.N. (1981). Myelinated Nociceptive Afferents Account for the Hyperalgesia That Follows a Burn to the Hand. Science, 213, 1527-1529.

Perl, E.R. (1968). Myelinated Afferent Fibres Innervating the Primate Skin and their Response to Noxious Stimuli. J. Physiol., 197, 593-615.

Perl, E.R., Kumazawa, T., Lynn, B., and Kenins, P. (1976). Sensitization of High-Threshold Receptors with Unmyelinated (C) Afferent Fibers. Prog. Brain Res., 43, 263-276.

Price, D.D., Hayes, R.L., Ruda, M. and Dubner, R. (1978). Spatial and Temporal Transformations of Input to Spinothalamic Tract Neurons and Their Relation to Somatic Sensations. J. Neurophysiol., 41, 933-947.

Thalhammer, J.G., and LaMotte, R.H. (1982). Spatial Properties of Nociceptor Sensitization Following Heat Injury of the Skin. Brain Res., 231, 257-265.

Torebjörk, H.E., LaMotte, R.H., and Robinson, C.J. Peripheral Neural Correlates of the Magnitude of Cutaneous Pain and Hyperalgesia: Simultaneous Recordings in Humans of Sensory Judgments of Pain and Evoked Responses in Nociceptors with C-Fibers. J. Neurophysiol., In Press.

Torebjörk, H.E., and Hallin, R.G. (1974). Identification of Afferent C Units in Intact Human Skin Nerves. Brain Res., 67, 387-403.

Torebjörk, H.E., and Hallin, R.G. (1977). Sensitization of Polymodal Nociceptors with C Fibres in Man. Proc. Int. Union Physiol. Sci., 13, 758.

Van Hees, J., and Gybels, J. (1981). C Nociceptor Activity in Human Nerve During Painful and Non Painful Skin Stimulation. J. Neurol. Neurosurg. Psychiatry, 44, 600-607.

Zotterman, Y. (1936). Specific Action Potentials in the Lingual Nerve of Cat. Scand. Arch. Physiol., 75, 105-120.

Zotterman, Y. (1939). Touch, pain and tickling: An electrophysiological investigation on cutaneous sensory nerve. J. Physiol., 95, 1-28.

DETERMINANTS OF SUBJECTIVE ATTRIBUTES OF NORMAL CUTANEOUS SENSATION AND OF PARESTHESIAE FROM ECTOPIC NERVE IMPULSE GENERATION

*JOSÉ OCHOA, **ERIK TOREBJÖRK, and †WILLIAM CULP

**Department of Neurology, University of Wisconsin Medical School, Madison, WI 53792, USA*
***Department of Clinical Neurophysiology, Academic Hospital, Uppsala, Sweden*
†Department of Biochemistry, Dartmouth Medical School, Hanover, NH 03756, USA

Fig. 1. Zotterman et al

Upon discovering the internal language of the nervous system, in the mid 1920's, Adrian and Zotterman (1926) immediately realized that the laconic messages transmitted as monotonous codes in which the only variable was impulse frequency, were probably decoded as the intensity of a particular sensory quality. Indeed, in their cats, intensity of a mechanical stimulus was reflected as frequency of afferent discharge, and so it took little imagination on the part of those introspective sensory physiologists to equate the receptor response feature with the subjective magnitude attribute.

This of course is well known. It is less well known that Zotterman was relatively disinterested in sensory magnitude. Like his viking ancestors he was rather fascinated by sensory quality (Figure 1) as betrayed in the title of his autobiography: "Touch, Tickle and Pain" (Zotterman, 1971).

But Adrian and Zotterman's experimental animals only gave reluctant and at best inconclusive indications of the quality of their sensations, and nevertheless Yngve convinced himself that the qualities touch, tickle, pain, cold, warm, etc... are determined not by the contents of the afferent messages, but by the nature of specific centrally connected afferent channels. Thus, there would exist specific touch channels, pain channels, tickle channels, etc... Earlier, and based on different evidence, von Frey had also advocated specificity to explain quality of cutaneous submodalities (for a brief, excellent review in English see von Frey, 1906). A keen observer, aware of the existence of microscopic end organs in the skin, von Frey had borrowed Johannes Müller's concept of "specific energy", whereby major sense modalities (vision, hearing, taste, etc.) would be determined by excitation of a specific kind of nerve by a specific stimulus energy. Von Frey extended the concept to embrace somatosensory submodalities and legitimately (after Blix) confirmed that in less densely innervated areas of human skin it is possible to identify sensitive spots whose activation would reproducibly evoke a distinct and pure sensation. A powerful argument by von Frey for specificity came from his observation that a same stimulus (i.e., minimal focally applied current, or heat) would evoke a different sensation (vibration vs. pricking pain, or cold vs. warmth) from independent sensory spots. "The most probable explanation of this fact is that the effect depends on the (specific) kind of terminal organ excited by the stimulus" (von Frey, 1906).

Not everybody accepted specificity, and as recently as the 1950's, the Oxford group came up with a rival theory to account for the basis of somatosensory qualities: the Pattern Theory. Weddell, Sinclair and their colleagues were fascinated with their own observation that in places like the cornea, which is supplied by one single kind of sensory receptor, one can feel a range of sensory qualities. The Oxonians then proposed that receptors and their afferent lines are uncommitted and that the particular nature of an adequate stimulus would cast a particular coded message which would be interpreted as a particular sensory quality upon decoding by the brain. Sinclair (1955) summarized the essence of the theory in a few eloquent lines: "But it seems more likely that one and the same type of ending can mediate different kinds of sensation according to the physical conditions to which it is exposed."

In retrospect it seems rather dangerous to build a neurophysiological theory on the grounds of neuroanatomical observations, and Nathan punishes that thinking ruthlessly: "All of us who accepted Weddell's view at the time were wrong in assuming that as two nerve fibers looked the same down a microscope they were functionally the same. Why should one judge nerve fibers on how they look? One might as well judge them on how they taste " (Nathan, 1976).

Thirty years later Weddell was of the opinion that the Pattern Theory "...does not hold water" (Weddell, written personal communication, 1981). Although today most workers are with Weddell against the pattern theory, definitive evidence in one or the other direction came forth very recently, because obtaining such evidence requires repeating Adrian and Zotterman's deed not in cats but in conscious humans. Indeed, it requires following up matters beyond correlation of kind of natural stimulus with kind of single fiber afferent message, to obtaining additional correlation between the latter and the subjective sensory experience evoked. Obviously this aim calls for an experiment in which afferent activity, initiated by natural cutaneous stimulation of identified sensory units, can be recorded in humans capable of reporting the subjective attributes of their sensations. In fact the microneurographic technique of Hagbarth and Vallbo (1967) fulfills the requirements. Unfortunately, when it comes to sorting out the enigma of what determines sensory quality, even microneurography falls short because, no matter how subtle or circumscribed, a natural stimulus coactivates a number of units with overlapping receptive fields. Thus there can be no certainty that the resulting sensation is contributed by the particular unit the intraneural microelectrode happens to be seeing: other coactivated units might be contributing quality to the subjective sensory blend or might be entirely responsible for it and yet their impulses might escape the recording electrode. Of course, a satisfactory compromise is reached through correlative experiments where both a human subject and an experimental animal receive identical natural stimuli; while man reports his subjective sensory experience, the animal's nerves are probed with microelectrodes to monitor afferent impulse activity. Although a great deal has been learned from this correlative animal-man strategy, the recorded afferent message does not quite belong to the experiencing subject, and this may be potentially fallacious. Recently the microneurographic technique was supplemented to allow both recording and microstimulation (INMS) of single sensory units through the same intraneural microelectrode (Torebjörk and Ochoa, 1980; Ochoa and Torebjörk, 1983). In this way it has become possible to administer a choice of afferent messages, in laconic trains or variously patterned, directly to single sensory fibers which can be individually identified by receptive field and classified by receptor-response characteristics. Amazingly, it is now possible to compare the contents of afferent messages, artificially delivered to the range of sensory units, with the subjective attributes of the sensations evoked in terms of quality and magnitude too.

When Zotterman was informed of preliminary results of studies of this kind during a meeting of the Scandinavian Physiological Society in the autumn of 1980, he gallantly acknowledged the step forward: with Adrian, Zotterman had shown how sensory units spell out messages destined to the brain; now one could use the human brain as a measuring instrument and directly ask how it reads incoming messages from single sensory units.

Emerging from the application of combined microneurography/ INMS, perhaps the most remarkable revelation has been insight into the exquisite discriminative capacity of the human brain to resolve subjective quality, magnitude and localization of somatic sensations evoked by impulse trains initiated in single primary sensory neurons (Torebjörk and Ochoa, 1980; Vallbo, 1981; Ochoa and Torebjörk, 1983; Schady, Torebjörk and Ochoa, 1983). For example, the quality attribute is normally decoded specifically from particular kinds of afferent channels: thus there is now final evidence in favor of specificity a la Blix, von Frey, Adrian and Zotterman.

The magnitude attribute of somatic sensation can be resolved in the total absence of spatial summation; that is to say, the contents of afferent messages in single units do carry information that the brain sizes quantitatively.

The cortical localization function (locognosia) is also precise, to the millimeter level, again in the absence of spatial summation. Indeed the brain "knows" the skin map of the hand at the single unit level of resolution.

It is only on the basis of the knowledge described above that one can begin to understand what determines the abnormal subjective attributes of certain positive sensory phenomena commonly described by patients suffering from dysfunction of the sensory pathways. Indeed, paresthetic sensations may have aberrant qualities, absurd magnitudes and chaotic projected localizations, and one wants to know why. Ideally one would like to be able to investigate this by implementing the same kind of experimental strategy described above: compare the contents of the afferent messages in single fibers of various types with the subjective sensory attributes verbalized....but during paresthesiae. Although there are few opportunities to stick electrodes into the nerves of neuropathic patients, it is quite easy to manipulate reversibly normal nerves into 'making' paresthesiae, for example in the postischemic state (Lewis, Pickering and Rothschild, 1931; Zotterman, 1933; Kugelberg, 1944; Weddell and Sinclair, 1947; Gordon, 1948; Merrington and Nathan, 1949). One can then readily record in volunteers unitary afferent impulses generated ectopically from nerve fibers previously subjected to ischemia (Torebjork, Ochoa and McCann, 1979; Ochoa and Torebjork, 1980). After 20 minutes or so of ischemia provoked by sphigmomanometer cuff inflated round the forearm above systolic pressure, the cuff is released and the cold blue hand becomes red, feels hot and then follows a stereotyped sequence of paresthesiae. It features "buzzing', 'pricking', 'pseudocramp' and ultimately 'tingling". There is no pain nor itch. There is also a period of muscle twitching, which is brief, probably reflecting the fact that motor nerve fibers retain better accommodation than the sensory fibers (see Culp, Ochoa and Torebjork, 1982).

Microneurographic recordings using intraneural Vallbo-Hagbarth electrodes disclose striking abnormal unitary discharges, often repeating in bursts. When "buzzing" is at its peak, the characteristic type of discharge is shown (Figure 2). Rather prolonged bursts lasting seconds repeat irregularly at relatively long intervals. When there is just intermittent "tingling," the typical discharges are brief bursts repeating rythmically (Figure 3).

So, here we have human sensory units misspelling patterned messages destined to the brain, and the subjective correlate evoked are sensations with abnormal attributes. A compelling question is: are the perverted attributes in paresthesiae the result of activity in specific afferent channels or are they determined by those unphysiological patterns of discharge.

In Torebjork's laboratory we used various patterns of discharge recorded during postischemic paresthesiae to drive the stimulator and we replayed them intraneurally onto identified single units of various types. It became clear that, as anticipated, modulation of the pattern of trains of impulses does not affect the specific quality of the sensation evoked by impulses initiated in particular types of primary sensory units. No "tingling" ever comes out of a nociceptor; no "pricking" comes out of a Pacini unit. In other words, these paresthesiae are

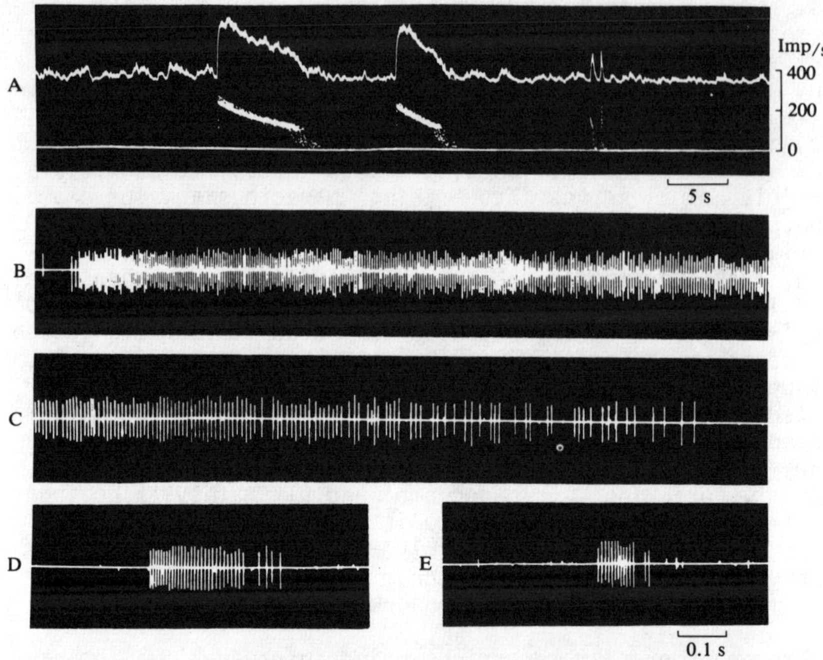

Fig. 2. Prolonged high-frequency discharges in a single unit recorded from the median nerve at elbow level. Unitary bursts appeared during the second minute after release of cuff compression round forearm, maintained for 25 min. A, integrated neurogram (upper trace) shows four abrupt deflections, representing single unit discharges, also displayed in B-E. 'Instantaneous' frequency plot (lower trace) shows initial frequency of about 220 Hz with exponential fall to about 150 Hz and subsequent break down. Duration of consecutive bursts diminished from an initial maximum of 7 s. B, displays beginning and C, the end of first unitary burst shown in A. Note regular firing frequency at the beginning, and missing beats towards the end. Last two bursts in A are displayed in D and E.

not new abnormal sensory qualities created by these new afferent patterns; they feature standard subjective attributes which have become distorted. Indeed, their high *magnitude* is disproportionate to their small area of projection due to high frequency of discharge in single units. Their *localization* is chaotic due to random recruitment order of intraneural generators with random projections in the cortical map. Most importantly, their *quality*

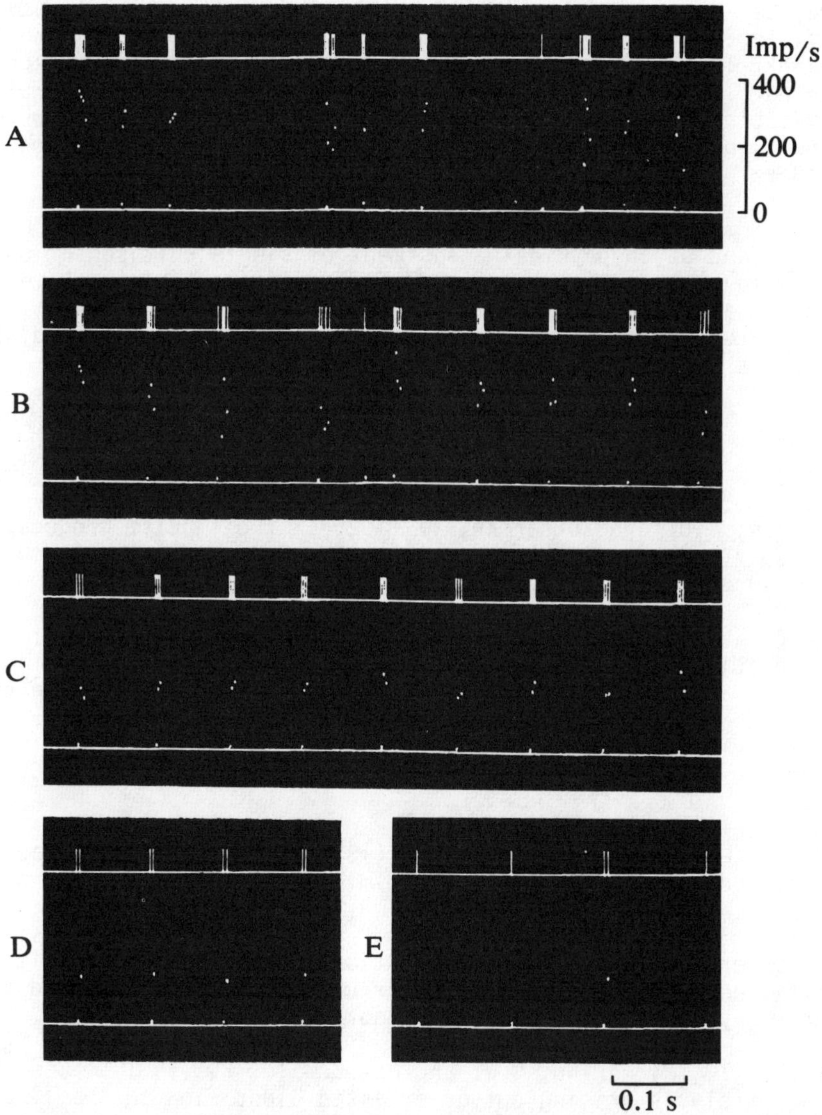

Fig. 3. Samples A-E cover a sequence from the sixth to fourteenth minute after ischaemia. Gallop rhythm in A evolved into a monotonous rhythm (B-E), together with a progressive reduction in number of spikes per burst. Unit ceased to fire at E, when subject reported cessation of rhythmic 'tingling' referred to index fingertip.

may be 'perverted' because the sensations evoked by activation of single units are pure, elementary, sensations (Ochoa and Torebjörk, 1983) that are not obvious in our normal sensory experiences. This is because excitation of skin by natural stimuli inevitably coactivates different types of sensory units, evoking the blends of pure sensations we normally experience (Ochoa, Torebjörk, Culp and Schady, 1982).

Future experiments employing combined intraneural recording, INMS, and psychophysical assessment of single unit sensation promise to provide important new information about basic elements in normal and abnormal sensation. It is a pleasure to acknowledge our intellectual and experimental indebtedness to the Swedish workers who pioneered in this field.

REFERENCES

Adrian, E.D. and Zotterman, Y. (1926). The impulse produced by sensory nerve endings. Part III. Impulses set up by touch and pressure. J. Physiol., 61, 465-483.

Culp, W.J., Ochoa, J. and Torebjörk, E. (1982). Ectopic impulse generation in myelinated sensory nerve fibers in man. In Abnormal Nerves and Muscles as Impulse Generators. (eds. W.J. Culp and J. Ochoa). Oxford Press, New York.

Gordon, G. (1948). 'Pins and needles'. Nature, London, 6, 742-743.

Hagbarth, K.E. and Vallbo, Å.B. (1967). Mechanoreceptor activity recorded percutaneously with semimicroelectrodes in human peripheral nerves. Acta Physiol. Scand., 69, 121-122.

Kugelberg, E. (1944). Accommodation in human nerves and its significance for the symptoms in circulatory disturbances and tetany. Acta Physiol. Scand., 8, Supplement 24, 1-105.

Lewis, T., Pickering, G.W. and Rothschild, P. (1931). Centripetal paralysis arising out of arrested blood flow to the limb, including notes on a form of tingling. Heart, 16, 1-32.

Merrington, W.R. and Nathan, P.W. (1949). A study of post-ischemic paraesthesiae. J. Neurol., Neurosurg., and Psychiatry, 12, 1-18.

Nathan, P.W. (1976). The gate control theory of pain: A critical review. Brain, 99, 123-158.

Ochoa, J.L. and Torebjörk, H.E. (1980). Paraesthesiae from ectopic impulse generation in human sensory nerves. Brain, 103, 835-853.

Ochoa, J. and Torebjörk, E. (1983). Sensations evoked by intraneural microstimulation of single mechanoreceptor units innervating the human hand. J. Physiol. (in press).

Ochoa, J.L., Torebjörk, H.E., Culp, W.J. and Schady, M.D. (1982). Abnormal spontaneous activity in single sensory nerve fibers in humans. Muscle and Nerve, 5, S74-S77.

Schady, W.J.L., Torebjörk, H.E. and Ochoa, J.L. (1983). Cerebral localisation function from the input of single mechanoreceptive units in man. Submitted to Acta Physiol. Scand.

Sinclair, D.C. (1955). Cutaneous sensation and the doctrine of specific energy. Brain, 78, 584-614.

Torebjörk, E. and Ochoa, J. (1980). Specific sensations evoked by activity in single identified sensory units in man. Acta Physiol. Scand., 110, 445-447.

Torebjörk, H.E., Ochoa, J.L. and McCann, F.V. (1979). Paresthesiae: Abnormal impulse generation in sensory fibres in man. Acta Physiol. Scand., 105, 518-520.

Vallbo, Å.B. (1981). Sensation evoked from the glabrous skin of the human hand by electrical stimulation of unitary mechanosensitive afferents. Brain Res., 215, 359-363.

vonFrey, M. (1906). The distribution of afferent nerves in the skin. J. Am. Med. Assc., 47, 645-648.

Weddell, G. and Sinclair, D.C. (1947). "Pins and Needles": Observations on some of the sensations aroused in a limb by the application of pressure. J. Neurol., Neurosurg., and Psychiatry, 10, 26-46.

Zotterman, Y. (1933). Studies in the peripheral nervous mechanism of pain. Acta Med. Scand., 80, 185-242.

Zotterman, Y. (1971). In Touch, Tickle and Pain. Pergamon Press, Oxford.

INDEX